System Validation and Verification

systems engineering series

series editor

A. Terry Bahill, University of Arizona

The Art of Systems Architecting
Eberhardt Rechtin, University of Southern California
Mark Maier, University of Alabama at Huntsville

Fuzzy Rule-Based Modeling with Applications to Geophysical, Biological and Engineering Systems
András Bárdossy, University of Stuttgart
Lucien Duckstein, University of Arizona

Systems Engineering Planning and Enterprise Identity
Jeffrey O. Grady, JOG System Engineering

Systems Integration
Jeffrey O. Grady, JOG System Engineering

Model-Based Systems Engineering
A. Wayne Wymore, Systems Analysis and Design Systems

Linear Systems Theory
Ferenc Szidarovszky, University of Arizona
A. Terry Bahill, University of Arizona

The Road Map to Repeatable Success: Using QFD to Implement Change
Barbara A. Bicknell, Bicknell Consulting, Inc.
Kris D. Bicknell, Bicknell Consulting, Inc.

Engineering Modeling and Design
William L. Chapman, Hughes Aircraft Company
A. Terry Bahill, University of Arizona
A. Wayne Wymore, Systems Analysis and Design Systems

The Theory and Applications of Iteration Methods
Ioannis K. Argyros, Cameron University
Ferenc Szidarovszky, University of Arizona

Systems Engineering Guidebook: A Process for Developing Systems and Processes
James N. Martin, Texas Instruments

Fuzzy Rule Based Computer Design
John Newport, Newport Systems Incorporated

IN.

System Validation and Verification

Jeffrey O. Grady
JOG System Engineering, Inc.
San Diego, California

CRC Press
Boca Raton Boston New York Washington London

Acquiring Editor: Robert Stern
Project Editor: Andrea Demby
Marketing Manager: Tim Pletscher
Cover design: Denise Craig
PrePress: Kevin Luong
Manufacturing: Carol Royal

Library of Congress Cataloging-in-Publication Data

Grady, Jeffrey O.
System validation and verification / Jeffrey O. Grady.
p. cm. -- (Systems engineering series)
Includes index.
ISBN 0-8493-7838-9 (alk. paper)
1. Systems engineering. 2. System analysis. I. Title.
II. Series.
TA168.G656 1997
620'.001'1--dc21 97-22214
CIP

International Standard Book Number 0-8493-7838-9
Library of Congress Card Number 97-22214
Printed in the United States of America 1 2 3 4 5 6 7 8 9 0
Printed on acid-free paper

Foreword

Writing about system engineering, once started, seems to be a pastime that is hard to set aside. This is my fourth book in the system engineering field. In each of the three previous books and this one as well, I have chosen consciously to focus on parts of the system engineering process rather than try to embrace the whole system engineering process in a single book because I concluded that the general books have not provided sufficient detail about the most important parts of the process. Previously, I had felt that there was not enough content to fill a book on verification, so I included one small chapter in my requirements analysis book. I changed my mind as a result of having to build a course for the subject and trying to find an adequate textbook.

As I have told several people, my *System Requirements Analysis*[1] book had its birth as a result of a chain of circumstances that I find almost comical now. As an engineering manager at General Dynamics Space Systems Division, I was encouraged to hire engineers directly out of college like all other department heads. Most system engineering department supervisors, including me, were not happy about this, but we followed orders and did our best to make it work. In my case, it dawned on me that my department, Systems Development, at the heart of the system engineering process, could only be successful under this policy if we partitioned the expansive job into smaller parts, specialized it, if I may use that term. As a result, we would be able to get new hires started on their career in system engineering and, through reassignments and in-house courses based on our written procedures, expand the narrow beginnings into full system engineering knowledge and very qualified system engineers.

As a result, I established a set of tasks in two tiers and developed a series of department procedures telling how to perform each of these detailed tasks. Since I had concluded that my division was most urgently in need of improvements in requirements analysis, I put the primary emphasis on that area. Soon after completing most of the department instructions, I had obligated myself to teach a class in this subject as the beginning class in the University of California San Diego Systems Engineering Certificate program. About the time I concluded that there was no adequate textbook in this field for the course, I had the misfortune to have a ladder come out from under me while painting my house.

The bad news was that I fractured both elbows. The good news was that I couldn't go to work for a few weeks, but I found that I could still operate a computer with two slings. So, I had the time, the physical capability, and the raw material at hand (department procedures) to write a text for the class I had agreed to teach.

Well, I had none of those advantages as I began this book in August 1994. I was very busy with other work building a new consulting business and I had to begin with an empty head in many of the areas of verification. But, I did have two good elbows. There was one common characteristic from my prior experience, however. I could find no one book that tied together several areas I felt should be connected as the basis for a course in verification. Also, the previous work on the *System Requirements Analysis* book was a plateau upon which to build since it covered the beginning of the verification story in the development of any system. Research that I had done for the *System Integration*[2] book also had exposed me to the important notion of product representations configuration management that relates directly to the validation process.

Another motivation for undertaking this work was a feeling of irritation over the confusing use of the words *verification* and *validation* in industry. My solution will not be universally accepted but at least I will lose my irritation in the process of exposing others to these ideas. No matter your cherished definitions of these words, I have defined them with some clarity so that you will be able to relate your views to those I have expressed. Perhaps this book should have been provided in electronic media with an entry screen allowing readers to fill in their preferred definitions with a resultant automatic book correction. If life were that simple, it might have been a good idea.

The problem is that everyone does not adhere to any universally accepted definition of these words, leading to continuing confusion between customers and contractors and within any one organization. The dictionary definitions are not very helpful either in resolving differences. Just prior to a lunch break in a requirements analysis course I presented at the Sandia National Laboratories I challenged the students to look up these words in a dictionary. One of the students came back to class with a printout from his computer dictionary for the V&V words, and it was not possible for any of us to distinguish between them relative to the activities discussed in this book.

I hope, by pulling together a complete package of work related to these two words, I think for the first time in a published form available to the general public, we can all look with some interest on properly naming the several components of that package. You may conclude that my choices are not optimum, but they represent my best efforts at the time this book was published.

The book is intended for practicing system engineers and others involved in validation and verification work on programs that solve complex problems entailing a need to prove that the original problem was, in fact, solved. The author intends to use the book as a basis for courses offered through his company in the V&V area and as a textbook when asked to teach a university course in verification. Other lecturers may find the content useful for similar purposes.

Jeffrey O. Grady
San Diego, California

References

1. Grady, J.O., *System Requirements Analysis,* McGraw-Hill, New York, 1993.
2. Grady, J.O., *System Integration,* CRC Press, Boca Raton, FL, 1994.

Acknowledgments

This book is dedicated to the memory of Mr. Max Wike, a retired naval officer and a system engineer in the department I managed at General Dynamics Space Systems Division in the mid-1980s. Before he passed away long before his time, he and his wife-to-be, Ms. Debbie Matsik, had succeeded in beginning my education in requirements verification and I very much appreciate that to this day.

It never fails that during every class on system requirements analysis that I have taught for University of California San Diego, UC Irvine, UC Berkeley, through the University Consortium for Continuing Education, and independently through my consulting firm at companies, students have offered me great insights and ideas about how this process can be done better. Many of these inputs find their way into future courses as well as new books like this one and, I hope, revisions of others. Unfortunately, I have not shown due diligence in keeping a list of these people by name and location so that I can properly thank them. The list would be very long. By the time this book is published it will likely include on the order of 400 or 500 engineers.

I benefited a great deal from a UC San Diego System Engineering Certificate Program course I attended while the ideas for this book were initially swarming in my mind. That course was titled "System Verification" taught by a great guy and fine system engineer, Mr. J. D. Hill, at the time a vice president at Scientific Applications International Corporation (SAIC) in San Diego, California. I very much appreciate the new knowledge I derived from his course.

Several people in industry provided much-appreciated valuable insights into modern techniques for validation and verification on items as small as integrated circuits and as large as a transport aircraft. Mr. David Holmes, a system engineer at Interstate Electronics Corporation in Anaheim, California very generously arranged a demonstration of how they develop application-specific integrated circuits (ASIC). Mr. David B. Leib, a Test Engineer on the McDonnell Douglas C-17 Flight Control System Simulator (Iron Bird), gave me a tour of the simulator at their facility in Long Beach, California that was very useful in describing a typical large-scale validation instrument.

Much of Section 6.6.4 and other areas touching on computer software verification was motivated by the work of Boris Beizer expressed in his

Software Testing Techniques[1] and *Black Box Testing*[2] to which the reader is directed for extensive and exhaustive treatment.

I am also very much indebted to Mr. Clarence Hicks who read the manuscript and offered valuable suggestions based on his many years of experience in verification work on aircraft, cruise missiles, and other things. I worked with Clarence on the General Dynamics Convair Advanced Cruise Missile program in the early 1980s and learned as a result that it was very important to verify your information. Through a strange chain of circumstances Clarence was reported to have passed away over a weekend. When he walked into the area late Monday morning just returning from vacation, we all thought it was the Second Coming. It turned out that there was another person by the same name employed by Convair and that person really had passed away over the weekend and the names and organizations had become confused.

While visiting my daughter, Kimberly, on one consulting trip, I went to dinner with her and her boyfriend, Barry Anderson, assuming we would have to talk about nontechnical things. I was surprised to find that Barry was an engineer at a large computer manufacturer, writing test scripts that were e-mailed to a worldwide manufacturing capability, stimulating a wider area of commentary. Thanks to Barry and my apologies to Kim for boring her to death.

I would like to again extend my thanks to Mr. Bernard Morais, President of Synergistic Applications in Sunnyvale, California. Barney read the manuscript and offered many suggestions for improvements. I very much value his opinions about our profession.

While I have benefited from the advice of those identified, they should not be held accountable for any errors of commission, omission, or wrongheaded thinking. They belong wholly to the author.

While writing a book, one has many down periods when he doesn't see how the end will ever be reached. Through all of those downs in writing this book my wife, Jane, has encouraged me to press on and without that encouragement, I doubt this book ever would have been completed. I recall several years ago when I was feeling sorry for myself about the possibility of being 50 years old before I completed a master's program. Jane told me, "Jeff, you can be fifty with a master's or fifty without a master's, but, with luck, you are going to be fifty." This kind of practical advice has sustained me in many situations.

I'd also like to add my appreciation for the friendship of Nicholas Grady which I will remember the rest of my life.

References

1. Beizer, B., *Software Testing Techniques,* International Thomson Computer Press, Boston, MA, 1990.
2. Beizer, B., *Black Box Testing,* John Wiley & Sons, New York, 1995.

Contents

List of illustrations

List of tables

chapter one

In the beginning

1.1 Requirements

This book covers two important parts of the systems approach for the development of solutions to complex problems. The systems approach to solving complex problems entails three fundamental steps: (1) define the problem in the form of requirements captured in specifications, (2) creatively synthesize the requirements into design solutions believed to satisfy those requirements, and (3) prove through test and other methods that the resultant product does in fact satisfy the requirements thereby solving the original problem. An organization that develops products to solve complex problems is right to apply this organized process to their work because it results in best customer value motivated by least cost and best quality and performance. This happens because the systems approach encourages consideration of alternatives and the thoughtful selection of the best solution from multiple perspectives as well as the earliest possible identification of risks and concerns followed by mitigation of these risks. At the center of this whole process are the requirements that act to define the problems that must be solved by designers or design teams.

The parts of the process covered in this book are referred to by the author as validation and verification. We will see that there is not universal understanding about the meaning of these words so, for the time being, let us say that V&V activities produce evidence that proves that it is possible to satisfy the requirements of an item or to prove that a particular design is compliant with the predetermined requirements. It is hoped that the reader is familiar with requirements analysis through experience or having read the author's earlier book *System Requirements Analysis*.[1] In the event the reader does not feel fully confident in his/her knowledge in this subject area, this first chapter provides a summary of the systems approach and the place of requirements in that process.

The techniques described in this book can be applied at any level of indenture in a system from the system level down through the component. So, when the author speaks of systems, the reader can understand that to mean whatever is his or her scope of interest. The content also applies to hardware and software alike, although the reader should be aware that the author's principal experience has been in hardware. Some of the content has been prepared specifically for hardware or software applications but most of the content applies to both.

The content also applies whether the product is a new weapons system for the Department of Defense (DoD), a new automobile model using electric power, or a new earth mover. This approach owes its emergence to the many military and space systems developed in the United States and other countries roughly during the period 1950 to 1990, but it is appropriate for any development effort intended to solve a very complex problem. This process may be viewed as a part of a risk management activity to encourage the identification of problems while there is still a lot of time to solve them or to mitigate their most adverse effects. If you choose to accept a lot of risk, you may not feel obligated to apply these techniques. But, you should steel yourself to deal with a lot of problems under stressful conditions resulting in unplanned cost, late delivery, and reduced product effectiveness. For risks are not just idle curiosities discussed in academic treatises. Many of them will be fulfilled during the life cycle of a system. There are ways to economize in the application of validation and verification and some of these are pointed out. One need not apply the full toolbox of techniques discussed in the book or to the level of intensity discussed to be at least reasonably effective in reducing program risk.

This book fills a void in the publicly available literature on two very important parts of the systems approach to the solution of complex problems. This systems approach entails the following principal steps:

a. Understand the customer's need, which is the ultimate system function for it tells what the customer wants. It defines the problem that must be solved at a fairly high level.
b. Expand the need into a critical mass of information necessary to trigger a more-detailed analysis of a system that will satisfy that need. This may entail customer surveys, focus groups, question-and-answer sessions, or mission analysis, in combination with a high-level functional analysis that exposes needed functionality, system design entailing allocation of the top-level functionality to things placed in a system physical architecture, and definition of requirements for the complete system (commonly captured in a system specification). Through this process, one will determine to what extent and for what elements solutions shall be attempted through hardware and software.
c. Further decompose the need, which represents a complex problem, within the context of an evolving system concept, into a series of related smaller problems described in terms of a set of requirements that must be satisfied by the solutions to the smaller problems.
d. Prior to design work it is sometimes necessary or desirable to improve one's confidence in one's understanding of the requirements or to

prove that it is physically possible to produce a design that is compliant. This is commonly accomplished through special tests and analyses, the result of which indicate the strength of the currently available related knowledge and technology base and the current state of readiness to accomplish design work using available knowledge.

e. Application of the creative genius of design engineers and market knowledge of procurement experts within the context of a supporting cast of specialized engineers and analysts to develop alternative solutions to the requirements for lower-level problems. Selections are made from the alternatives studied based on knowledge to identify the preferred design concept that will be compliant with the requirements.
f. Integration, testing, and analysis activities are applied to designs, special test articles, preproduction articles, and initial production articles that prove that the designs actually do satisfy the requirements.

This whole process is held together, especially during the initial theoretical stage, by requirements. These requirements are statements that define the needed characteristics of a solution to an engineering problem prior to the development of the solution to the problem. This book is about requirements, proving that it is possible to satisfy them, and proving that they have been satisfied in designs. Requirements are defined as a prerequisite to accomplishing design work because of the nature of complex systems and the need for many people to work together to develop complex systems.

1.2 The two Vs

The reader is forewarned that the meanings of the two words in the title beginning with the letter V are not universally agreed upon. There are at least three different meanings attached to these two words. The word validation will be used in this book to mean a process carried out to demonstrate that one or more requirements are clearly understood and that it is possible to satisfy them through design work within the current technological state of the art. This is a part of the process of challenging the need for particular requirements and specific values tagged to those requirements prior to the development of solutions or as part of a process of developing alternative concepts, one of which may be the optimum solution in satisfying the requirements. Many system engineers prefer to express the meaning of validation as proving that the right system is being or has been built. The author truncates this meaning to accept only the first part of it, that is, proving that the right system is being or is going to be built, and refers to all postdesign V&V work as verification that requirements have been complied with.

The word verification will be used to mean a proof process for unequivocally showing that a particular design will or does satisfy the corresponding requirements upon which it is based. Verification is done after the design process to develop evidence of design solution requirements compliance. A shorthand statement of the meaning of verification is commonly voiced as, "Did I build the system right?"

In software development, the word verification is often taken to mean a process of ensuring that the information developed in phase n of a development process is traceable to information in phase n – 1. Validation, in this view, is the process of proving compliance with customer requirements. The latter is the essentially the same definition used for verification in this book.

Validation, as used in this book, therefore precedes the design solution and verification follows the design in the normal flow of events in the system development process. It should be pointed out that these words are used in an almost exactly opposite way in some circles, as noted above. The professional system engineer and program manager must be very careful to understand early in a program the way their customer or contractor counterparts are using these two words to avoid later confusion and unhappiness. People from different departments in the same company may even have different views initially on the V words and these should be sorted out, before the company consciously seeks to standardize its process and vocabulary.

These V words are also used in technical data development in the DoD community. The contractor validates technical data by performance of the steps in the manual commonly at the company facility using company personnel. Technical data is then verified by the customer, generally at the customer's facility or area of operations, by demonstrating accomplishment of content using customer personnel. When the author was a field engineer for Teledyne Ryan Aeronautical, a gifted technical data supervisor and former Navy Chief Warrant Officer, Mr. Beaumont, informed him, in a way that has not yet been forgotten, how to remember the difference between validation and verification of technical data. He said, "A comes before E! Now, even a former Marine can remember that, Grady!" This simple statement also applies to requirements validation and verification in this book.

Before we proceed in detail about how to perform validation and verification work, we need to build a more complete common understanding as a basis for discussion. In order to do so, we must build a clear understanding of requirements, what they are, why they are necessary, how they are captured, and how they are applied in the development process. The author's book *System Requirements Analysis*[1] provides that background, but the remainder of this chapter offers a brief, stand-alone piece to give a proper background to those who do not have access to that book.

1.3 The foundation of system engineering

Complex systems flow from a complex process that converts organized human thought, materials, and time into descriptions of organized sets of things that function together to achieve a predetermined purpose. This complex process is called the systems approach, the system development process, or the exercise of system engineering. Why must it be a complex undertaking and how do the validation and verification activities fit into it?

There are some simple truths about system engineering that, once appreciated, make everything else fall into place and the V words with them. The amount of knowledge that humans have exposed is far in excess of the amount of knowledge that any one person can master and efficiently apply in solving problems. Developing systems is a problem-solving process that transforms a simple statement of a customer need into a clear description of a preferred solution to the problem implied in the need. Companies that are able to bring to bear more of the knowledge base related to a given complex problem will have the best chance of success in solving complex problems. Over time, this ability can be translated into the basis for a successful business venture in the development of complex systems.

Most of the simple problems faced by humankind have been solved somewhere in the world. Many of the problems we now conceive represent fairly complex new problems or extensions or combinations of previous solutions phrased on a grander, more-general, problem space. Complex problems commonly require an appeal to multiple technologies and a broader knowledge base than any one person can master. In order to accumulate the knowledge breadth necessary to deal with a complex problem, we must team more than one person, each a relatively narrow specialist, together and organize their work environment to ensure that synergism occurs. Synergism, a key systems term, is the interaction of elements that when combined produce a total effect greater than the sum of the individual elements. Synergism is necessary to weld together the individual specialized knowledge spaces into an aggregate knowledge base that the team needs to solve the problem. At the same time, a team may fall into a lucky condition of serendipity resulting from the application of creative genius or dumb luck which should not be cast aside by systems purists.

We are clear on what the specialized engineers do within process systems in the development of product systems. They ply their specialized trade in the definition of product requirements and development of design features related to their specialty. While success in this work is a necessary condition for the success of the overall project, it is also true that these specialized engineers cannot among themselves accomplish the development of a complex product system, because of their specialization, without the cooperation of generalists which we may call system engineers.

Specialists have a great deal in common with Sherlock Holmes who told Watson in the first of these unforgettable stories that he had to guard against admitting extraneous information into his mind for it would crowd out information important to his profession. If an enterprise were to choose to expand and generalize the knowledge base of their specialized engineers to eliminate the need for any overlying coordinating function, they will, in general, become less competitive because their specialists will not, over time, be as sharp in their special fields as their more finely focused counterparts in competing firms.

Some time during 1968 in the officer's club at Bien Hoa Air Force Base, South Vietnam, the author once debated the relative merits of specialization

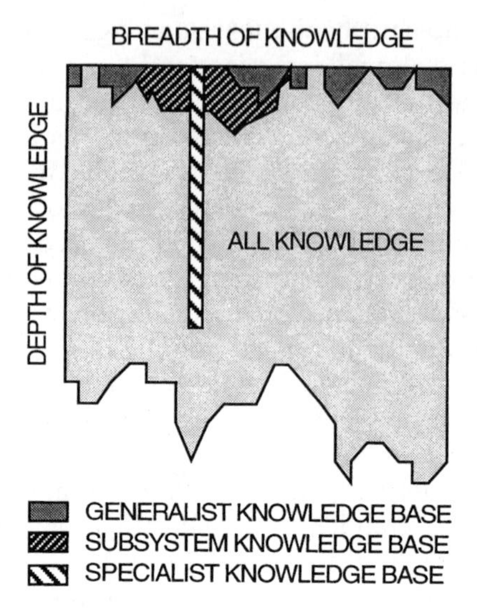

Figure 1-1 We are all specialists.

and generalism with an Air Force captain for several hours. At the time the author was a field engineer representing Teledyne Ryan Aeronautical serving with the Strategy Air Command which was operating a series of unmanned photo reconnaissance aircraft out of Bien Hoa. His job, at the time 10,000 miles from the plant and often alone, was to support the customer for all systems in all active models and related launch aircraft equipment for maintenance and operations purposes so, by necessity, he drifted into an appreciation for generalism. The captain ended the conversation that evening with the comment, "Now let me see if I've got this straight, Jeff. You are specializing in generalism." As deflating as this comment was at the time, the author has since grudgingly concluded that the captain was absolutely correct. We are all specialists with a different balance point between knowledge breadth and depth as illustrated in Figure 1-1.

We have our choice of knowing a great deal about one or a few things or a little about many things. This is the basis for the oft-quoted statement about the ignorance of system engineers relative to specialists, "They are a mile wide and an inch deep." Yes, it is true, and it is necessary. The alternatives, qualitatively clear from Figure 1-1, are that individuals may apply their knowledge capacity in either direction but not both in a comprehensive way. In the development of systems, we need both the narrow specialist to develop ideas appropriate to the solution of specific problems and generalists skilled in fitting all of the specialized views of a system into a whole. In very complex problems and systems, this hierarchy of knowledge granularity may require three or four distinct levels.

Over time, humans have devised this very organized process, the systems approach, for solution of complex problems that orchestrates a progressive refinement of the solution derived by a group of specialists, none of whom is individually capable of a complete solution. A complex problem is broken into a set of smaller problems each of which can be efficiently attacked by small teams of specialists. To ensure that all of the small problem solutions will combine synergistically to solve the original complex problem, we first predefine solution characteristics called requirements that are traceable to the required characteristics for the larger problem, all of which are derived from the customer need, the ultimate requirement.

The resultant logical continuity gives us confidence that the small problem solutions will fit together to solve the large problem. Validation is applied during concept development and early design work to ensure that specific requirements are necessary, optimally valued, and that it is possible to satisfy them. Verification work is applied subsequent to the design work on test articles and early production items to produce evidence that the design solutions do, in fact, satisfy the requirements.

The rationale for decomposing large problems into many related smaller ones is based on two phenomena. First, as noted above, a very large problem will commonly appeal to many different technologies. Complex problem decomposition results in smaller problems each one of which appeals to a smaller number of different technologies than necessary for the large problem. Therefore, fewer different specialists are needed and the coordination problem is reduced from what it would be for direct solution of the larger problem. Second, breaking large problems into smaller ones encourages an organization size and granularity consistent with our span of control capabilities. The systems approach is, therefore, driven by perfectly normal individual human problems of knowledge and span of control limitations. It is a method for organizing work of a complex nature for compatibility with human capabilities and limitations.

To master all knowledge, we have found it necessary to specialize. Because we specialize, we need an organized approach to encourage cooperative action on problems so complex that no one person can solve them. System engineering work is very difficult to do well because it does require cooperative work on the part of many people, whereas specialists generally have the luxury of focusing on their own specialized knowledge base. System engineering work entails an outgoing interest direction from the practitioner while specialized work tends toward an inward-looking focus. System engineering promises to become more difficult to perform well in the future because our individual knowledge limitation is fixed and our knowledge base continues to expand. This growing problem is made more serious because more-refined specialization is relatively easy to implement, but the compensating improvements in our ability to integrate and optimize at higher levels is more difficult to put in place. The former is oriented toward the individual. The latter requires groups of people and is thus more difficult.

It is this difficulty based on many people cooperating in a common responsibility that has caused some development organizations to evolve a very rigorous, serial, and military-like or dictatorial environment within which the work of humans is forced into specific pigeonholes clearly defined in terms of cost and schedule allocations and the humans themselves devalued. The original intent for the systems approach was to encourage cooperative work among many specialists concurrently, but in practice the process has not been applied well in many organizations focused on functional departments and a serial work pattern. This book encourages what is now called the concurrent development process performed through cross-functional teams on programs with functional department support in terms of trained personnel, good written practices, and good tools all incrementally improved in a coordinated way over time.

It is understood that these people, once assigned to the program, will be organized by the program into product-oriented teams and managed on a day-to-day basis by the program and not by functional management. This is necessary in order to avoid the single most difficult problem in the development of systems, cross-organizational interface responsibility voids. We organize by product to cause perfect alignment between the interfaces between items in the system hierarchy and the human communication patterns that must take place to resolve these interfaces resulting in clear interface development accountability.

The work described in this book requires the cooperation of many specialists involved in different technology disciplines and company functional areas. This work requires the capture of a tremendous amount of information that must be accessible to and reviewed by many people. The amount of information involved and the broad need for access suggests the use of distributed databases within a client-server structure and that conclusion will be reinforced throughout the book.

1.4 System development phasing overview

Customers for very large and complex systems like the DoD and the National Aeronautics and Space Administration (NASA) have, over a period of many years, developed very organized methods for acquisition of those systems. Commonly, these methods entail a multiphased process that partitions complex problem solving into a series of steps or phases each ending in a major review where it is determined whether or not the planned progress has been achieved before committing to expenditure of additional funding to carry on into the next phase where each phase involves increasing funding demands and potential risk of failure. The purpose of the phasing is to control program risk by cutting the complete job into manageable components with intermediate decision points that allow redirection at critical points in the development process.

Since these two agencies are government bodies, they must function within a growing volume of laws, rules, and regulations intended to preserve a level playing field for businesses seeking to satisfy their needs, maximize the benefits to the country from procurements, and discourage failed procurements. The use of public funds properly carries with it burdens with which commercial enterprises need not contend. Therefore, commercial enterprises should properly avoid some of the extensive oversight and management-oriented methods applied to government programs, but they should think carefully before they throw the whole process out the window including the technical processes useful in both military and commercial product development.

Figure 1-2 offers a generic phasing diagram suitable for companies with a varied customer base with overlays for DoD and NASA procurements. Each phase is planned to accomplish specific goals, to answer specific questions, and control future risks. Each phase is terminated by a major review during which the developer and, in the case of a DoD or NASA customer, the customer reach an agreement on whether or not phase goals have been satisfied.

Generally, the information developed in a phase is approved as a package and used to define a baseline serving as a platform for subsequent phase work. In this way, controlled progress is made toward solving the original complex problem represented by the customer need.

The development of commercial products has commonly been accomplished by companies in a very different environment. There appears to be, among engineers and managers in commercial enterprises, a great dislike of the traditional DoD-inspired systems approach to complex problem solving. The author is a persistent student of these attitudes but has not mastered what he believes to be the real basis for them. In talking to people in the commercial development world about their product development process, one finds one of two possibilities to be in effect: (1) they cannot explain their process, suggesting with which it is an *ad hoc* process that just happens fueled by the personalities at work and the thoughts of the moment or (2) they explain a process that most anyone experienced in the DoD approach could identify, possibly with some semantic differences.

The degree to which the process is predefined aside, the commercial world commonly does not have the benefit of customers who speaks to them clearly about their needs. Some would argue that this does not often happen with large government customers either, especially as it relates to the clarity of the message. But it is true that DoD and NASA developments proceed from a customer position on the need. In commercial developments, the customer is seldom a homogeneous whole with a single spokesman. Developers must make a real effort to uncover who the customer is and what will sell to that customer. They must play the part of the customer as well as the producer. They must have a built-in voice of the customer that accurately reflects real customer interests.

The author maintains that the program spark may be different between the DoD and commercial worlds but that what evolves thereafter need not be so very different. The fundamentals of the DoD-inspired approach are valid no matter the situation so long as we are interested in solving complex problems that require the knowledge of many people in specialized disciplines. The most efficient way to solve such problems that we currently understand is to decompose them into a related series of smaller problems

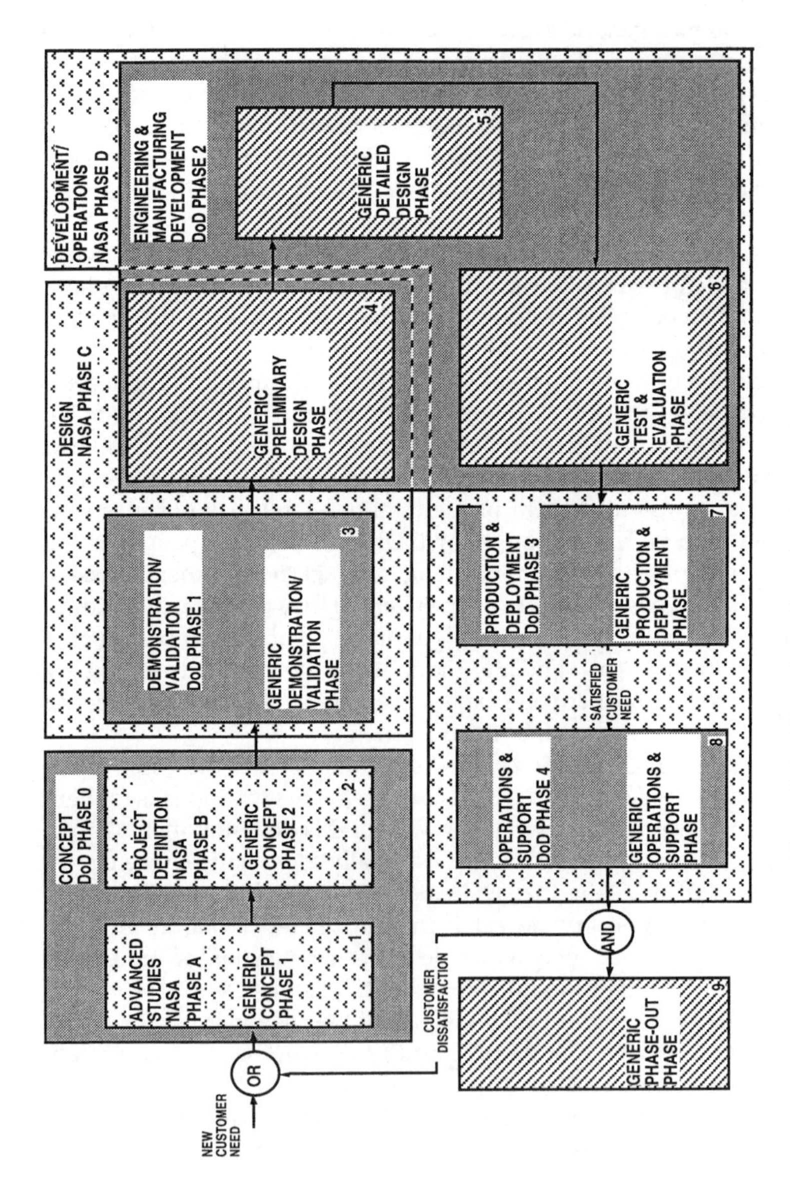

Figure 1-2 Generic system development process.

as discussed above. So, it is problem complexity and not the nature of the customer that defines the need for an effective system engineering process.

We could also dig deeper and inquire just how different the beginning is between commercial and government programs. In a commercial enterprise, particularly in the consumer products area, a good deal of effort must be made to try to understand potential market opportunities in terms of what will sell and who might buy their products. This is done through product questionnaires received with the product, focus groups, market studies, analysis of statistical data from government agencies and polling firms, historical trends, and lucky guesses.

If you scratch the surface of a successful DoD contractor, you will find activities that you might expect to be going on in a commercial firm to understand the customer. You will find the marketing department and its supporting engineering predesign group working interactively with one or more customer organizations, either cooperatively or with some separation, to understand the customer's needs, offer suggestions for solving those they perceive, and to remain abreast of customer interests. This is the basis of the often repeated saying that, "If you first become aware of a program in the *Commerce Business Daily* (CBD), it is too late to bid." Federal government agencies are required to announce any procurement over a certain dollar value in the CBD, which comes out every workday. The rationale behind the saying is that the future winner of the program noted in the CBD probably helped the customer understand that the need existed through months or years of close cooperative work and related independent research and development (IRAD) work to control technical risks and demonstrate mastery of needed technologies.

So, the fundamental difference between DoD and commercial development may be as simple as the degree of focus and leadership within the two customer bases. In the commercial world, especially for consumer products and less so for large commercial projects like large buildings, commuter rail line construction, and oil exploration, the voice of the customer is scattered and uncoordinated. In DoD procurements, the voice of the customer is focused at a point in the government program manager. Yes, it is true that there are many different expressions of the DoD customer voice in the form of the procurement agency, government laboratories, and multiple users, but the customer voice is focused in the request for proposal or contract under the leadership of the program manager. Not so in commercial industry.

1.5 Toward a standard process

Enterprises evolve over time to serve a particular customer base offering a relatively narrow product line. In most enterprises, even those with multiple kinds of customers (government and commercial, for example) a single generic process can be effective. This is a valid goal because the employees will progress toward excellent performance faster through repetition than when they must apply different techniques on different programs. It is also difficult to engage in continuous improvement when you are maintaining more than one fundamental process. A single process encourages development and improvement of personnel training programs, documented practices, and tools keyed to the training and practices. It is hard to imagine why one enterprise would want to maintain multiple processes, but some do, reinventing themselves for each new proposal based on bad feelings about their previous program performance.

Figure 1-2 provides one view of a generic process diagram emphasizing the product and process development activity. This process follows the traditional sequential, or waterfall model, at a very high level involving requirements before design and verification of design compliance with those requirements. One must recognize that on any given program, at any given time, some items may have completed the requirements analysis process and have moved to the design process while other requirements work continues for other items.

Figure 1-3 provides a standard phasing diagram within which a standard process functions to achieve program phase goals. Figure 1-4 illustrates how these two views, generic process and program phasing, correlate. In each phase we accomplish a degree of development work by applying our standard process for a particular system depth and span. The next phase may require some iteration of previous work based on new knowledge suggesting a different approach providing for better system optimization. In each phase we repeat some of our generic process steps mixed with some new processes not previously accomplished.

1.6 Development environments

While a single process is encouraged, that does not mean that the organization should have no flexibility in choosing the most effective development environment for the given situation. As the reader will find below, the author encourages developing organization skill in several models and techniques from which they can choose but each of these should have an evolving and improving standard associated with it for that company.

Up through the 1960s and 1970s, there was only one accepted system development model. That was what we now call the waterfall model. Since then, the software world has evolved alternative models proven useful in the development of software under some circumstances. These include the spiral and the V models. The author also offers a variation on the V model called the N model. In all of these cases, one should note, that at the microlevel the same cycle of activities is taking place, that being: requirements analysis, design, verification. This cycle is at the heart of the systems approach. In some models discussed below, we intend that we will complete each of these steps in one pass. In others, the understanding is that we will have more than one opportunity to complete any one of the steps. But in all cases, we will pass through each step in the same order.

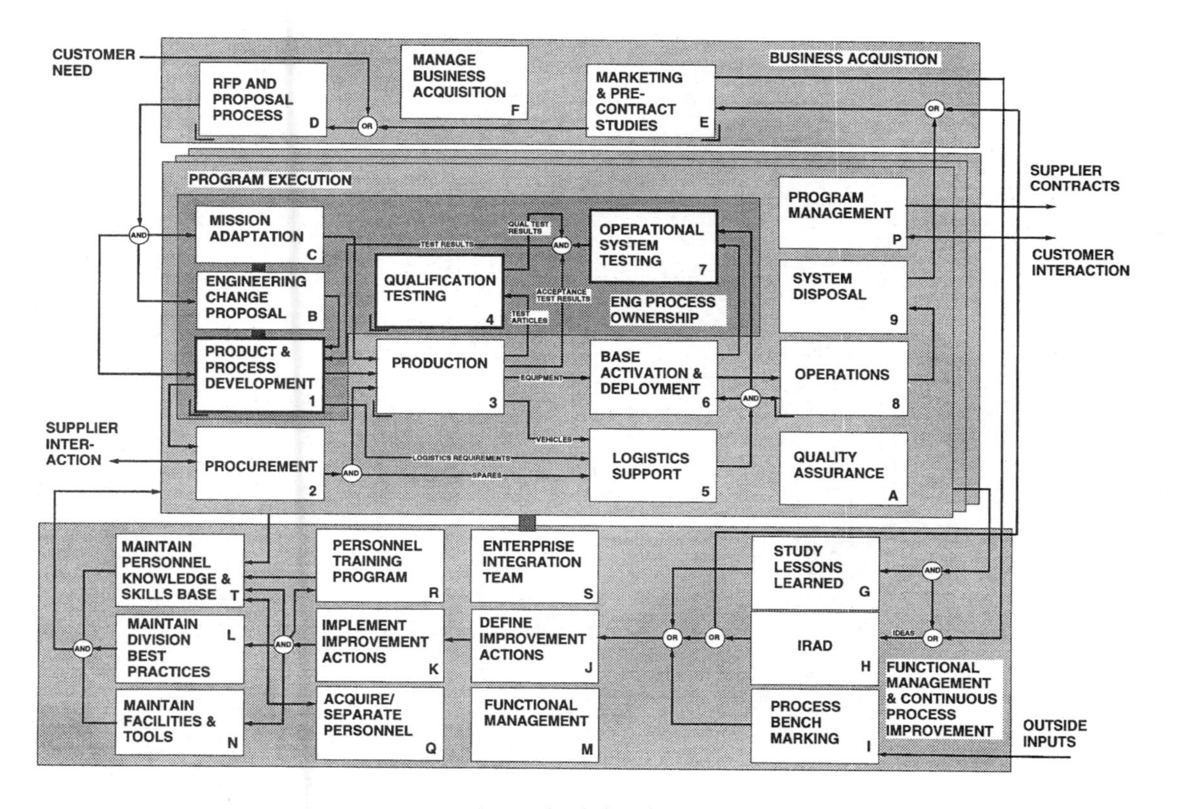

Figure 1-3 A standard development process.

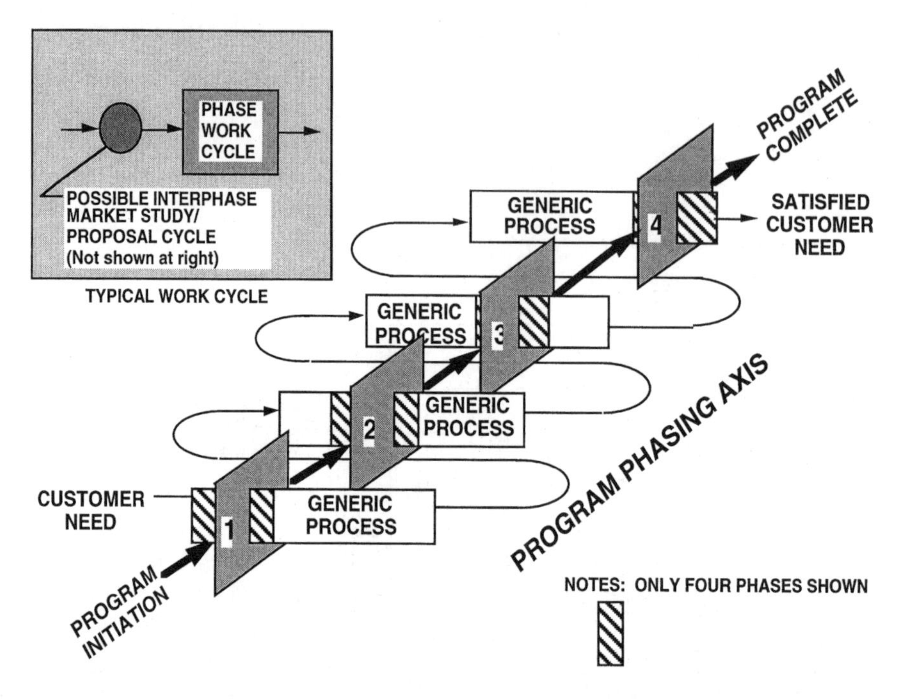

Figure 1-4 Correlation of a process model and program phasing. Generic process work accomplished in the indicated phase.

In applying these models we can choose to apply the same model throughout a single system development activity. Or, we can choose to apply one model at the system level and one or more others at the lower tier in the same development activity. So, for a given development activity, the model may be a composite, created from the models discussed below, that is very difficult to illustrate in two dimensions, but imaginable based on the pictures we will use.

1.6.1 The waterfall development model

The most commonly applied development model is called the waterfall model because of the analogy of water falling from one step to the next in Figure 1-5. In its pure configuration, each step ends abruptly coinciding with the beginning of the next step. In reality, these steps overlap because when

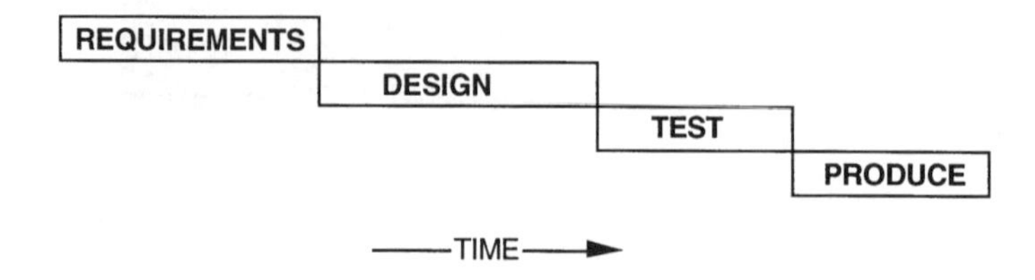

Figure 1-5 Waterfall model.

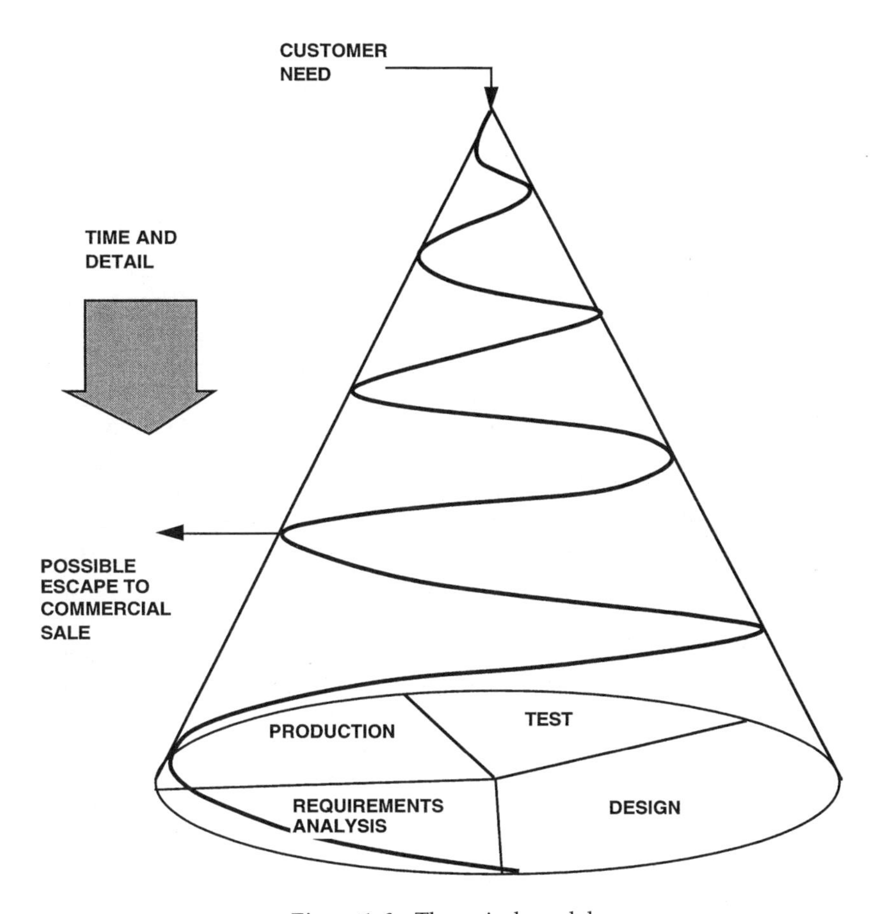

Figure 1-6 The spiral model.

the requirements are approved for one item its design work may begin while other requirements analysis work continues for other items. This overlap condition exists throughout the development span. Some people prefer to invert the sequence such that the water falls uphill, but the effect is the same.

1.6.2 The spiral development model

The spiral model, shown in Figure 1-6, illustrates the growth of information about the system between the beginning, expressed in the customer need statement, and delivery by virtue of the expanding diameter of the spiral in time. For any given item being developed, the development team first defines requirements to the extent they are capable, does some design work based on those requirements, and builds a prototype. Testing of that prototype reveals new information about appropriate requirements and design features not previously exposed. This process may be repeated several times until the team is satisfied that the product design is satisfactory.

This model is attacked by some as leading to point designs without benefit of an adequate requirements analysis activity and clear definition of the problem. That may be a valid criticism, but there are situations where we have difficulty faithfully following the structured approach suggested in this chapter. One of these situations involves the development of computer software that must be very interactive with the actions of a human operator. It is very difficult to characterize an appropriate set of requirements for such a system because of the subjective nature of the human responses to particular stimulus. The development will flow much more rapidly if the team has an opportunity to experience physically one or more alternative concepts and feed back the results into the next development iteration.

This model is also useful in describing software development through a sequence of versions. The team may conclude at some point in the development that the product is competitive relative to the products in development by rival firms and, even though it may be imperfect, it can be released to beat the competition to market. Subsequent to a version release, work continues to evolve improvements and fixes for known bugs that the customer may or may not be expected to encounter with the present release. Between product releases, customer service provides phone help to persons unfamiliar with the residual bugs and how to prevent them from becoming a bother. A product may remain in this development process for years as in the case of the word processor used in the creation of this book.

1.6.3 The V development model

The development process can be described in terms of a development downstroke and an upstroke. On the downstroke, we decompose the need into a clear definition of the problem in terms of what items the system shall consist of and the needed characteristics of those items. At the bottom of that stroke, we design the product in accordance with the requirements defined on the downstroke. Sufficient product articles are manufactured to test in accordance with test requirements defined during the downstroke. The test results qualify the product for use in the intended application.

Figure 1-7 illustrates the V model which is especially effective in exposing the close relationship between the requirements analysis process and the verification activity to prove that the design satisfies the requirements. This is a very simple view of the development process. In actuality the system expands on the second level into two or more end items or configuration items each of which expands into multiple subsystems and components. So, the V actually involves a shredout on the downstroke and an integrating stream on the upstroke that is not clear from the simplistic one-dimensional diagram shown below. For a more expansive display of this view, see a paper by Kevin Forsberg and Harold Mooz titled "The Relationship of Systems Engineering to the Project Cycle" in *A Commitment to Success*, the proceedings from a joint symposium sponsored by the National Council on Systems

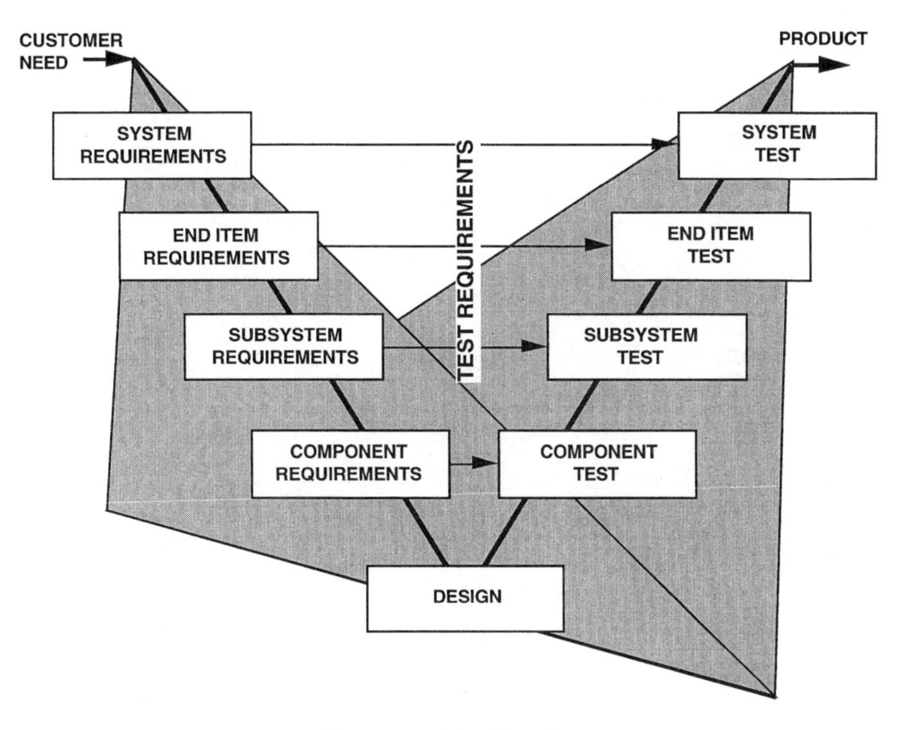

Figure 1-7 The V model.

Engineering and the American Society for Engineering Management in Chattanooga, Tennessee in October 1991.

1.6.4 The N development model

The author prefers to split the downstroke of the V model into two components to emphasize the fundamental difference between the functional and physical models useful in characterizing systems under development. This results in a forked V or an N model illustrated in Figure 1-8. The functional plane corresponds to the functional analysis process through which we gain an insight into what the system must do. As explained below, we allocate exposed functionality to physical things and assemble them into an architecture in the physical plane. The allocated functionality provides the transform between the functional model and physical model, and for each allocation we are obligated to define one or more performance requirements for the item to which we allocated the functionality. The design constraints, we shall describe shortly, are defined in the physical plane once it has been determined what kind of items shall be included in the system. The functional model terminates when we have completed our architecture definition to the degree that we understand the problem in terms of a physical architecture all items of which at the lowest tier of each branch will either yield

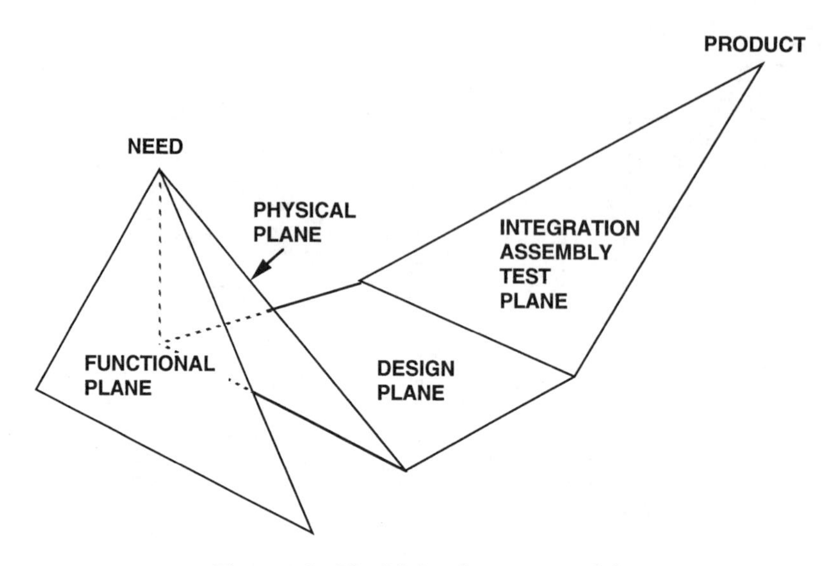

Figure 1-8 The N development model.

to purchase from vendors or yield to design actions by a team of people in our own enterprise.

The completion of the functional and physical model plane work clearly defines the problem that must be solved. We assign responsibility for development of the products represented by the blocks of the architecture diagram to teams at some level of granularity to accomplish the integrated design work for those items. These are the small problems we defined as a result of decomposing the large problem represented by the need. As the design team product designs become available, we manufacture test articles and subject them to planned tests and analyses to determine if they conform to the requirements previously defined. The final result is a product that satisfies the original need.

1.6.5 Development environment integration

Figure 1-9 is an attempt to identify every conceivable development environment from which one could select the desired environment for a particular development activity. It does so through a three-dimensional Venn diagram showing combinations of different sequences (waterfall, spiral, and V), different phasing possibilities (rapid prototyping vs. rigorous phasing), and one of three development attitudes: grand design, incremental, and evolutionary.

In the grand design approach, the team develops the product in a straight-through process from beginning to end in accordance with a well-defined, predetermined plan. This attitude is normally well matched to the waterfall model. In the incremental approach, the final requirements are defined before design work begins but foreseen problems, such as immature technology, prevent a straight-line approach to solution. Therefore, a series

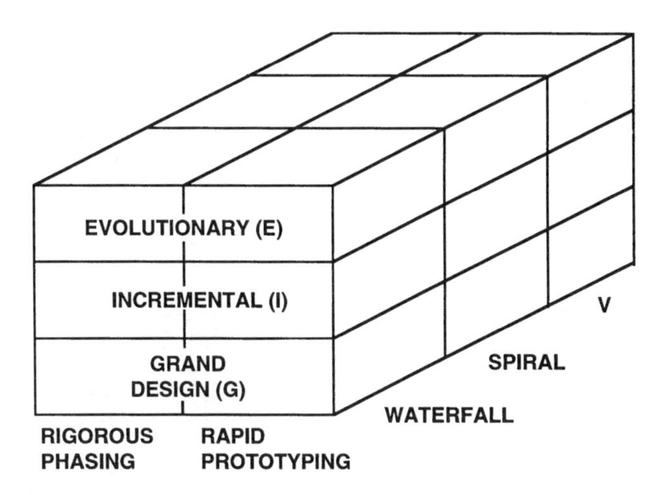

Figure 1-9 Development environment space.

of two or more builds are planned where the first satisfies the most-pressing requirements for the product. Each incremental build permits the developer and customer to gain experience with the problem and refinement of the next design cycle, ever working toward the final design configuration.

In the evolutionary approach, we may be unclear about the requirements for the final configuration and conclude that we can only acquire sure knowledge of our requirements through experience with the product. This is, of course, a chicken-and-egg kind of problem or, what others might call, a bootstrap problem. In order to understand the problem space, we have to experience the solution space. The overall evolutionary program is structured in a series of builds and each build helps us to understand the final requirements more clearly as well as a more-fitting design solution. Note the unique difference between the incremental and evolutionary approaches. In the incremental we understand the final requirements and conclude that we cannot get there in one build (grand design). In the evolutionary case, we do not understand the final requirements and need experience with some design solution to help us understand them.

Earlier we said that a unique enterprise should have a generic process used to develop product and should repeat that process with incremental improvements as one element of their continuous improvement method. We could choose to close our minds to flexibility and interpret this to mean that we should pick one of the spaces of Figure 1-9 and apply only that environment in all possible situations. Alternatively, we can choose to allow our programs and development teams to apply an environment most suited to the product and development situation as a function of their unlimited choice. Neither of these extremes is the best choice.

Figure 1-10 offers one way to provide guidance between these extremes. Our internal system engineering manual could include this sketch as a means to define alternatives available without prior approval from enterprise

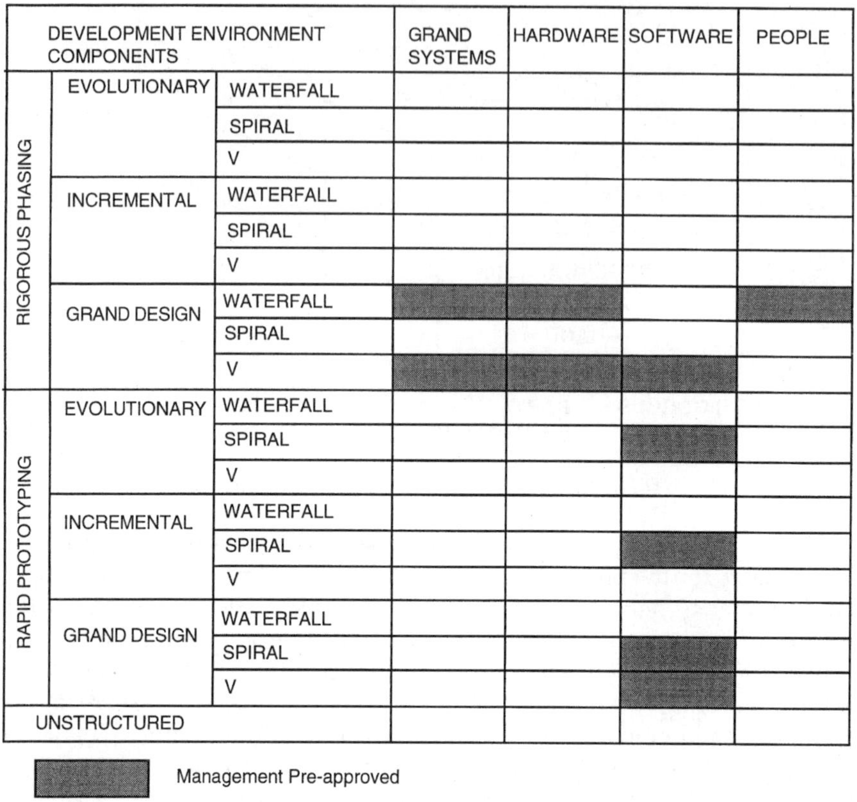

DEVELOPMENT ENVIRONMENT COMPONENTS			GRAND SYSTEMS	HARDWARE	SOFTWARE	PEOPLE
RIGOROUS PHASING	EVOLUTIONARY	WATERFALL				
		SPIRAL				
		V				
	INCREMENTAL	WATERFALL				
		SPIRAL				
		V				
	GRAND DESIGN	WATERFALL				
		SPIRAL				
		V				
RAPID PROTOTYPING	EVOLUTIONARY	WATERFALL				
		SPIRAL				
		V				
	INCREMENTAL	WATERFALL				
		SPIRAL				
		V				
	GRAND DESIGN	WATERFALL				
		SPIRAL				
		V				
UNSTRUCTURED						

Management Pre-approved

Figure 1-10 Preferred spaces example.

management. We might ask why programs should be restricted in this fashion. The reason is that we need to maintain proficiency of our workforce in a particular process. If each program is allowed to apply any combination desired and for which the working force is not already skilled, then the enterprise may have great difficulty evolving a very effective process and skilled workforce.

Figure 1-10 forms a matrix of the 18 different environments displayed in Figure 1-9, plus it recognizes the unstructured or *ad hoc* possibility (noted for completeness but discouraged) for a total of 19 environments. The other matrix axis is for the kind of element involved. Grand systems are composed of two or more of the kinds of things listed or, at the time, we do not yet know how the system will be implemented. This figure is marked up with one possible permission set indicating that company teams may apply the Rigorous Phasing, Grand Design, or Waterfall environment for the development of Grand Systems, Hardware, and Procedural (People) elements. It also approves five different development environments for computer software. A particular enterprise may find a different set of environments better matches their needs.

The number of different environments preapproved by management in an enterprise should have something to do with at least three factors. As the number of programs in the house at any one time increases we can probably afford to preapprove a greater number of environments. This is because on any one program it is more likely that the staff will include people with experience in the environments selected for that program from a simple statistical perspective. Second as the diversity of the program technologies increases in the firm, we need to tolerate a broader range of possibilities. If, for example, we have been only developing hardware in the past and now conclude that we will also develop much of our own software related to that hardware, we will need to adopt one or more environments useful in software development rather than try to force fit our software development activities into the same mold used for hardware. Finally, the difficulty and degree of precedence in the problems we are called upon by customers to solve has a bearing as well. For example, if in the past we have only solved problems where a direct, structured approach will work, we may have settled on the Rigorous Phasing, Grand Design, Waterfall environment. If now we find ourselves involved with customers with problems about which little is known, we may have to apply an environment that permits us and the customer to gain experience as we work out their solution.

Figure 1-10 exposes us to a degree of complexity that we may not be comfortable with. We may be much happier with our ignorance and prefer relying on our historical work patterns. Some companies will reach this conclusion and find themselves in great trouble later. If there is anything we know about the future, it is that the pace of change will increase. A healthy firm interested in its future will evolve a capability that blends a generic process encouraging repetition of a standard process with a degree of flexibility with respect to allowable development environments.

These goals are not in conflict. In all of the environments discussed, we rely on requirements analysis as a prerequisite to design. In some cases, we know we may not be successful in one pass at the requirements analysis process but the same techniques described in the next paragraph will be effective in any of these environments.

1.7 System requirements analysis

This book is about requirements that play a key role in the development process as explained above. In the English-speaking world, requirements are phrased in English-sentences that cannot be distinguished structurally from English sentences constructed for any other purpose. It is relatively easy to write requirements once you know what to write them about. A requirement is simply a statement in the chosen language that clearly tells an expectation placed on the design process prior to implementing the creative design process. Requirements are intended to constrain the solution space to solutions which will encourage small problem solutions that synergistically work together to satisfy the large problem solution. Requirements are formed from the words and symbols of the chosen language. They include all of the standard language components arranged in the way that a good course in that language, commonly studied in the lower grades in school, specifies. Those who have difficulty writing requirements experience one of three problems: (1) fundamental problems expressing themselves in the language of choice that can be corrected by studying the language, (2) technical knowledge deficiency that makes it difficult to understand the technical aspects of the subject about which the requirements are phrased which can be corrected through study of the related technologies, and (3) difficulty in deciding what to write requirements about.

The solution to the latter problem is the possession, on a personal, company, and program basis, of a complete toolbox of effective requirements analysis tools and the knowledge and experience to use the tools effectively. Providing the toolbox is the function of the system requirements analysis process. This toolbox must encourage the identification of product characteristics that must be controlled and selection of numerical values appropriate to those characteristics. So, this toolbox must help us to understand what to write requirements about. It gives us a list of characteristics that we should control to encourage synergism within the evolving system definition. If the analyst had a list of characteristics and a way to value them for specific items, language knowledge, and command of the related technologies, he/she would be in a very good position to write a specification.

One approach to a solution to this problem on what to write requirements about is boilerplate and this is essentially what specification standards such as MIL-STD-490A provided for many years and its replacement in MIL-STD-961D, Appendix A is likely to do for many more within the context of military systems. In this approach, you have a list of paragraph numbers and titles and you attempt to define how each title relates to the item of interest. This results in a complete requirement statement that becomes the text for the paragraph stating that requirement. The problem with this approach is that there are many kinds of requirements that cannot be specifically identified in such a generic listing. One could create a performance requirements boilerplate that covered every conceivable performance requirement category with some difficulty only to find that it was more difficult to weed out those categories that did not apply to a specific item than it would have been to have determined the appropriate categories from scratch. This is why one would find no lower level of detail in a specification standard than performance even though there may evolve 50 pages of performance requirements during the analysis. Interfaces are another area where boilerplate is not effective at a lower level of detail.

Many design engineers complain that their system engineers fail to flow down the customer requirements to their level of detail in a timely way. Their solution is to begin designing at their level of detail without requirements because their management has imposed a drawing release schedule they dare not violate. These engineers are often right about the failure of their system engineers but wrong to proceed in a vacuum as if they know what they are doing. The most common failure in requirements analysis is the gap between system requirements defined in a system specification given to the contractor and the component level requirements in the performance requirements area. In these organizations in trouble, the goal in the engineering organization is often stated as getting the system performance requirements flowed down to the design level or deriving design requirements from system requirements. Unfortunately, no one seems to know quite how to do that and avoidable errors creep into the point design solutions in the void that develops.

Flowdown is a valid requirements analysis strategy, as is boilerplate (which the author refers to as one form of cloning); it just is not a completely effective tool across all of the kinds of requirements that must be defined and particularly not effective for performance requirements. Flowdown is composed of two subordinate techniques: allocation or partitioning and derivation. Allocation is applied where the child requirement has the same units as the parent requirement and we partition the parent value into child values in accordance with some mathematical rule. Where the child requirement has different units from the parent requirement, the value may have to be derived from one or more parent requirements through parametric analysis, use of mathematical models, or use of simulation techniques.

The problem in the flowdown strategy is that you must have a set of parent requirements in order to perform requirements analysis. One may ask, where did the parent requirements come from? This difficulty is solved by applying some form of modeling that leads to the identification of characteristics that the item must have. The model should be extensible with respect to the problem represented by the system being developed such that it can be used to identify needed functionality and be iteratively expanded for lower tiers of the problem. Functional analysis (described later in this chapter) is an example of a means to identify performance requirements. The "flowdown" occurs through the functional analysis activity as an integral part of accomplishing the analysis. Once the physical hierarchy of the product structure is established through functional analysis and allocation, other models can be applied to identify design constraints and appropriate values in interface, environmental, and specialty engineering areas. In the latter two cases, requirements allocation can be very effective in a requirements hierarchy sense. The N development model explained above is intended to draw attention to this phenomenon.

Our toolbox should help us understand what to write requirements about, what characteristics we should seek to control. Once we have this list of characteristics, we must pair each with an appropriate numerical value and weave words around them to make the meaning very clear. We need to quantify our requirements because we will want to be able to determine whether or not a particular design solution satisfies the requirement. This is very hard to do when the requirement is not quantified. As we shall see, this

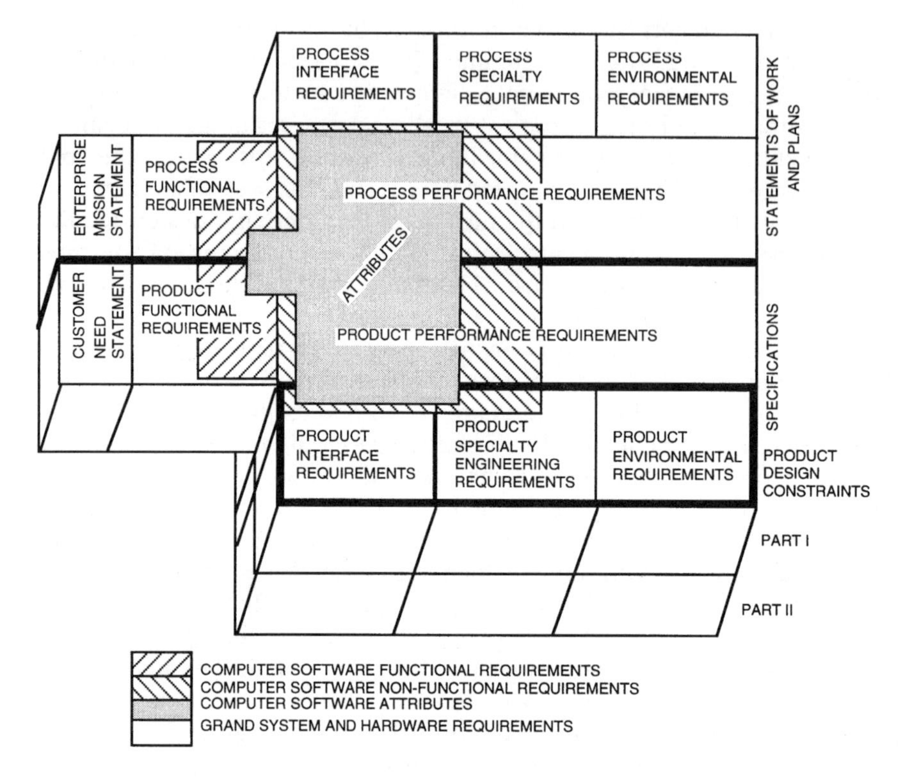

Figure 1-11 Requirements taxonomy.

is the reason the author encourages simultaneous definition of requirements and the companion test and analysis, or verification, requirements for proving the solution satisfies them.

Our system requirements analysis process should clearly tell us what kinds of characteristics we should seek to write requirements about and provide a set of tools to help us add characteristics to our list. We also need to have the means to quantify them. We seek a list of characteristics that is complete, in that it does not omit any necessary characteristics, and minimized, since we wish to provide the maximum possible solution space for the designer. We would like to have a more specific understanding of how this requirements analysis solution is characterized by completeness and minimization, but there is no easy answer. The best prescription the author can offer is to apply an organized approach that connects specific tools up with specific requirements categories and apply the tools with skill based on knowledge derived in practice or through training.

Figure 1-11 offers an overall taxonomy of every kind of requirement that one would ever have to write in the form of a three-dimensional Venn diagram. The top layer, Part I in MIL-STD-490A or the performance specifications in the context of MIL-STD-961D, corresponds to development requirements, often called *design-to* requirements that must be clearly understood before design. The lower layer, Part II in 490A context or detailed specifications under 961D, corresponds to product requirements, commonly called *build-to* requirements. The portion of the construct below the heavy middle line is for product specifications in which we are primarily interested in this book. The requirements above this line correspond to process requirements captured in statements of work and plans. Many of the same techniques discussed in this book for specifications apply to process requirements but most of the book specifically concerns product specifications and proving that the design satisfies their content.

Development requirements can be further categorized as shown in Figure 1-11 as performance requirements and design constraints of three different kinds. Performance requirements tell what an item must do and how well it must do it. Constraints form boundary conditions within which the designer must remain while satisfying the performance requirements. The toolbox offered in this book evolves all performance requirements from the customer need statement, the ultimate system function, through a product functional analysis process that identifies functional requirements. These functional requirements are allocated to things in the evolving system physical model, commonly a hierarchical architecture, and expanded into performance requirements for that item. Functional analysis provides one set of tools. We will identify three other sets for the three kinds of constraints.

As noted in the figure, verification requirements are not illustrated in the diagram because they are paired with the item requirements in our imagination. One could construct another dimension for design and verification requirements but it would be very hard to envision and draw in two dimensions. This solution corresponds to the fundamental notion that flows through this whole book that a design requirement should always be paired with a verification requirement at the time the design requirement is defined. The reason for this is that it results in the definition of much better design requirements when you are forced to tell how you will determine whether or not the resultant design has satisfied that design requirement. If nothing else, it will encourage quantifying requirements because you will find it very difficult to tell how it will be verified without a numerical value against which you can measure product characteristics.

Before moving on to the toolbox for the other categories illustrated in Figure 1-11, we must agree on the right timing for requirements analysis. Some designers reacted to the rebirth of the concurrent design approach with acceptance because they believed it meant that it was finally okay to develop requirements and designs simultaneously. That is not the meaning at all, of course. The systems approach, and the concurrent development approach that has added some new vocabulary to an old idea, seeks to develop all of the appropriate requirements for an item prior to the commitment to design work. We team the many specialists together to first understand the problem phrased in a list of requirements. We then team together to create a solution to that problem in the form of a product design and coordinated process (material, manufacturing, quality, test, and logistics) design. The concurrency relates to the activity within each fundamental step in the systematic development of product accomplished in three serial steps, requirements, design, and verification, not to all of these steps simultaneously. Even in the spiral sequence, we rotate between these steps serially repeatedly.

It is true that we may be forced by circumstances to pursue design solutions prior to fully developing the requirements (identifying the problem) as discussed under development environments, above, but requirements before design is the intent. A case where this is appropriate is when we wish to solve a problem before we have sufficient knowledge about the problem and it is too hard to define the requirements in our ignorance. A physical model created in a rapid prototyping activity can provide great insights into appropriate requirements that would be very difficult to recognize through a pure structured approach.

1.7.1 The need and its expansion

The very first requirement for every system is the customer's need which is a simple statement describing the customer's expectations for the new, modified, or reengineered system. Unfortunately, the need is seldom preserved once a system specification has been prepared and thus system engineers fail to appreciate a great truth about their profession. That truth is that the development process should be characterized by logical continuity from the need throughout all of the details of the system.

The tools discussed in this section are designed to expand upon the need statement to characterize the problem space fully. The exposed functionality is allocated to things and translated into performance requirements for those things identified in the physical plane. We then define design constraints appropriate to those things. All of our verification actions will map back to these requirements.

The need statement seldom provides sufficient information by itself to ignite this analytical process so may be preceded by efforts to understand the customer's need and the related mission or desired action or condition. In military systems this may take the form of a mission analysis. In commercial product developments we may seek to understand customer needs through surveys, focus groups, and conversations with selected customer representatives.

This early work is focused on two goals. First, we seek to gain expanded knowledge of the need. In the process we seek to ignite the functional decomposition process targeted on understanding the problem represented by the need and making decisions about the top-level elements of the solution by allocating top-level functionality to major items in the architecture. Some system engineers refer to this process as requirements analysis but it is more properly thought of as the beginning of the system development process involving some functional analysis, some requirements analysis, and some synthesis and integration. Second we seek to validate that it is possible to solve the problem stated in the need. We are capable of thinking of needs that are not possible to satisfy. The problem we conceive may be completely impossible to solve or it may only be impossible with the current state of the art. In the latter case, our study may be accomplished in parallel with one or more technology demonstrations that will remove technological boundaries permitting development to proceed beyond the study.

This early activity should include some form of functional analysis as a precursor to the identification of the principal system items. These top-level allocations may require one or more trade studies to make a knowledge-driven selection of the preferred description of the problem. This process begins with the recognition that the ultimate function is the need statement and that we can create the ultimate functional flow diagram (one of several decomposition techniques) by simply drawing one block for this function F that circumscribes a simplified version of the need statement. The next step is a creative one that cannot be easily prescribed. Given this ultimate function, we seek to identify a list of subordinate functions such that the accomplishment of the related activities in some particular sequence assures that the need is satisfied. We seek to decompose the grand problem represented by the need into a series of smaller, related problems and determine what kinds of resources could be applied to solving these smaller problems in a way that their synergism solves the greater problem.

If our need is to transport assigned payloads to low Earth orbit, we may identify such lower order problems, or functions, as: F1 = Integrate and Prepare For Transport, F2 = Launch, F3 = Transport, and F4 = Release Payload. This particular sequence of lower-tier functions is keyed to the use of some kind of rocket launched from the surface of the Earth. If we chose to think in terms of shooting payloads into orbit from a huge cannon, our choice of lower-tier functionality would be a little different, as it would be if we elected the Orbital Science Corporation's Pegasus solution of airborne launch.

If our need was the movement of 500,000 cubic yards of naturally deployed soil per day, we might think in terms of past point design solutions involving a steam shovel and dump trucks or open up our minds to other alternatives thus inventing the earthmover.

So, the major functions subordinate to the need are interrelated with the mission scenario and we cannot think of the two independently. The thought process that is effective here is to brainstorm several mission scenarios and develop corresponding top-level functional flow diagrams (called master flow diagrams or zero-level diagrams by some) at the highest level for these scenarios. We may then determine ways of selecting the preferred method of stating the problem by identifying key parameters which can be used to examine each scenario. This is essentially a requirements analysis activity to define quantified figures of merit useful in making selections between alternative scenarios.

As noted above, we may have to run trade studies to select the most effective problem statement as a precursor to allocation of top-level functionality to things. We may also have to accomplish trade studies as a means to derive a solution based on facts for the most appropriate allocation of particular functions. Finally, we may have to trade the design concept for particular things conceived through the allocation process. Trade studies are but organized ways to reach a conclusion on a very difficult question that will not yield to a simple decision based on the known facts. All too often in engineering, as in life in general, we are forced to make decisions based on incomplete knowledge. When forced to do so, we should seek out a framework that encourages a thorough, systematic evaluation of the possibilities, and the trade study process offers that.

The customer need does not have to be focused on a new unprecedented problem as suggested in the preceding comments. The customer's need may express an alteration of a problem previously solved through earlier delivery of a system. The early study of this kind of problem may involve determining to what extent the elements of existing systems can be applied and to what extent they must be modified or elements replaced or upgraded. This is not a radically different situation. It only truncates the development responsibilities by including within the system architecture things that do not require new development. Those items that do require new development will respond to the structured approach expressed in this chapter at their own level of difficulty.

This early analytical process may start with no more than an expression of the customer need in a paragraph of text. The products with which we should conclude our early mission analysis work include:

a. A master functional flow diagram (or equivalent diagrammatic treatment) coordinated with a planned mission scenario briefly described in text and simple cartoonlike graphics or a simple event list.
b. An architecture block diagram defining the physical model at the top level.
c. A top-level schematic block diagram or n-square diagram that defines what interfaces must exist between the items illustrated on the architecture block diagram.
d. Design concept sketches for the major items in the system depicted on the architecture block diagram.
e. A record of the decision-making process that led to the selection of the final system design concept.
f. A list of quantified system requirements statements. These may be simply listed on a single sheet of paper or captured in a more formal system or operational requirements document.

So, this mission analysis activity is but the front end of the system development process employing all the tools used throughout the development downstroke. It simply starts the ball rolling from a complete stop and develops the beginning of the system documentation. It also serves as the first step in the requirements validation activity. Through the accomplishment of this work we either gain confidence that it is possible to satisfy this need or conclude that the need cannot be satisfied with the available resources at our disposal. We may conclude, based on our experience, that we should proceed with development, await successful efforts to acquire the necessary resources, focus on enabling technology development for a period of months or years, or move on to other pursuits that show promise. In the case of commercial products, this decision process may focus on marketing possibilities based on estimates of consumer demand and cost of production and distribution.

1.7.2 Structured decomposition

Structured decomposition is a technique for decomposing large complex problems into a series of smaller related problems. We seek to do this for the reasons discussed earlier. We are interested in an organized or systematic approach to doing this because we wish to make sure we solve the right problem and solve it completely. We wish to avoid, late in the development effort, finding that we failed to account for part of the problem that forces us to spend additional time and money to correct and brings into question the validity of our current solution. We wish to avoid avoidable errors because they cost so much in time, money, and credibility. This cost rises sharply the farther into the development process we proceed.

The understanding is that the problems we seek to solve are very complex and that their solution will require many people each specialized in a particular technical discipline. Further, we understand that we must encourage these specialists to work together to attain a condition of synergism of their knowledge and skill and apply that to the solution of the complex problem. This is not a field of play for rugged individuals except in the leadership of these bands of specialists. They need skilled and forceful leadership by a person possessed of great knowledge applicable to the development work and able to make sound decisions when offered the best evidence available.

During the development of several intercontinental ballistic missile (ICBM) systems by the U.S. Air Force, a very organized process called functional analysis came to be used as a means to thoroughly understand the problem, reach a solution that posed the maximum threat to our enemies consistent with the maximum safety for America, and make the best possible choices in the use of available resources. We could argue whether or not this process was optimum and successful in terms of the money spent and public safety, but we would have difficulty arguing with the results following the demise of the Soviet Union as a threat to our nation's future.

At the time these systems were in development, computer and software technology were also in a state of development. The government evolved a very organized method for accumulating the information upon which development decisions were made involving computer capture and reporting of this information. Specifications were prepared on the organized systems that contractors were required to respect for preparing this information. Generally, these systems were conceived by people with a paper mind-set within an environment of immature computer capability. Paper forms were used based on the Hollerith 80-column card and intended as data entry forms. These completed forms went to a keypunch operator. The computer generated poorly crafted report forms. But, this was a beginning and very successful in it final results.

This process included at its heart a tool called functional flow diagramming discussed briefly above. The technique uses a simple graphical image created from blocks, lines, and combinatorial flow symbols to model or illustrate needed functionality. It was no chance choice of a graphical approach to do this. It has long been well known that we humans can gain a lot more understanding from pictures than we can from text. It is true that a picture is worth ten to the third words. Imagine for a moment the amount of information we take in from a road map glanced at while driving in traffic and the data rate involved. All of the structured decomposition techniques employ graphical methods that encourage analytical completeness as well as minimizing the time required to achieve the end result.

While functional flow diagramming was an early technique useful in association with the most general kind of problem, computer software analysis has contributed many more-recent variations better suited to the narrower characteristics of software. U.S. Air Force ICBM programs required adherence to a system requirements analysis standard and delivery of data prepared in accordance with a data item description for functional flow diagramming. While functional flow diagramming is still a very effective technique for grand systems and hardware, it is not as effective for computer software analysis as other techniques developed specifically for it. So, our toolbox of analytical techniques should have several tools in it including functional flow diagramming and several others.

We should now return to a discussion past. Encouragement has been offered for a single process repeated each time a program is undertaken in the interest of repetition and personnel capability improvement over time. Yet, here is another chance for process divergence to add to the different development environments offered earlier. As expressed in Figure 1-10 earlier, we have our choice on the degree of flexibility we will permit our programs. We can lock them down into the code, "functional flow diagramming or die," as a way of encouraging continual improvement. Alternatively, we can permit them to select from a wider range of approved decomposition practices in the interest of program efficiency as a function of product type and team experience. Figure 1-12 offers a diagram similar to Figure 1-10 for decomposition practices preapproval marked up for a particular set of choices.

The reality is that in 1995, as this book was being written, the system development process was not a static, unchanging entity. Improvements were being made and they will continue far into the future until a dramatic insight permits someone to describe how the application of fuzzy logic, mathematics of chaos, new research into the workings of the human mind,

DECOMPOSITION PRACTICE	GRAND SYSTEMS	HARD-WARE	SOFT-WARE	PEOPLE
FUNCTIONAL FLOW DIAG.				
FUNCTIONAL HIERARCHY				
IDEF0				
BEHAVIORAL DIAG.				
FLOW CHARTING				
YOURDON-DEMARCO				
HATLEY-PIRBHIA				
OBJECT ORIENTED				
IDEF1X				

PRE-APPROVED FOR USE ON PROGRAMS

Figure 1-12 Sample decomposition tool approval matrix.

and hypermedia techniques revolutionizes the development of complex man-made systems for the better. Until then, we must be alert to opportunities to plug in proven new techniques and integrate them into our toolbox and the skill base of our people.

As in the case of development environments, it is not necessary, or even desirable, that we use exactly the same decomposition method throughout the evolving system architecture. In the beginning, we may choose to apply functional flow diagramming. As we begin allocating exposed functionality to things, we may allocate particular functionality to software. At that point the assigned team or team leader may choose to apply the Hatley–Pirbhai model knowing that this will have to be developed as real-time software. Another team may select behavioral diagramming because they will apply the Ascent Logic requirements tool RDD-100 in their requirements analysis work. Some of these techniques are more appropriate to hardware and others to software. But, the adherents of most of them will make a convincing argument why their preferred technique will work equally well for either product type.

1.7.2.1 Grand systems and hardware approaches

Figure 1-12 lists several techniques for decomposition of large problems, one of which, functional flow diagramming, has been discussed earlier. It is the author's preferred approach for grand systems because of its simplicity and generality. This process starts with the need as function F which is expanded into a set of next-tier functions which are all things that have to happen in a prescribed sequence (serial, parallel, or some combination) to result in function F being accomplished. One draws a block for each lower-tier activity and links them together in a sequence using directed line segments to show sequence. Logical OR and AND symbols are used on the connecting lines to indicate combinatorial possibilities that must be respected. This process continues to expand each function, represented by a block, into lower-tier functions. Figure 1-13 sketches this overall process for discussion.

A function statement begins with an action verb that acts on a noun term. The functions exposed in this process are expanded into functional requirements statements that numerically define how well the function must be performed. This step can be accomplished before allocation of the function statement to a thing in the physical system architecture or after. But, in any case, the allocation of the function or functional requirement obligates the analyst to write one or more performance requirements based on the allocated function for the thing to which it is allocated. This is the reason for the power of all decomposition techniques. They are exhaustively complete when done well by experienced practitioners. It is less likely that we will have missed anything compared with an *ad hoc* approach.

This process begins with the need and ends when the lowest tier of all items in the physical architecture in each branch satisfies one of these criteria: (1) the item will be purchased from another entity at that level or (2) the developing organization has confidence that it will surrender to detailed design by a small team within the company.

There are two extreme theories on the pacing of the allocation relative to the functional decomposition work. Some system engineers prefer to remain on the problem plane (refer to Figure 1-8) as long as possible to ensure a complete understanding of the problem. This may result in a seven-tier functional flow diagram before anything is allocated. In the other extreme, the analyst expands a higher-tier function into a lower-tier diagram and immediately allocates the exposed functionality. This selection is more a matter of art and experience than science, but the author believes a happy medium between the extremes noted above is optimum.

If we accumulate too much functional information before allocation, we run the risk of a disconnect between lower-tier functions and the design concepts associated with the higher-order architecture that results. If, for example, we allocate a master function for Transport Dirt to an earthmover, we may have difficulty allocating lower-tier functionality related to moving the digging and loading device (which, in our high-level design concept, is integrated with the moving device). Allocation accomplished too rapidly can lead to instability in the design concept development process because of continuing changes in the higher-tier functional definition.

The ideal pacing involves progressive allocation. We accumulate exposed functionality to a depth that permits us to analyze thoroughly system performance at a high level, possibly even run simulations or models of system behavior under different situations with different functionality combinations and performance figure-of-merit values in search of the optimum configuration. Allocation of high-order functionality prior to completing these studies is premature and will generally result in a less than optimum system and many changes that ripple from the analysis process to the architecture synthesis and concept development work. Throughout this

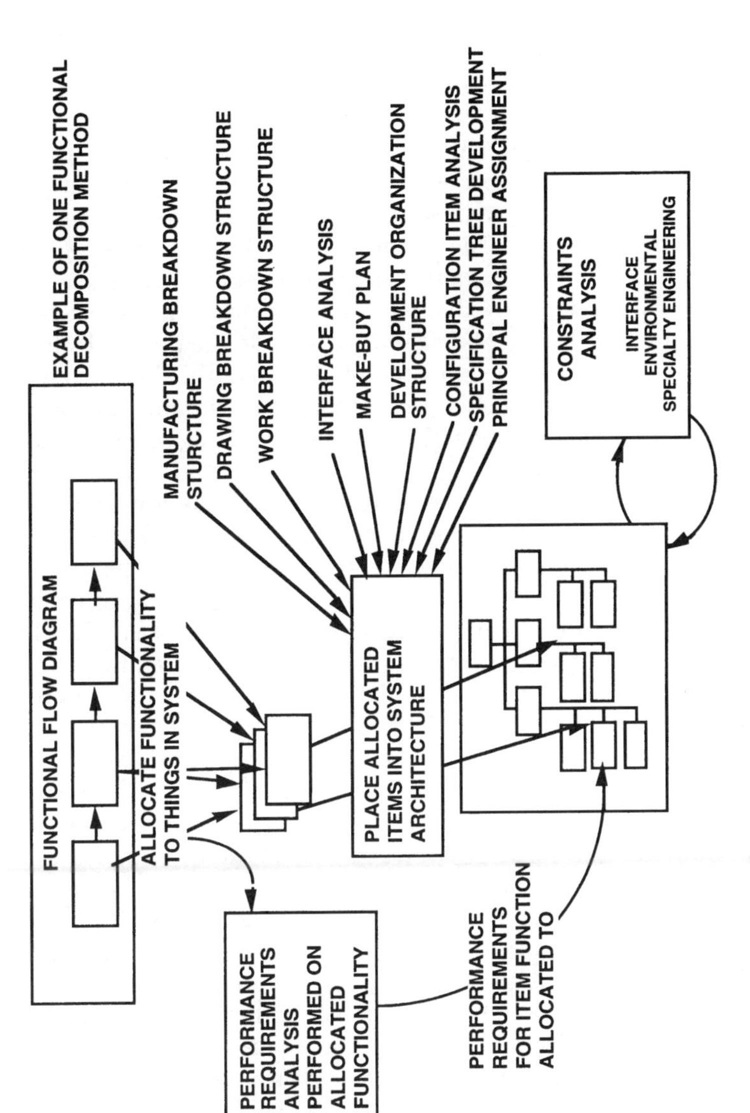

Figure 1-13 Structured decomposition for grand systems and hardware.

period we have to deal with functional requirements rather than raw function statements so, when we do allocate this higher-order functionality, it will be as functional requirements rather than raw function names. Before allocating lower-tier functionality we should allocate this higher-order functionality and validate them with preliminary design concepts. These design concepts should then be fed back into the lower-tier functional analysis to tune it to the current reality. Subsequent allocations can often be made using the raw functions followed by expansion of them into performance requirements after allocation.

Some purists would claim that this is a prescription for point designs in the lower tiers. There is some danger from that and the team must be encouraged to think through its lower-tier design concepts for innovative alternatives to the status quo. The big advantage, however, to progressive tuning of the functional flow diagram though concept feedback is that at the lower tiers the functional flow diagram takes on the characteristics of a process diagram where the blocks map very neatly to the physical situation that the logistics support people must continue to analyze. This prevents the development of a separate logistics process diagram with possible undesirable differences from the functional flow diagram. Once again, we are maintaining process continuity.

The author believes the difference between functional flow and process diagrams is that the former is a sequence of things that must happen whereas the latter is a model of physical reality. When we are applying the functional flow diagram to problem solving, we do not necessarily know what the physical situation is nor of what items the system shall consist. The blocks of the process diagram represent actual physical situations and resources.

The U.S. Air Force developed a variation of the functional flow or process diagram called the IDEF diagram. IDEF is a compound acronym with the meaning ICAM DEFinition, where ICAM = integrated computer-aided manufacturing analysis. In addition to the horizontal inputs and outputs that reflect sequence, these diagrams also have inputs at the top and bottom edges that reflect controlling influences and resources required for the steps. This diagrammatic technique was developed from an earlier SADT diagramming technique developed for software analysis and applied to the development of contractor manufacturing process analysis. It does permit analysis of a more complex situation than process diagramming, but the diagram developed runs the risk of totally confusing the user with the profusion of lines. Many of the advantages claimed for IDEF can be satisfied through the use of a simpler functional or process flow diagram teamed with a dictionary. These diagrams present a simple view that the eye and mind can use to acquire understanding of complex relationships, and the dictionary presents details related to the blocks that would confuse the eye if included on the diagram. The IDEF technique has evolved into an IDEF0 for process analysis, IDEF1X for relational data analysis, IDEF2 for dynamic analysis, and IDEF3 for process description, and several others.

Some system engineers, particularly in the avionics field, have found it useful to apply what can be called hierarchical functional analysis. In this technique, the analyst makes a list of the needed lower-tier functionality in support of a parent function. These functions are thought of as subordinate functions in a hierarchy rather than a sequence of functions as in flow diagramming. They are allocated to things in the evolving architecture generally in a simple one-to-one relationship. The concern with this approach is that it tends to support a leap to point design solutions familiar to the analyst. It can offer a very quick approach in a fairly standardized product line involving modular electronics equipment as a way to encourage completeness of the analysis. This techniques also does not support time line analysis as does functional flow diagramming since there is no sequence notion in the functional hierarchy.

Whatever techniques we use to expose the needed functionality, we have to collect the allocations of that functionality into a hierarchical architecture block diagram reflecting the progressive decomposition of the problem into a synthesis of the preferred solution. The peak of this hierarchy is the block-titled system, which is the solution for the problem (function) identified as the need. Subordinate elements, identified through allocation of lower-tier functionality, form branches and tiers beneath the system block. The architecture should be assembled recognizing several overlays to ensure that everyone on the program is recognizing the same architecture: work breakdown structure (finance), manufacturing breakdown structure (assembly sequence), engineering drawing structure, specification tree, configuration or end item identification, make–buy map, and development responsibility.

As the architecture is assembled, the needed interfaces between these items must be examined and defined as a prerequisite to defining their requirements. These interfaces will have been predetermined by the way that we have allocated functionality to things and modified as a function of how we have organized the things in the architecture and the design concepts for those things. During the architecture synthesis and initial concept development work, the interfaces can be defined for the physical model using schematic block or n-square diagrams.

Ascent Logic has popularized another technique called Behavioral Diagramming that combines the functional flow diagram arranged in a vertical orientation on paper with a data and interface flow arranged in the horizontal orientation. The strength of this technique is that we are forced to evaluate needed process and data needs simultaneously rather than as two separate and, possibly disconnected, analyses. The inclusion of architecture synthesis and interface analysis under the functional analysis umbrella does mitigate the differences somewhat. The tool RDD-100 uses this analysis model leading to the capability to simulate system operation and output the system functionality in several different views including functional flow, IDEF0, or n-square diagrams.

1.7.2.2 Computer software approaches

It is not possible for a functioning system to exist that is entirely computer software, for software requires a machine medium within which to function. Systems that include software will always include hardware, a computing instrument as a minimum, and most often will involve people in some way. Software is to the machine as our thoughts, ideas, and reasoning are to the gray matter making up our mind. While some people firmly believe in out-of-body experiences for people, few would accept a similar situation for software. A particular business entity may be responsible for creating only the software element of a system and, to them, what they are developing could be construed a system, but their product can never be an operating reality by itself. This is part of the difficulty in the development of software; it has no physical reality. It is no wonder then that we might turn to a graphical and symbolic expression as a means to capture its essence.

We face the same problem in software as hardware in the beginning. We tend to understand our problems first in the broadest sense. We need some way to capture our thoughts about what the software must be capable of accomplishing and to retain that information while we seek to expand upon the growing knowledge base. We have developed many techniques to accomplish this end over the period of 40 to 50 years during which software has been a recognized system component.

The earliest software analytical tool was flow diagramming which lays out a stream of processing steps similar to a functional flow diagram (commonly in a vertical orientation rather than horizontal probably because of the relative ease of printing them on line printers) where the blocks are very specialized functions called computer processes. Few analysts apply flow diagramming today, having surrendered to data flow diagramming (DFD), the Hatley–Pribhai extension of this technique, or object-oriented analysis. Alternative techniques have been developed that focus on the data that the computer processes. The reasonable adherents of the process and data orientation schools of software analysis would today accept that both are required and some have made efforts to bridge this gap. The most recent round of tool development focuses on a merger of these two orientations in what is called object-oriented software development. Earlier attempts included input process output (IPO) which was the basis for behavioral diagramming used by Mack Alfred in the development of RDD, and integration of entity relationship diagramming with structured analysis.

All software analysis tools (and hardware-oriented ones as well) involve some kind of graphical symbols (bubbles or boxes) representing data or process entities connected by lines, generally directed ones. Most of these processes begin with a context diagram formed by a bubble representing the complete software entity connected to a ring of blocks that correspond to external interfaces that provide or receive data. This master bubble corresponds to the need, or ultimate function, in functional analysis, and its allocation to the thing called a system. The most traditional technique was developed principally by Yourdon, DeMarco, and Constantine. It involves expansion of the context diagram bubble into lower-tier processing bubbles that represent subprocesses just as in functional analysis. These bubbles are connected by lines indicating data that must pass from one to the other. Store symbols are used to indicate a need to store temporarily a data element for subsequent use. These stores are also connected to bubbles by lines to show source and destination of the data. Since the directed lines represent a flow of data between computer processing entities (bubbles), the central diagram in this technique is often referred to as a data flow diagram.

In all software analysis techniques, there is a connection between the symbols used on the diagrammatic portrayal to text information that characterizes the requirements for the illustrated processes and data needs. In the traditional line-and-bubble analysis approach, referred to as DFDs, one writes a process specification for each lowest-tier bubble on the complete set of diagrams and provides a line entry in a data dictionary for each line and data store on all diagrams. Other diagrams are often used in the process specification to explain the need for controlling influences on the data and the needed data relationships. All of this information taken together becomes the specification for the design work that involves selection of a specific machine upon which the software will run, a language or languages that will be used, and an organization of the exposed functionality into "physical" modules that will subsequently be implemented in the selected language through programming work. A good general reference for process- and data-oriented software analysis methods is Yourdon's *Modern Structured Analysis.*[2] Tom DeMarco's *Structured Analysis and System Specification*[3] is another excellent reference for these techniques.

Much of the early software analysis tool work focused on information batch processing because central processors, in the form of large mainframe computers, were in vogue. More recently, distributed processing on networks and software embedded in systems have played a more prominent role revealing that some of the earlier analysis techniques were limited in their utility to expose the analyst to needed characteristics. Derek Hatley and the late Imtiaz Pirbhai offer an extension of the traditional approach in their *Strategies for Real-Time System Specification*[4] to account for the special difficulties encountered in embedded, real-time software development. They differentiate between data-flow needs and control flow needs and provide a very organized environment for allocation of exposed requirements model content to an architecture model. The specification consists of information derived from the analytical work supporting both of these models.

Fred McFadden and Jeffrey Hoffer have written an excellent book on the development of software for relational databases in general and client-server systems specifically titled *Modern Database Management.*[5] With this title, it is understandable that they would apply a data-oriented approach involving entity–relationship (ER) diagrams and a variation on IDEF1X. The latter is explained well in a Department of Commerce Federal Information Processing Standards Publication (FIBS PUB) 184. McFadden and Hoffer also explain a merger between IDEF1X and object-oriented analysis.

The schism between process-oriented analysis and data-oriented analysis, which had been patched together in earlier analysis methods, has been joined together more effectively in object-oriented analysis about which there have been many books written. A series that is useful and readable is by Coad and Yourdon (Volumes 1 and 2, *Object Oriented Analysis* and *Object Oriented Design*,[6] respectively) and Coad and Nicola (Volume 3, *Object Oriented Programming*[7]). Two others are James Rumbaugh et al., *Object Oriented Modeling and Design*[8] and Grady Booch, *Object-Oriented Analysis and Design with Applications.*[9] At the time this book was written, the object-oriented approach was still in the creative phase and had not become standardized.

1.7.3 Performance requirements analysis

Performance requirements define what the system or item must do and how well it must do those things. Precursors of performance requirements take the form of function statements or functional requirements (quantified function statements). These should be determined as a result of a structured analysis process that decomposes the customer need as noted above using an appropriate hardware or software technique.

Many organizations find that they fail to develop the requirements needed by the design community in a timely way. They keep repeating the same cycle on each program and fail to understand their problem. This cycle consists of receipt of the customer's requirements or approval of their requirements in a specification created by the contractor followed by a phony war on requirements where the systems people revert to documentation specialists and the design community creates a drawing release schedule in response to management demand for progress. As the design becomes firm, the design people prepare an in-house requirements document that essentially characterizes the preexisting design. Commonly, the managers in these organizations express this problem as, "We have difficulty flowing down system requirements to the designers."

The problem is that the flowdown strategy is only effective for some specialty engineering and environmental design constraints. It is not a good strategy for interface and performance requirements. It is no wonder these companies have difficulty. There is no one magic bullet for requirements analysis. One needs the whole toolbox described in this chapter. Performance requirements are best exposed and defined through the application of a structured process for exposing needed functionality and allocation of the exposed functionality to things in the architecture. You need not stop at the system level in applying this technique. It is useful throughout the hierarchy of the system.

Performance requirements are traceable to (and thus flow from) the process from which they are exposed much more effectively than in a vertical sense through the product architecture. In the context of Figure 1-11, they trace to the problem or functional plane. Constraints are generally traceable within the product or solution plane. This point is lost on many engineers and managers and thus they find themselves repeating failed practices indefinitely.

Given that we have an effective method for identifying valid performance requirements as described under functional analysis above, we must have a way to associate them with quantitative values. In cases where flowdown is effective, within a single requirement category, such as weight, reliability, or cost, a lower-tier requirement value can be determined by allocating the parent item value to all its child items in the product architecture. This process can be followed in each discipline creating a value model for the discipline. Mass properties and reliability math models are examples. In the case of performance requirements, we commonly do not have this clean relationship, so allocation is seldom effective in the same way.

Often the values for several requirements are linked into a best compromise and to understand a good combination we must evaluate several combinations and observe the effect on selected system figures of merit like cost, maximum flight speed in an aircraft, automobile operating economy, or land fighting vehicle survivability. This process can best and most quickly be accomplished through a simulation of system performance where we are allowed to control certain independent parameters and observe the effects on dependent variables used to base a selection upon. We select the combination of values of the independent variables that produces the best combination of effects in the dependent variables.

Budget accounts can also be used effectively to help establish sound values for performance requirements. For example, given a need to communicate across 150 miles between the Earth's surface and a satellite in low Earth orbit, we may allocate gain (and loss) across this distance, the antenna systems on both ends, connecting cables, receiver and transmitter. Thus, the transmitter power output requirement and the receiver sensitivity requirement are determined through a system-level study of the complete communications link.

This work involved in establishing appropriate values for performance requirements is part of the requirements validation process to the extent that it gives us confidence that the selected values are achievable within the time and money limits established for development.

1.7.4 Design constraints analysis

Design constraints are boundary conditions within which the designer must remain while satisfying performance requirements. All of them can be grouped into the three kinds described below. Performance requirements can be defined prior to the identification of the things to which they are ultimately allocated. Design constraints generally must be defined subsequent to the definition of the item to which they apply. Performance requirements provide the bridge between the problem and solution planes through allocation. Once we have established the architecture, we can apply three kinds of constraints analysis to these items. In the case of each constraint category, we need a special tool set to help us understand in some organized way what characteristics we should seek to control.

1.7.4.1 *Interface requirements analysis*

Systems consist of things. These things in systems must interact in some way to achieve the need. A collection of things that do not in some way interact is a simple collection of things, not a system. An interface is a relationship between two things in a system. This relationship may be completed through many different media, such as wires, plumbing, a mechanical linkage, or a physical bolt pattern. These interfaces are also characterized by a source and a destination, that is, two terminals each of which is associated with one thing in the system. Our challenge in developing systems is to identify the existence of interfaces and then to characterize them, each with a set of requirements mutually agreed upon by those responsible for the two terminals.

Note the unique difference between the requirements for things in the system and interfaces. The things in systems can be clearly assigned to a single person or team for development. Interfaces must have a dual responsibility where the terminals are things with different responsibilities. This complicates the development of systems because the greatest weakness is at the interfaces and accountability for these interfaces can be avoided. The opportunities for accountability avoidance can be reduced by assignment of teams responsible for development as a function of the product architecture rather than the departments of the functional organization. This results in perfect alignment between the product cross-organizational interfaces (interfaces with different terminal organizational responsibilities) and the development team communication patterns that must take place to develop them. Responsibility and accountability is very clear.

The reader is encouraged to refer to the author's *System Requirements Analysis*[10] or *System Integration*[11] for a thorough discussion of schematic block and n-square diagramming techniques. As a result of having applied these techniques during the architecture synthesis of allocated functionality, the system engineer will have exposed the things about which interface requirements must be written. Once again, the purpose of our tools is to do just this, to help us understand what to write requirements about. The use of organized methods encourages completeness and avoidance of extraneous requirements that increase cost out of proportion to their value.

Once it has been determined what interfaces we must respect, it is necessary to determine what technology will be used to implement them, such as electrical wiring, fluid plumbing, or physical contact, for example. Finally, the resultant design in the selected media is constrained by quantified requirements statements appropriate to the technology and media. Each line on a schematic block diagram or intersection on an n-square diagram must be translated into one or more interface requirements that must be respected by the persons or teams responsible for the two terminal elements. The development requirements for the two terminal items may be very close to identical such as a specified data rate, degree of precision, or wire size. The product requirements, however, will often have an opposite nature to them, such as male and female connectors, bolt hole or captive bolt and threaded bore hole, or transmitter and receiver.

1.7.4.2 Environmental requirements analysis

One of the most fundamental questions in system development involves the system boundary. We must be able to unequivocally determine whether any particular item is in the system or not in the system. If it is not in the system, it is in the system environment. If an item is in the system environment, it is either important to the system or not. If it is not, we may disregard it in an effort to simplify the system development. If it is important to the system, we must define the relationship to the system as an environmental influence.

We may categorize all system environmental influences in the five following classes:

a. *Natural Environment* — Space, time, and the natural elements such as atmospheric pressure, temperature, and so forth. This environment is, of course, a function of the locale and can be very different from that with which we are familiar in our immediate surroundings on Earth as in the case of Mars or the Moon.
b. *Hostile Systems Environment* — Systems under the control of others that are operated specifically to counter, degrade, or destroy the system under consideration.
c. *Noncooperative Environment* — Systems that are not operated for the purpose of degrading the system under consideration but have that effect unintentionally.
d. *Cooperative Systems Environment* — Systems not part of the system under consideration that interact in some planned way. Generally, these influences are actually addressed as interfaces between the systems rather than environmental conditions because there is a person from the other system with whom we may cooperate to control the influences.
e. *Induced Environment* — Composed of influences that would not exist but for the presence of the system. These influences are commonly initiated by energy sources within the system that interact with the natural environment to produce new environmental effects.

As noted above, cooperative environmental influences can be more successfully treated as system interfaces. Hostile and noncooperative influences can be characterized through the identification of threats to system success and the results joined with the natural environmental effects. The induced environment is best understood through seeking out system energy sources and determining if those sources will interact with the natural environment in ways that could be detrimental to the system.

The natural environment is defined in standards for every conceivable parameter for Earth, space, and some other bodies. The challenge to the system engineer is to isolate on those parameters that are important and those that are not and then to select parameter ranges that are reasonable for those parameters that will have an impact on our system under development. The union of the results of all of these analyses forms the system environmental requirements. It is not adequate to stop at this point in the analysis, however.

Systems are composed of many things that we can arrange in a family hierarchy. Items in this hierarchy that are physically integrated in at least some portions of their system operational use, such as an aircraft in an aircraft system or a tank in a ground combat system, can be referred to as end items. We will find that these end items in operational use will have to be used in one or more environments influenced by the system environment but, in some cases, modified by elements of the system. For example, an aircraft will have to be used on the ground, in the air through a wide range of speed and altitude, and in hangars. These are different environments that we can characterize as subsets of the system environment definition. The best way to do this is first to define the system process in terms of some finite number of physical analogs of system operation. We may then map the system end items to these process steps at some level of indenture. Next, we must determine the natural environmental subsets influencing each process step. This forms a set of environmental vectors in three space. In those cases where a particular end item is used in more than one process, we will have to apply some rule for determination of the aggregate effect of the environments in those different process steps. The rule may be worst case or some other one. This technique is called environmental use profiling.

The final step in environmental requirements analysis involves definition of component environmental requirements. These components are installed in end items. The end items can be partitioned into zones of common environmental parameters as a function of the end item structure and energy sources that will change natural environmental influences. We define the zone environments and then map the components into those zones. In the process, we may find that we have to provide an environmental control system in one or more of the zones to reduce the design difficulty of some components. We may also conclude, given that we must have at least one such system, that we should relocate one or more components into the space thus controlled. So, the environmental requirements for a component are predetermined by the zone in which it is located and we may derive its environmental requirements by copying (cloning) the zone requirements. Component environmental requirements may have to be adjusted for shipping and other non-installed situations.

In summary, this set of three tools (standards, environmental use profiling, and end item zoning) may be used to characterize the environmental requirements for all system items fully from the system down through its components. In all cases, we must be careful to phrase these requirements in terms that recognize our inability to control the ultimate natural environment.

The discussion above focuses on the effects of the environment on our system. We must also consider the effects of our system on the natural environment. This is most often referred to as environmental impact analysis. It has not always been so, but today we must be sensitive to a bidirectional nature in our environmental specification. Our efforts in the distant past were very small compared with the tremendous forces of nature. Today, the scope of our access, control, and application of energy and toxic substances is substantial, and potential damage to local and regional natural environments is a real concern. Environmental impact requirements are commonly defined in laws and regulations. We can determine in what ways the environment will be exposed to system-generated stress by evaluating all system energy sources, toxic substances used, and any exhaust products. The author prefers to classify this topic under the banner of environmental requirements analysis but it could be thought of as a specific specialty engineering discipline.

These effects must be considered both for the life of the system during its use and, perhaps more importantly, for its disposition following its useful life. There can be no better example of how difficult and dangerous this can be than the case of the nuclear weapons cleanup process seriously begun in the 1990s after the end of the nuclear confrontation between the U.S.S.R. and the United States lasting 40 years. Throughout the confrontation, both sides were so concerned with survival and the urgency of development that little thought was given to the problems that they may be causing for the future. By the time the problem could no longer be ignored, it was so substantial that a solution was more difficult than the complex scientific and technical problems that had to be solved to create it. One can be very unpopular expressing interest in system disposition during the heady system development times, especially with one's proposal or program manager. But, customers will increasingly be concerned with this matter as the problems we attempt to solve become increasingly more general and their scope more pervasive.

Computer software environmental requirements are limited. The natural environment impinges on the computer hardware within which the software functions but does not in any direct way influence the software itself in the absence of hardware deficiencies that cause computer circuits to fail due to environmental stresses. The context diagram used to expose the boundary between the software and its environment is the best place to start in determining the appropriateness of any environmental influences directly on the software. One such requirements category of interest commonly is the user environment describing how the user will interact with the software. These influences could alternatively be covered under the heading of interface requirements. One must first determine whether the operator is within or outside the system, however, and this is no small question.

The principal element in the software environment is in the form of cooperative systems exposed through interface analysis. What other software and hardware systems will supply information or use information created by the system in question? The noncooperative system component may exist in very complex software systems but is hard to characterize in terms of an example. A very common and increasingly important environmental factor is the hostile environment. This matter will more commonly be addressed as the specialty engineering discipline called security engineering, however, than as an environmental consideration.

1.7.4.3 Specialty engineering requirements analysis

The evolution of the systems approach to development of systems to solve complex problems has its roots in the specialization of the engineering field into a wide range of very specialized disciplines for the very good reasons noted earlier. Our challenge in system engineering is to weld these many specialists together into the equivalent of one all-knowing mind and applying that knowledge base effectively to the definition of appropriate requirements followed by development of responsive and compliant designs and assessment of those designs for compliance with the requirements as part of the verification activity.

There are many specialized disciplines recognized, including reliability, maintainability, logistics engineering, availability, supportability, survivability and vulnerability, guidance analysis, producibility, system safety, human engineering, system security, aerodynamics, stress, structural dynamics, thermodynamics, and transportability. For any specific development activity, some or all of these disciplines will be needed to supplement the knowledge pool provided by the more general design staff.

Specialty engineers apply two general methods in their requirements analysis efforts. Some of these disciplines use mathematical models of the system as in reliability and maintainability models of failure rates and remove-and-replace or total repair time. The values in these system-level models are extracted from the model into item specifications. Commonly, these models are built in three layers. First, the system value is allocated to progressively lower levels to establish design goals. Next, the specialty engineers assess the design against the allocations and establish predictions. Finally, the specialists establish actual values based on testing results and customer field use of the product.

Another technique applied is an appeal to authority in the form of customer-defined standards and specifications. A requirement using this technique will typically call for a particular parameter to be in accordance with the standard. One of these standards may include a hundred requirements and they all flow into the program specification through reference to the document unless it is tailored as discussed in Section 1.7.6. Specialty engineers must, therefore, be thoroughly knowledgeable about the content of these standards; their company's product line, development processes, and customer application of that product; and the basis for tailoring standards for equivalence to the company processes and preferred design techniques.

1.7.5 Verification requirements

The verification requirements are paired with the requirements classes included on Figure 1-11 and thus not specifically addressed as a separate class of requirements. The requirements classes illustrated in that figure will normally appear in the requirements section of your specification. The verification requirements will normally appear in a quality assurance or verification section of your specification by whatever name. For every performance requirement or design constraint included in the specification, there should be one or more verification requirements that tell how it will be determined whether or not the design solution satisfies that requirement.

The first step in this process is to build a verification traceability matrix listing all of the requirements in the left column by paragraph number and title followed by a definition of verification methods. The latter can be provided by a column for each of the accepted methods that may include test, analysis, examination, demonstration, and none required. An X is placed in this matrix for the methods that will be applied for each requirement (more than one method may be applied to one requirement). The matrix is completed by a column of verification requirement paragraph numbers. There should be one or more verification requirements defined for each X in the matrix.

We must also identify at what level of system hierarchy the requirement will be verified. For example, if the requirement for an aircraft altitude control unit requires that it maintain aircraft barometric altitude to within 100 feet, we could require at the system level that a flight test demonstrate this capability with actual altitude measured by a separate standard and that it not deviate by more than 100 feet from the altitude commanded. At the avionics system level, this may be verified through simulation by including the actual altimeter in an avionics system test bench with measured altitude error under a wide range of conditions and situations. At the black box level, this may be stated in terms of a test that measures an electrical output signal against a predicted value for given situations. Subsequent flight testing would be used to confirm the accuracy of the predictions.

The requirements for the tests and analyses corresponding to proving that the design solution satisfies the requirements must be captured in some form and used as the basis for those actions. In specifications following the military format, Section 4, Quality Assurance, has been used to do this but in many organizations this section is only very general in nature with the real verification requirements included in an integrated test plan or procedure. Commonly, this results in coverage of only the test requirements with analysis requirements being difficult to find and to manage. The best solution to this problem is to include the verification requirements (test and analysis) in the corresponding specification, to develop them in concert with the performance requirements and constraints, and to use them as the basis for test and analysis planning work that is made traceable to those requirements.

Later in this book we will expand the simple verification traceability matrix discussed here into a requirements compliance matrix used in the management of the whole verification process. This matrix will tie together the information provided for each specification with the verification tasks planned to create evidence of the degree of compliance and the evidence resulting from those tasks. This matrix will be one of the most important documents in the life of the program.

1.7.6 Applicable documents

Requirements come in two kinds when measured with respect to their scope. Most of the requirements we identify through the practices described in this section are specific to the product or process we are seeking to define. Other requirements apply to that product or process by reference to some documentation source external to the program prepared for general use on many programs by a large customer organization, government agency, or industry society. These external documents commonly take the form of standards and specifications that describe a preferred solution or constrain a solution with preferred requirements and/or values for those requirements.

The benefit of applicable documents is that they offer proven standards and it is a simple process to identify them by a simple reference to the document containing them in the program specification. The downside is that one has to be very careful not to import unnecessary requirements through this route. If a complete document is referenced without qualification, the understanding is that the product must comply with the complete content. There are two ways to limit applicability. First, we can state the requirement such that it limits the appeal and therefore the document applies only to the extent covered in the specification statement. The second approach is to tailor the standard using one of two approaches. The first tailoring technique is to make a legalistic list of changes to the document and include that list in the specification. The second technique is to mark up a copy of the standard and gain customer acceptance of the marked-up version. The former method is more commonly applied because it is easy to embed the results in contract language but it can lead to a great deal of difficulty where the number of changes is large and their effect is complex.

1.7.7 Process requirements analysis

The techniques appropriate to product requirements analysis may also be turned inwardly toward our development, production, quality, test, and logistics support processes. Ideally, we should be performing true cross-functional

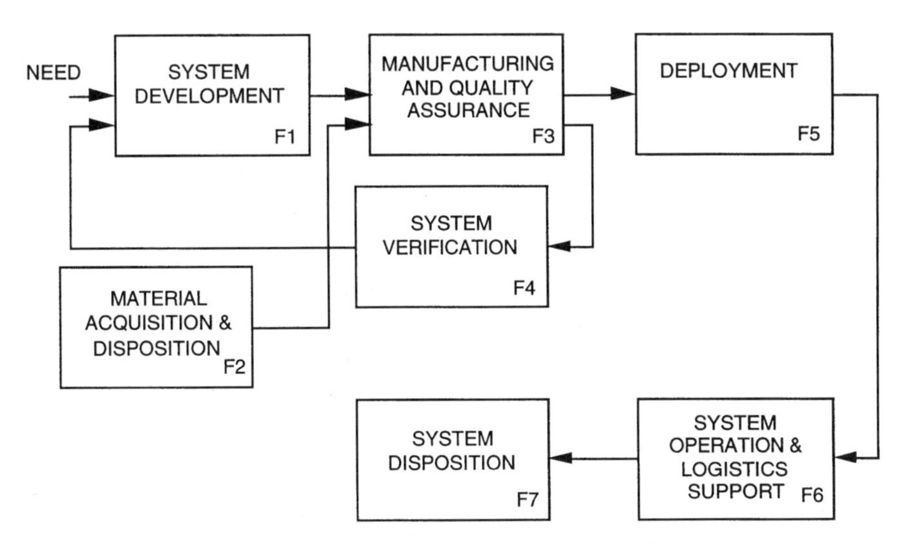

Figure 1-14 Life cycle functional flow diagram.

requirements analysis during the time product requirements are being developed. We should be optimizing at the true system level involving all product functions but process functions as well. We should terminate this development step with a clear understanding of the product design requirements and the process design requirements.

This outcome is encouraged if we establish our top-level program flow diagram at a sufficiently high level. We commonly make the mistake of drawing a product functional flow diagram only focused on the operational mission of the product. Our top-level diagram should recognize product development and testing, product manufacture and logistic support, and product disposition at the end of its life. This should truly be a life cycle diagram.

Figure 1-14 is an example of such a total process functional flow diagram. System development (F1), material acquisition and disposition (F2), and integrated manufacturing and quality assurance (F3) functions can be represented by program evaluation review technique (PERT) or critical path method (CPM) diagrams using a networking and scheduling tool. The deployment function (F5) may entail a series of very difficult questions involving gaining stakeholder buy-in as well as identification of technical product-peculiar problems reflecting back on the design of the product. At least one ICBM program was killed because it was not possible to pacify the inhabitants of several western states where the system would be based. Every community has local, smaller-scale examples of this problem in the location of the new sewage treatment plant, dump, or prison. It is referred to as the not-in-my-backyard problem.

The traditional functional flow diagram commonly focuses only on function F6 and often omits the logistics functions related to maintenance and support. This is an important function and the one that will contribute most voluminously to the identification of product performance and support requirements. Expansion of F6 is what we commonly think of as the system functional flow diagram. The system disposition function (F7) can also be expanded through a process diagram based on the architecture that is identified in function F1. During function F1, we must build this model of the system and related processes, expand each function progressively and allocate observed functionality to specific things in the system architecture and processes to be used to create and support the system.

All of these functions must be defined and subjected to analysis during the requirements analysis activity and the results folded mutually consistently into product and process requirements. Decisions on tooling requirements must be coordinated with loads for the structure. Product test requirements must be coordinated with factory test equipment requirements. Quality assurance inspections must be coordinated with manufacturing process requirements. There are, of course, many, many other coordination needs between product and process requirements.

This book is focused primarily on work that occurs within F1 and F4 of Figure 1-14. All of the validation work will normally occur in F1 and much of the planning work for verification as well. Some of the verification work accomplished by analysis will occur in F1 but F4 provides the principal machinery for proving that the product satisfies its requirements through testing. In order to accomplish this work it is necessary to manufacture product which can be tested. Some of this product may not be manufactured on a final production line but in an engineering laboratory so the separation of development (F1) and verification (F4) by the manufacturing (F3) function does not clearly reflect every possible development situation. In any case, where the verification process indicates a failure to satisfy requirements, it may stimulate a design change that must flow back through the process to influence the product subjected to verification testing.

1.8 Validation overview

Chapters 2 and 3 cover requirements validation. Chapter 2 describes several validation techniques and how they can be applied as a part of the program risk management activity. It is very important during this work that the development team maintain control over the many product representations used during validation and subsequent design and verification work to model or stand in place of the actual product designs, which in all likelihood will not exist during the early development work. Chapter 3 identifies these many representations and offers a method for simple configuration management of the items during the dynamic program period that commonly is not served by a formal configuration management activity.

Use of the wrong product representation for analysis or development testing will commonly result in the wrong answer, and the seriousness of the resulting error can be compounded in other work amplifying the amount of time and cost required to correct the problem when finally discovered. To prevent these kinds of occurrences, we should clearly define a list of representations and establish the configuration of them relative to the current development baseline. This condition must be tracked throughout the development period.

1.9 Verification overview

Chapter 4 introduces verification and covers the planning work needed for two kinds of verification activities: qualification and acceptance. Chapter 5 offers a top-down expansion of the verification process definition. Chapters 6 through 9 cover qualification process requirements verification at the component, end item, and configuration item level. Chapter 10 extends the qualification process to the system level. Chapters 11 through 13 cover the same territory for acceptance verification.

In both cases, we are interested in creating evidence of design compliance with the previously defined requirements. One method of creating evidence is to subject an item to tests specifically designed to expose item capabilities with respect to its requirements and record the results of those tests, which are the enduring evidence of compliance. We wish to have convincing evidence either supporting or contradicting the premise that the items satisfies its requirements. As the reader might conclude from this statement of purpose, logic should play an important role in crafting the verification planning so as to produce best evidence. We might ask, "What is best evidence?" It is information that is persuasive about the truth of a statement. The statement can be characterized as, "Item X has features and capabilities that support the conclusion that it satisfies requirement Y." Ideally, it should be possible to construct a deductive string of the form, "If A_1, then A_2," "If A_2, then A_3," … "If A_{n-1} then A_n." In this string, A_1 is the final results of the test and A_n is the statement that the item satisfies the requirement in question. If there is a logically correct string of statements connecting these two pieces of information, then we may conclude that the requirement is verified.

It should be pointed out that the verification effort staged on a program cannot focus totally and only on testing. This book, therefore, covers all forms of verification actions. These include test, analysis, demonstration, and examination. As commendable as it is to prepare an integrated test plan on a program covering all testing, it is not enough. An integrated validation and verification plan would be a superior goal. The author, like so many other system engineers, has had those moments of inspiration on programs to do just that, but the difficulty in coordinating the many analysis activities, acquiring the resulting information, and storing, processing, and reporting it makes this a very difficult task. Some encouragement may be provided in this book for those not yet in burn-out from previous attempts at integrated validation and verification planning.

1.10 Process V&V

While this book is primarily focused on product V&V, we should not completely neglect the application of these techniques to processes used to develop product. The requirements for these processes will normally, when written down, be included in statements of work, plans, and process specifications. Chapter 14 offers methods for applying V&V techniques explored in Chapters 2 through 13 to program processes.

References

1. Grady, J.O., *System Requirements Analysis,* McGraw-Hill, New York, 1993.
2. Yourdon, E., *Modern Structured Analysis,* Prentice-Hall, Englewood Cliffs, NJ, 1989.
3. De Marco, T., *Structured Analysis and System Specifications,* Yourdon Press, Englewood Cliffs, NJ, 1979.
4. Hatley, D. and Pirbhai, I., *Strategies for Real-Time System Specification,* Dorset House, New York, 1988.
5. McFadden, F. and Hoffer, J., *Modern Data Base Management,* 4th ed., Benjamin/Cummings, Redwood City, CA, 1994.
6. Coad, P. and Yourdon, E., *Object Oriented Analysis,* and *Object Oriented Design,* Vols. 1, 2, Yourdon Press, Englewood Cliffs, NJ, 1990.
7. Coad, P. and Nicola, J., *Object Oriented Programming,* Vol. 3, Yourdon Press, 1993.
8. Rumbaugh, J. et al., *Object Oriented Modeling and Design,* Prentice-Hall, Englewood Cliffs, NJ, 1991.
9. Booch, G., *Object-Oriented Analysis and Design with Applications,* Benjamin/Cummings, Redwood City, CA, 1994.
10. Grady, J.O., *System Requirements Analysis,* McGraw-Hill, New York, 1993.
11. Grady, J.O., *System Integration,* CRC Press, Boca Raton, FL, 1994.

chapter two

Validation before and during synthesis

2.1 Validation and risk

Requirements validation is a subset of program risk management. It is that component of the risk activity that is focused on the quality of the requirements defined for items. Figure 2-1 illustrates this relationship. The requirements that flow out of the requirements analysis process should be evaluated in the requirements validation process to determine if they are necessary and if so whether or not we have confidence that we can synthesize them, that is, create a compliant design. The answer to the latter question is a function of the nature of the requirements, the requirement values, the technology available to us, and our skill as designers. Therefore, the answer to the question cannot be provided strictly as a requirements issue. Rather, it must be answered in the context of the skill with which we can synthesize the requirements. We may have at least to define a credible design concept to gain confidence that we can satisfy the requirement. So, validation can be thought of as part of the transform between requirements and design development.

There are several techniques that can be used within the validation process to gain confidence that we can satisfy the requirements, including test, analysis, demonstration, and inspection. Where it promises to require a considerable period of time to gain confidence, we should apply technical performance measurement (TPM), which is a metric approach for tracking in time the required value relative to the current demonstrated capability requiring a clear plan of action to drive toward compliance.

Since every program must marshal its scarce resources and selectively apply them to first priorities, it is useful to evaluate all risks, including the potential problems we may have in satisfying requirements, in the context of the two parameters illustrated in Figure 2-2. Those requirements that pose the problems most likely to occur with the most severe consequences should rank high on our validation list and be pursued with the greatest energy.

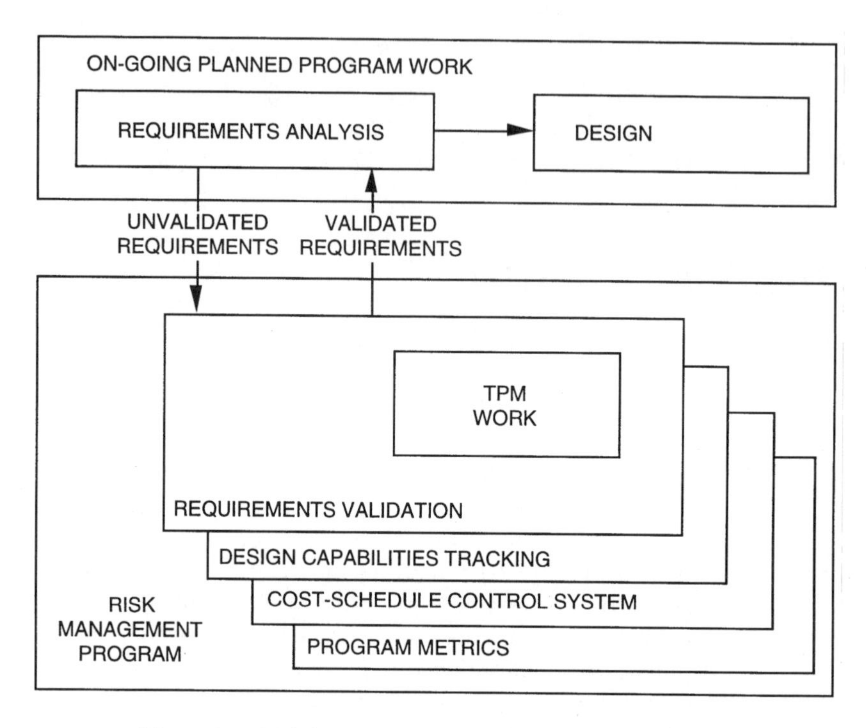

Figure 2-1 Validation is embedded in the risk program.

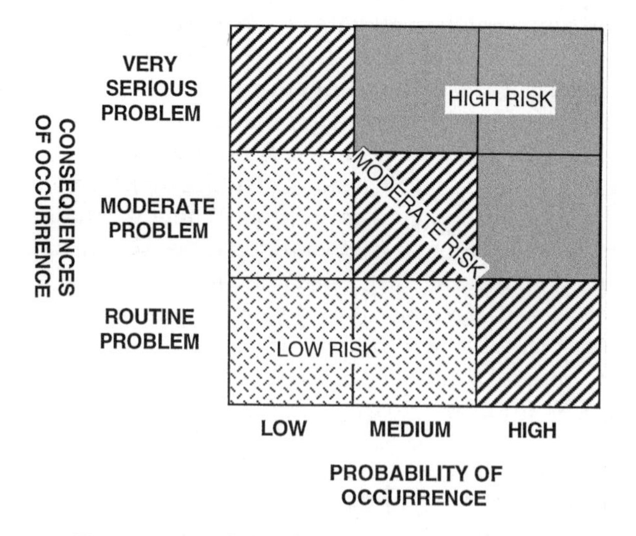

Figure 2-2 Risk-level assignment and display.

There are methods to translate these risk categories into numerical probabilistic values. Some program managers are comfortable with numerical values but many are not. It is generally true that the consequences and occurrence probabilities are subjectively determined so at the heart of all numerical systems is a subjective underpinning, perhaps with a finer granularity than suggested by the Venn diagram offered by Figure 2-2.

All risks, including requirements needing validation, can be partitioned into three program parameters: cost, schedule, and performance. While everything can be related to cost, it is useful to relate a risk to a finer granularity than that, and these three parameters are very commonly used. Most requirements pose performance risks but a broader view of the problem would suggest that most anything can be accomplished given enough money and time. If a customer absolutely needs a product satisfying a particularly difficult requirement and it is only a matter of piling up enough time and money, the risk of failing to satisfy the requirement may be more meaningfully related to a cost or schedule risk associated with success in satisfying the requirement. Some people working on advanced programs that push available technology in several directions at once prefer to include technology as a fourth risk parameter.

Since risk is future oriented, it is very difficult to accurately characterize related data numerically, or even qualitatively sometimes. This problem, however, is insufficient grounds upon which to base hostility to risk management. The intent is to identify potential problems so that we can avoid future surprises and to characterize those problems as accurately as possible as a basis for work focused on avoiding the adverse consequences of those problems should they occur. We may make mistakes in initially characterizing the probability of occurrence or consequences of occurrence but if we continue to concentrate on avoiding particular outcomes, our estimates will become more accurate over time. If we simply disregard future possibilities and let things happen, we will most often be disappointed with the results.

Risks are realities that good managers should avoid by identifying and mitigating them as early as possible. As a cross-check on our decision-making process, we should find no requirements risks that are not embraced by our validation process. The planned validation work for the requirements for an item essentially becomes a major piece of the risk mitigation plan for the item. It is true that some programs must accept considerable risk of failure in order to press the state of the art but this can be done with as much safety as possible by consciously trying to foresee the nature of problems that may occur.

Risks come in several varieties and sometimes one of these risks is more important to avoid than the others, resulting in increasing one risk in order to reduce another. For example, a system may have a very pressing delivery date requirement and a very difficult technical requirement for navigation accuracy. In the interest of mitigating the delivery and technical risks, the developer may elect to procure two different navigation units in a parallel procurement in the hope that one of them will satisfy the demanding accuracy requirement in time to satisfy the delivery requirement. In pursuing this course, the developer may build a cost risk into the program of less concern than the other risks. Both of these sets may have to be carried as alternative installations for some time until the accuracy and delivery date issues become more clear. Throughout this period, the developer will have to pursue validation and verification of the requirements for both of these equipments.

2.2 *The validation time span*

Validation properly takes place over the period beginning with the identification of system requirements and ending with a credible preliminary design solution for all of the elements of the design. For most system elements, the end will correspond with an approved conceptual or preliminary design and this is a good goal in general. For other items characterized by major concerns and challenges, the validation process may stretch further into the design process. We should understand that the longer it takes to gain confidence in our ability to implement a design compliant with the requirements, the longer we must carry with us the burden of risk that we may not be able to arrive at a sound solution or that we will select a bad design that later either must be changed or accepted with regret by our customer. All of the requirements validation work should absolutely be complete by the time management and/or the customer concludes that the design is complete. In the case of DoD programs, this occurs at the milestone called critical design review (CDR) commonly coinciding with 95% of all planned engineering drawings having been released.

Some readers, especially those with engineering design experience, will feel that the wrong slant has been placed on the validation process in this chapter. They will feel that validation is a design process and not a requirements-related process. Clearly, it is both. It is through what the author calls validation that we gain confidence that it is possible to synthesize the requirements into a viable design. So, many would say that this is a process of validating the design. For those who feel this way, let us agree that we are both right. Validation is one component of a sound risk management process that embraces both the requirements definition and design solution processes. Validation applies to the transform between the requirements defined for an item and the design concept created in response.

Between the requirements and the design solution there exists a chasm with the rims separated by the creative design process that takes place within the mind of one or more gifted engineers. The system engineer must not become depressed over the lack of traceability between these two, rather should seek to build the requirements definition as carefully and quickly as possible and, in concert with the designer, designers, or design team, reach a conclusion about their ability to translate these requirements into a compliant design. The validation process spans this chasm between the ordered world of requirements and the chaos (no offense intended) of creative design work.

As requirements for the system and its elements are identified, they should be partitioned into those that will require formal validation and those where, in our good engineering judgment, the cost of formal validation is not warranted by the likely results. So, there is a requirements-oriented component to validation. But, the validation process also extends into the design process because we wish to gain assurance that we are capable of satisfying the requirement. The design concepts conceived by designers may have to be subjected to careful scrutiny through test and analysis to prove that they can be effectively employed. If, for example, the needed system functionality requires a computer speed faster than any machine yet developed, we would be remiss not to build a breadboard/brassboard model based on new technology and test it to show that the technology was sufficiently mature and effective to solve the problem.

There is a natural reluctance to submit requirements or a design concept to validation action in any public way because it implies that they might be wrong. The ego of the designer and management can be caught up in the decision logic leading to a foolish feeling of infallibility. This is very easy to fall prey to because it occurs early in a program where you have just finished building a winning proposal and everything appears new and possible. You begin to believe the optimistic statements you have been feeding to the potential customer through your proposal and subsequent meetings. The systems engineering community must remain firm at this point in its encouragement of the need for validation where it is appropriate. The system engineer interesting in pursuing validation should recognize that it may be career limiting, where the practitioner appears to be an obstacle to program progress. So, the need for validation should be sold to the program through the proposal or business planning process and not left under the surface until program implementation. Successful enterprises will make it a fundamental part of their way of doing business.

2.3 *Avoiding a null solution space*

Many system engineers believe that it is impossible to overdo the job of requirements definition because the function of requirements is to "specify" or state in full and explicit terms. This mind-set is encouraged when a system engineer creates the specification and a designer must implement it; that is, two different people do the work. This attitude is also more pronounced in system engineers who have never served as a design engineer. We must always keep foremost in our mind that the whole systems approach to the development of systems to solve complex problems is driven by a need to protect the specialized design engineer or team from knowledge deficit while providing him/her/them with the widest possible solution space. A requirements analyst should seek a condition of requirements completeness and sufficiency not an overpowering number of requirements flowing from excessive zeal. The best specification is not the heaviest possible specification.

All requirements do not constrain the design solution space as they should. Too many requirements constrain the design solution space too much, possibly preventing the designer from selecting an alternative that could be superior to at least some of those allowed by a more reasonable set of requirements. At the time we are doing requirements analysis there is no simple technique for limiting the requirements we identify to only those that are absolutely essential. There is no way that computers can do this for us no matter the tools that we bring to bear.

The best protection against overspecification is to apply structured requirements analysis processes, such as those described in Chapter 1, that encourage us to identify only the necessary requirements. These methods also encourage completeness and therefore sufficiency. A second way to prevent requirements excesses is to perform an ongoing traceability analysis during the requirements analysis process. Where it is not possible to establish a traceability link from one item requirement to a parent item requirement or to a structured analysis process, we should ask whether or not the requirement is really necessary. The requirements should also be devoid of solutions. They should be design independent. We should ask ourselves if each of the requirements defined for an item is necessary. If the requirement were not included would it be possible to create a design satisfying the remaining requirements that failed to interact synergistically with all other items to satisfy higher-order requirements all the way to the customer need? If the answer is yes, then the requirement should be retained.

Finally, we should supplement a structured approach to requirements analysis, a watchfulness for excluding design content, and an appeal to the discipline of traceability with a selective formal validation activity. Validation, in this book, refers to a process of proving the need for particular requirements, the validity of their values, and that it is possible to synthesize them into a design solution. As noted above, it is possible to define requirements for an item such that they unnecessarily limit the solution possibilities. It is also possible to limit the solution space to a null such that it is impossible within the laws of science to solve the problem.

Ideally, we would never stray from the norm of only capturing the necessary and sufficient requirements. But, requirements are identified by humans and humans can make mistakes. Therefore, it is possible that we will fail to meet this high standard. Since it is better to discover a null solution space or one unnecessarily restricted by requirements earlier than later, we should seek to uncover these conditions as early as possible. An effective validation activity can be used to detect and correct this condition during the requirements definition process.

It is probably true that the validation work should include at least one person who was not deeply involved in creating the specification content. A fresh mind not contaminated by past wrong thinking will generate many questions that would not occur to the participants due to the limitations imposed by group think. These questions may sting the egos of those who developed the requirements more than a little so the process should be implemented by secure, mature people. But, even if it results in fist fights, it is a valuable process that will prevent many program problems later in the program when there is less flexibility to handle them.

2.4 Validation process description

2.4.1 Overview

Figure 2-3 illustrates an overall process for validation as it can be applied to a single item for which the requirements are being determined. This process fits within the Requirements Validation box of Figure 2-1. If the terms are new to you, for the time being simply try to absorb the process flow alternatives. We will discuss each path in some detail in this section. Validation should be applied to requirements as we identify them and early in the period when the designer or design team is working to synthesize them into a design solution. Figure 2-3 focuses on a single requirement for one item and this process must be applied to hundreds or even thousands of requirements for many different items on a program. Think of this process as a sieve through which we pour the requirements for an item. The decision logic steers the requirement through the chains of activity.

On a large program, we may have a number of sets of requirements in various stages of the validation process at any one time and the status of these requirements with respect to the validation process will be changing over time as well. So, we need a means to track the status of the validation process throughout the requirements analysis and design synthesis processes. We will introduce matrices for this purpose, but the ideal method is to have this capability built into the tool used to capture the requirements, accomplish traceability on them, and print specifications containing them. While many computer tools supported verification, few supported validation and requirements maturity tracking at the time this book was written. It is hoped that will change.

As shown in Figure 2-3, we must first check for completeness and screen identified requirements for correctness and necessity in the Evaluate Requirements task. Surviving requirements must then be screened for validation action based on our evaluation. We may conclude that there is no need to validate a particular requirement formally in accordance with a specific criteria, in which case the requirement can be passed on through to validation complete without any formal validation activity. Some requirements may appear very difficult to satisfy because they have never before been satisfied by our organization and they entail a great deal of risk. Our first act should be to challenge the need for them and the difficult value assigned. In some cases we may conclude the requirement is unnecessary, in which case it should be deleted. Otherwise, it may be possible to gain customer approval or internal agreement on a change to the value that gives us more confidence.

Requirements with surviving concerns should be passed through one of two remaining channels both of which are essentially the same and are differentiated by the period of time we expect it to take to complete the validation process. A requirement that will surrender to a near-term validation action, such as a design evaluation test (DET), an analysis, an inspection,

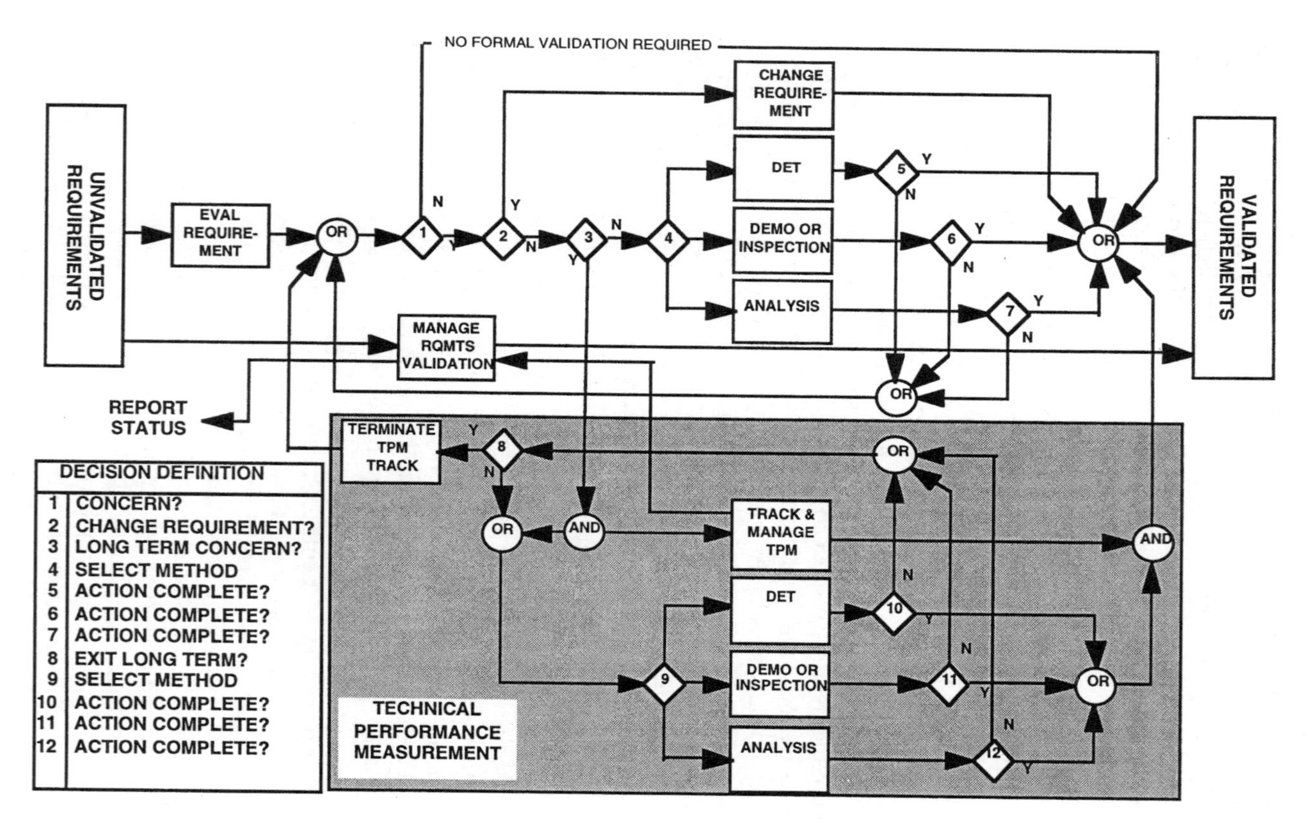

Figure 2-3 Item requirements validation process.

or demonstration can be passed through the upper channel on Figure 2-3. If the validation action is going to take a considerable period of time, we should track the work through a TPM program to be explained in more detail later in this chapter. If we choose TPM, we have essentially the same situation as far as validating the requirement is concerned with the difference being that we are tracking the situation more intensely over a longer period of time.

As noted by the recycling possibilities, we may have to apply several techniques in serial over an extended period of time in parallel with preliminary design efforts. The use of decision symbols in Figure 2-3 implies a single channel selection process but we could carry a requirement in several channels simultaneously. Perhaps inclusive OR symbols rather than decision symbols would have been more appropriate for this reason. For example, we may conclude that the value for a requirement will have to be validated through simulation (a combination of DET and analysis). We may also have a concern for our ability to synthesize the requirements with our current technology leading to a technology demonstration. The customer may also direct that we track the value as a TPM parameter regardless of how long we believe it will require to resolve the validation issue. Our challenge is to determine what combination of paths to apply for each requirement that should be validated and to extract ourselves from validation as quickly as good judgment dictates.

Validation work may all be initiated during the requirements analysis process or later when, during the preliminary design process, members of the design team become concerned about their understanding of the requirements or their ability to synthesize them. In either case, the validation process is concerned with requirements for items and should not be disconnected from them. The more general risk management process may involve risks that are not product requirements related, so we should think of the requirements validation activity as only a component of a larger activity. One could conceivably run TPM tracking on programmatic or functional management parameters but these will be referred to as programmatic and functional metrics, respectively. TPM commonly is thought of as only related to product requirements value tracking and reporting but it is a subset of a larger metric family as are the techniques of cost/schedule control. Figure 2-4 illustrates the relationship between these several activities in a multilayer Venn diagram. The double crosshatched area represents the overall risk management activity for a program while the requirements validation activity space relates to the material discussed in this book. You will note that TPM is completely a subset of the validation process, while validation is but a part of the overall risk activity.

TPM tracks both the required value but also the current capability of the design solution over time. As the design solution matures, the design people should maintain an understanding about the capabilities of the design solution not only corresponding to those parameters selected for TPM tracking but for all design features. Ideally, the designers should be working to satisfy

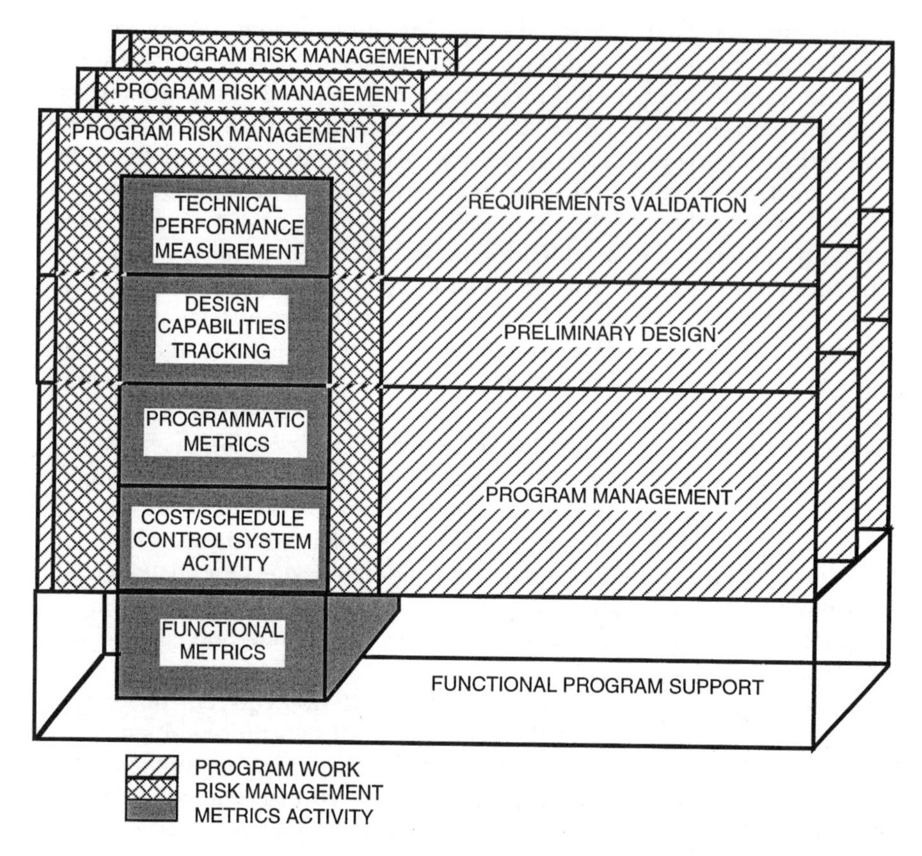

Figure 2-4 Correlation of validation with metrics and the program risk universe.

the requirements in the design capabilities but not exceed them significantly. To the extent that design requirements are exceeded, overdesign is introduced into the design generally along with unnecessary cost. This is an area that systems people should focus on in doing integration and optimization work on the evolving design and is not related to the validation process.

Figure 2-3 illustrates a situation where there are three programs in development by the company at one time. Each must manage its own risk program and the related validation work as well as the other activities that derive information from the metrics work indicated. All programs may also contribute information to functional metrics that should provide useful management information back to the programs individually as well as comparative between programs that may help a program gain insight into things that could be done better.

Upon completion of the planned validation process shown in Figure 2-3, the requirements can be considered validated. The possible outcomes for a given requirement as a result of the validation work are delete the requirement, change the requirement (its value or context), or retain it as currently written. The decision should be based on facts derived through the validation process. The validation process produces evidence of validity of the requirements or a rationale for changes.

2.4.2 *Initial screening of the requirements for validation*

Figure 2-5 expands on the Evaluate Requirements task of Figure 2-3. The first filter applied in the validation process must be carried through the program until the specification is approved and it asks whether or not we have identified all of the appropriate requirements. This is not an easy question to answer for it inquires into things not then known. There is no magic potion that can be applied here but the structured analysis process offers the best assurance that nothing important has been missed. So, two criteria are advanced as follows:

a. If we have applied structured processes, have we missed anything in those processes? Are our models complete? If we did not use a structured process, there is little hope of finding unidentified necessary requirements.
b. Are there parent item requirements that do not currently flow down to this item? All need not, but those that do not should be consciously reviewed for whether or not they should.

The first filter to be applied to identified requirements is whether the requirement is necessary. If it is not necessary, it should be rejected or deleted. These requirements may have to be identified in a transitory way as requirements subject to validation, but they should rapidly fade from the scene as they are proved to be unnecessary. An unnecessary requirement invariably adds unnecessary cost and makes it more difficult to derive a compliant design. A criterion for necessity is offered as follows:

a. If the requirement is not recognized, can a design solution be derived for the system that fails to satisfy the need or for the item such that it fails to interact synergistically with other items to the end that the need is not satisfied.
b. The requirement is not contained in another requirement for the item.
c. The requirement is traceable to a parent item requirement or to an entity in an approved structured analysis model from which it was derived.
d. The requirement should be design independent rather than simply narrowing the design solutions directly.

If the requirement is necessary, we must also ask if it has been properly characterized. These questions should include the following:

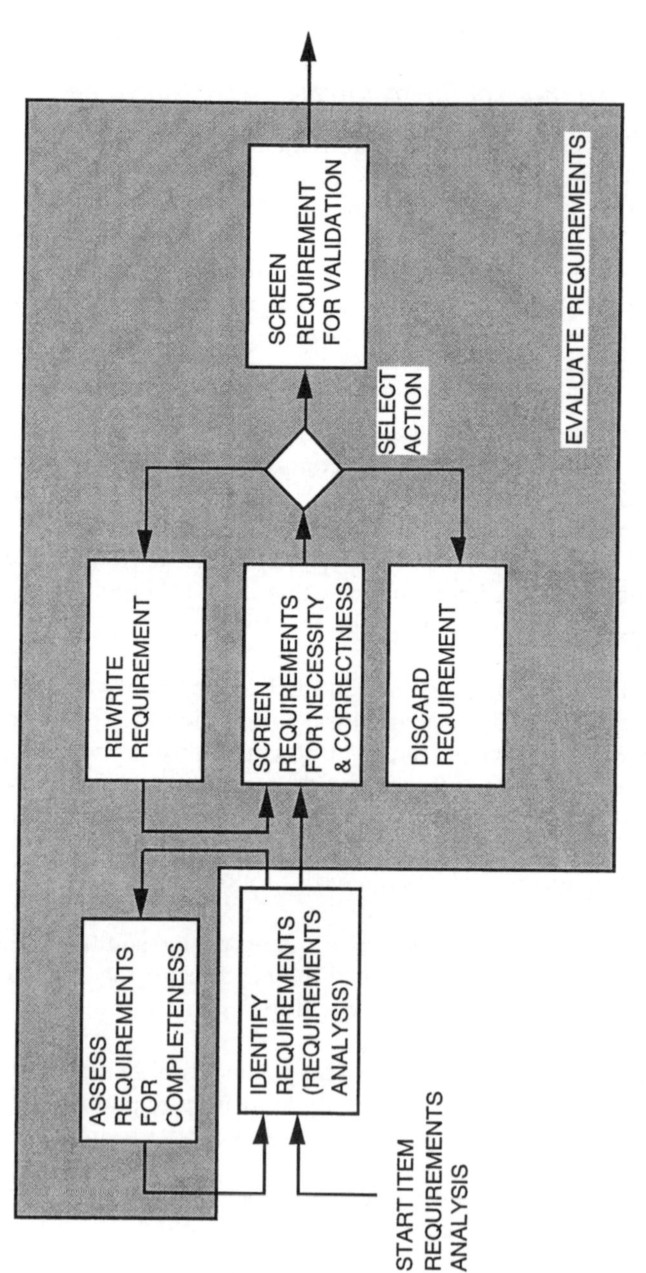

Figure 2-5 Evaluate requirements activity.

a. Is the value correct? We can gain confidence in requirements values that were derived from a system math model for the parameter in question because it enforces discipline on value assignment. Simulations properly constructed and operated may also give us confidence. We should be concerned about values for which the person who assigned it cannot give a reasonable rationale for that value.
b. Is the requirement stated correctly? Aside from correct use of the language, the requirement should avoid unnecessarily difficult and very specialized words. The requirement should be stated in positive terms rather than what to avoid.
c. Is it possible to obtain more than one meaning for the requirement? The requirement should be perfectly clear so that two or more people will not derive different meaning. One could also ask if the statement makes any sense in a very practical way.

If the requirement is not perfectly stated, it must be rewritten to overcome the critical findings and dumped back into the evaluation pool. This process may require several rounds before the requirement clears the evaluation. To the extent that requirements are commonly found poorly written, the principal engineer, team leader, or other person responsible for the item should find out why and take steps to correct the problem as it is parasitic in nature and will reduce the resources of time and money available later in the program.

Figure 2-6 summarizes the different subsets that a particular requirement may fall into. Up to this point we have discussed requirements that are not in the current specification that should be and requirements that are in the current specification that should not be. Figure 2-6 breaks the former into two subsets, those previously deleted and those never contained. In both of these cases, it is possible that the specification should not contain them, but it is also possible that a requirement could be deleted in error or not recognized during the analysis. All of the proper content not included and improper content included represents risk and the validation process should encourage these subsets (identified as a crosshatched space) to go to null.

2.4.3 Validation intensity selection

Next, we wish to determine whether the surviving requirements, in the current proper subset of Figure 2-6, should be subjected to formal validation. Your organization should have some predefined criteria for selection of the appropriate validation action. If you do not now have such criteria, you may wish to consider the following suggested criteria:

a. Is the value known? If the value is not known is there a known process in work that will result in a credible value?
b. Does the development organization have confidence, based on data, that it will be possible to satisfy the requirement?

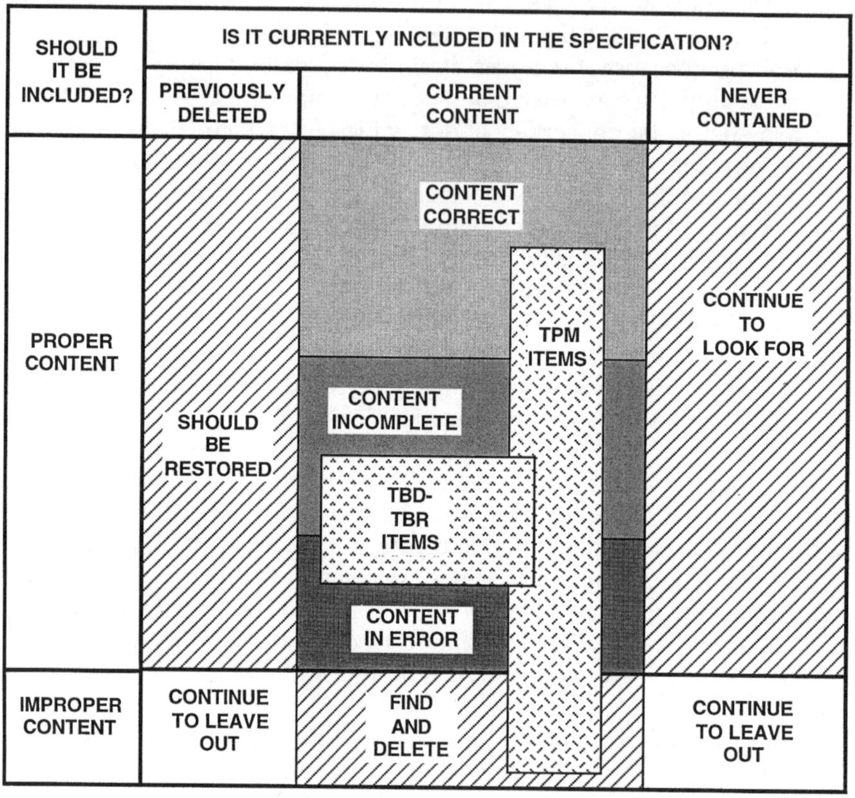

SHOULD IT BE INCLUDED?	IS IT CURRENTLY INCLUDED IN THE SPECIFICATION?		
	PREVIOUSLY DELETED	CURRENT CONTENT	NEVER CONTAINED
PROPER CONTENT	SHOULD BE RESTORED	CONTENT CORRECT TPM ITEMS CONTENT INCOMPLETE TBD-TBR ITEMS CONTENT IN ERROR	CONTINUE TO LOOK FOR
IMPROPER CONTENT	CONTINUE TO LEAVE OUT	FIND AND DELETE	CONTINUE TO LEAVE OUT

Figure 2-6 Requirements validation intensity hierarchy.

c. Is there a history of success in satisfying this kind of requirement by the development organization?
d. Will the development for the item in response to this requirement entail only technologies with which the development organization has experience and knowledge?

If the answer to all of these questions for a given requirement is "yes," then there is little need to accomplish formal validation action for the requirement. If the answer to all of these questions is "no," it is likely that some kind of formal validation action is needed. In between these two extremes, one must rely upon experience and judgment to reach a conclusion about the need for formal validation.

Ideally, these questions should pass through our mind as a matter of habit as we are initially conceiving each requirement and thus avoid identifying some requirements in the first place. But we should also pass through a formal requirements evaluation step as the specification content takes on a nearly complete form and before it is initially released. The results should be reviewed at a requirements and design concept review prior to the team being authorized to begin preliminary design. Requirements databases should be structured to capture the validation status and to track that status to closure. In the best of all worlds, the analyst would make the binary validation decision at the time the requirement is initially entered into the database. The principal engineer or team leader for the item might review that input and make a final decision and define the validation method and responsibility.

Requirements validation does cost time and money. Therefore, formal validation should be applied with selectivity. For all of the requirements identified for an item, we should first partition them into two subsets, those for which there is little concern and those which should be subjected to some form of validation. As a result of this binary partitioning, conducted initially early in the program and maintained throughout the requirements identification and preliminary design processes, we should focus on one or two principal paths through Figure 2-3. The top channel is appropriate where it will be possible in fairly short order to complete the validation action entailing test, demonstration, inspection, or analysis work. The lower channel is appropriate where we will have to track our requirements maturation and design solution process over an extended period of time and provide management with current and historical data on our movement toward completion of the validation work.

The requirements to be formally validated may be included in one or more of four management streams: risk management, TPM, TBD-TBR items tracking, and formal validation. All of the requirements excluded from these four subsets will have essentially been informally validated by a cursory analytical exclusion. A purist may feel that this will open us to risk since not all requirements are being formally validated. The problem is that we have to obey the fundamental tenants of economics as well as engineering. We have limited resources on every program and must set priorities recognizing a condition of balance between the different kinds of risk. If we attempt to validate every requirement formally through test and analysis, we may relieve the program of all performance risk except that we may have exceeded the available program funding and realized a cost risk. Program management entails balancing risks and recognizing that it is not possible to study every possibility in complete detail. Wherever possible we have to make system development decisions that exclude the need for work based on our experience and history and focus available funding on the most serious issues facing us.

2.4.4 Formal requirements validation management

Where the conclusion is drawn that formal validation action is needed, we should create a tracking matrix to identify the responsible parties and needed validation actions clearly, summarize current status, and offer reference to a

ITEM	REQUIREMENT	METHOD	VALIDATION ACTION	PRINCIPAL	PLANNED COMPLETE
A1342	3.2.5.7	Analysis	Validate value in simulator	Jones	10-30-96
A1342	3.2.5.9	Test	Tech demo our capability	Burns	11-15-96

Figure 2-7 Requirements validation tracking matrix.

simple plan for future validation work. Figure 2-7 is structured for system-level tracking of all validation actions. If you are using a database system with validation tracking capability, it may be possible to follow this model but it is uncommon for commercially available tools to provide this function. An alternative is to maintain a separate spreadsheet or database for this data but this violates the concept of data integrity in that identical data items will be located in more than one place. This will require extreme vigilance to avoid divergence of data entered in different systems. If you are using word processors to create specifications, you could include a validation table in Section 6 (Notes in a DoD specification) while the specification was being prepared. This matrix would not need the item column and the matrix should disappear when you have completed all of the planned validation actions. This condition should occur before final release of the specification.

A better place to put this table is in an integrated validation and verification document where it could be used to offer current status of the validation effort. You might also simply refer, in that integrated document, to this matrix published separately in memorandum report format periodically with revision letters applied to the memo number.

The matrix is adequate to identify and track all of the validation actions but each validation principal engineer should be required to maintain a current action plan and schedule defining planned work and giving current status. This can be done in a very simple way and does not require extensive documentation even in a DoD program. A final report should describe results of the validation work. If the work involves several actions, it may be necessary to link up multiple test and analysis reports under the cover of a primary report

2.4.5 Validation through risk management

One or more of the requirements that have been caught by the validation screening activity described above may pose sufficient concern on the part of management to be selected for inclusion in the risk management program. We are only concerned here with requirements risks but all risks identified should be listed in a risk list for the program and assigned to specific contractors, teams, and persons. It should be clear that the responsible person and organization has the responsibility to understand the risk fully, report those findings, find ways to mitigate the risk (ideally to zero), maintain records of risk mitigation work, prepare and maintain a risk action plan, periodically brief status, and implement approved action plan elements. There should be a definite time when planned mitigation actions must be complete. The action plan may call for development evaluation tests, analyses, inspections, or demonstrations as noted on Figure 2-3.

Generally, the risk associated with requirements will be carried as a technical risk where difficulty is expected in synthesizing the requirement. But, if the problem is stated in terms of the cost or schedule to achieve a result that is known to be possible, then it may be carried in one of those subsets. Likewise, if a program uses technology as a fourth risk category and the risk is perceived as unavailability of the technology to achieve the requirement, it may be carried in that category. So, the work that will be undertaken to mitigate a requirements risk will have to be colored as a function of the risk category chosen. It may focus on ways to get cost or schedule time out of the design and development process. Alternatively, it may call for one or more technology demonstrations that are structured to obtain access to new technologies that support design solutions or manufacturing techniques that are capable of satisfying the requirement.

One of the most difficult design problems the author ever observed was the design of the wings for the Advanced Cruise Missile in the mid 1980s at General Dynamics Convair (part of Hughes Missile in Tucson at the time this book was written). The task fell to a terrific mechanical engineer named Bill Doane and equally skilled team of people from aerodyunamics, stress, materials manufacturing, and radar signature groups. Bill had to satisfy all of the normal aerodynamic requirements that resulted in sufficient lift to sustain normal flight within the context of the wings being initially contained within the fuselage prior to launch. That is, the wings had to reliably deploy after launch from a B52, B1, or B2 aircraft. This requirement placed obvious limitations on the size and configuration of the wings. In addition, the missile had to be unobservable by hostile ground or airborne radar and wings can be excellent reflectors of radar energy. The missile required long range so low missile empty weight was very critical to provide for a large fuel load. A successful wing design matured over a very long period within the context of numerous technology demonstrations, risk mitigation sequences, trades studies, and design concept developments balanced against ongoing design work on many other parts of the missile often in conflict.

Since not all program risks are driven by product requirements, even if your requirements database includes validation and risk tracking, it will not satisfy the total program need for risk management reporting. Ideally, one would be using a relational database system with linkage between the requirements database and the risk database. The only way this could be done at the time this book was written would be to build the database system yourself. It is hoped, as time progresses, that the requirements tool makers will begin to see the benefits to themselves in a standardized and open tool interface architecture that permits one to hook up their internal relational databases to their commercially available tools.

2.4.6 Technical performance measurement

A further subset of the requirements validation activity should entail TPM items. As in the case of risk items, we must be selective in picking TPM parameters because each does entail small but real budget and schedule impacts. The program for a very complex system may have only 25 parameters at the system level. The Lockheed Martin/U.S. Air Force F22 program at one point included over 100 parameters at several levels. The number of parameters should be selected for consistency with the way the program is going to be managed. Each organizational structure (program, team, principal engineer) may have assigned TPM parameters chosen from the requirements contained in the corresponding specification.

The TPM process reports the historical evolution of a requirement value keyed to past and planned program events as a requirement value vs. time comparing the required value (which may change over the development period) with the projected or achieved value. Figure 2-8 illustrates an example of TPM reporting documentation. The first chart tracks parameter value in time and the second chart provides the action plan for future work to reach closure. The utility in doing this is that management gets insight into how well the program is doing in terms of a limited number of "test points." We should select the TPM parameters with this view in mind; that is, what are the best test points to use in determining the health of the evolving design solution?

As in the case of risks, all TPM parameters should be listed in a management matrix similar to the validation tracking matrix shown in Figure 2-7. This matrix can be extended to include TPM parameter identification and management information but, a better solution, once again, is to link a relational TPM parameter database to the validation database (in turn linked to the requirements database). This summary report should identify each TPM parameter, indicate the current status (see Table 2-1 for one set of status designations), and name the principal engineer. The database should be structured to print out the TPM documentation charts and status matrix.

2.4.7 Requirements maturity control

Throughout the early requirements definition period, we are faced with the gradual maturation of requirements knowledge. Figure 2-6 identifies requirements for which an appropriate value has not been determined as TBD-TBR items. TBD means "to be determined" and TBR means "to be resolved." Some people prefer the TBD designation and others TBR. Some organizations use both terms with slightly different meanings. In those cases, TBD is used to mean that no value has been determined and TBR means that a best guess or challenge value has been included while we work toward resolving the final achievable value.

As noted in Chapter 1, values are commonly defined through either an allocation model or an analytical derivation possibly involving modeling

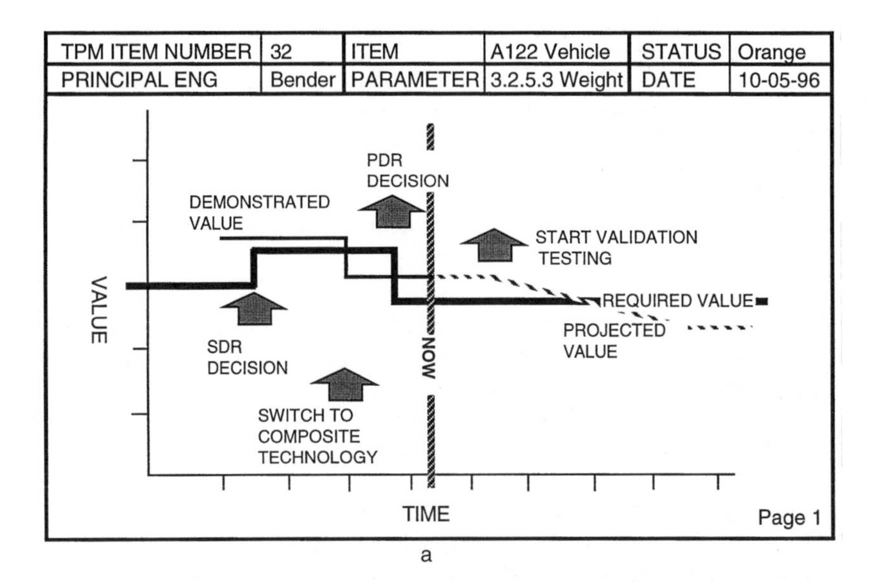

a

TPM ITEM NUMBER	32	ITEM	A122 Vehicle	STATUS	Orange
PRINCIPAL ENG	Bender	PARAMETER	3.2.5.3 Weight	DATE	10-05-96

CONCERN: Current weight exceeds required value by 10% even after a switch from aluminum to composite center section structure.

PLANNED ACTIONS:

1. Continue to search for ways to reduce weight in other structural sections and on-board equipment. Complete by 11-15-96.

2. During validation testing of the structure, be alert for opportunities for mass removal without adversely affecting strength. Testing will run between 12-01-96 and 04-30-97.

3. Investigate composite outer wing panels and tail surfaces. Complete by 01-15-97.

CRITICAL POINT: CDR is 08-15-97 and required value must be attained.

Page 2

b

Figure 2-8 TPM parameter documentation. (a) Parameter value chart; (b) action planning chart.

and simulation. An appropriate value for most requirements can readily be determined concurrently with identification of the requirement using flow-down based on the parent item value and a logical approach to defining all peer item requirements. Weight and reliability are examples that follow this pattern. Many of the most-challenging requirements, however, offer considerable difficulty in value determination through the derivation approach. For example, on a transport aircraft program, we may agonize over optimum gross weight, fuel load, engine power/thrust, and lift and drag figures. These

Table 2-1 TPM Parameter Status Designations

Status	Explanation
RED	Serious problem exists that must be solved
ORANGE	Parameter marginally not under control but there is a credible plan to get there
GREEN	Parameter under control
BLUE	Parameter well under control. Possible excess capability, growth potential, or suboptimization

are all interrelated and cannot be easily determined in isolation. Very often, customer needs are not finely determined in early work leading to some uncertainty about these kinds of requirements.

There is a wide range of opinions on how to proceed with these requirements with value maturity difficulty including the following:

a. Identify an initial value and see what happens.
b. Identify goals or targets and see what happens.
c. Initially identify the value as TBD and actively work toward closure.
d. Identify a goal but leave the value TBR and actively work toward closure.
e. Refrain from identifying a requirement until its value has been determined.

The urgency in defining all of the appropriate requirements and their values is at least in part a function of the contractual relationship associated with the specification within which those requirements will appear. If we are dealing with a customer specification (whether written by the contractor or customer), we cannot base a clear contractual relationship on unknown or incorrect values of key requirements. There will be a lot of pressure exerted to release the specification with values, no matter how ill conceived, in order to meet data delivery schedules. This is true of supplier, vendor, or subcontract specifications as well in order to "get the show on the road." This same situation can evolve in an associate relationship where the two parties and the common customer seek to reach agreement on the contents of an interface control document.

The author would encourage one of the compromise positions "c" or "d" within an environment that requires timely closure. The other three alternatives are just not very smart. Approaches "a" and "b" are essentially the same. Good luck may be the only salvation. Once the proper value is determined, and it is different from the initial value or goal, there will have to be a cycle of corrective work that simply costs a lot of money. This is a perfect example of faulty system engineering work. The skilled application of the systems approach should lead to a condition of no surprises. The last solution listed above has the same result as the first except that the developer may totally ignore an important aspect of the final product rather than respond to an incorrect value for that parameter.

PARAGRAPH NUMBER	PARAGRAPH TITLE	PRINCIPAL	ACTION PLAN	DUE DATE	TBD	TBR
3.2.5.7	Engine Thrust	Logan	Simulation Run	10-30-96		005
3.2.6.8	Fuel Interface	Burnston	Fuel Vol. Study	11-02-96	010	
3.3.5	Lube Oil Pressure	Stone	Max RPM	10-10-96		006

Figure 2-9 TBD/TBR closure matrix.

Alternative "d" can be implemented as shown in the following example of an engine thrust requirement in an engine procurement specification:

> 3.2.5.7 Engine Thrust. Maximum engine thrust under standard day conditions at sea level shall be equal to 10,000 pounds (preliminary value, final value **TBR-05**) plus or minus 500 pounds.

In the specification, we would have to explain the meaning of preliminary values and how they are identified and that all requirements not so identified are final. We would also have to explain in the contract how changes in value of these preliminary requirements will be covered from a financial perspective. And it is in the contract language that these problems must be addressed. If there are one or more key requirements for which a firm and validated value has not been determined, the contract has to address how the contractor and customer will share in the risk. A fixed price approach will not protect the contractor where the available technology will not support the value. A cost-plus approach will not protect the customer from a contractor ill prepared to solve the problem. Regardless of the contract type, there should be recognized a class of troublesome requirements and the contract must tell how requirements may be included in that class, removed from it, and how related issues will be resolved.

One of the fundamentals in this resolution process is to have a clear list of those requirements. An example of this is shown in the TBD/TBR closure matrix in Figure 2-9. This matrix should establish accountability, note a plan of action, and time constraints. All of the TBD/TBR items in the specification should be numbered as in the example noted above and coordinated in the matrix. Both notations could be numbered in the same string or in independent strings so long as their notation is unique.

If specifications are prepared directly using a word processor, the matrix could be placed in the preliminary specification notes section and eliminated by the time the specification is initially released. It can also be maintained separately. The ideal method would, once again, entail integration into the requirements tool set but all commercially available tools do not provide this functionality. The tool should be able to provide a list for a particular specification as well as a global program list or count. This is a useful system

FIELD NAME	TYPE	NUMBER OF CHARACTERS	SAMPLE DATA
ARCHITECTURE _ID	CHARACTER	16	A1233
RQT_ID	CHARACTER	3	034
FRAG_ID	CHARACTER	1	
PARAGRAPH_NBR	CHARACTER	16	3.2.5.7
TITLE	CHARACTER	60	Engine Thrust
ATTRIBUTE	CHARACTER	120	maximum thrust standard day conditions at sea level
TYPE_REQUIREMENT	CHARACTER	1	Q (Quantified)
RELATION	CHARACTER	1	E (Equals With Tolerance)
VALUE	NUMBER	8, 8	10,000
UNITS	CHARACTER	30	pounds
MATURITY_STAT	CHARACTER	1	R (TBR)
MATURITY_ID	CHARACTER	2	13
TOLERANCE	NUMBER	8, 8	500
CAPABILITY	NUMBER	8,8	9,200
TEXT	MEMO		

Figure 2-10 Database structure subset supporting TBD/TBR.

engineering metric that is very tedious to compute without the automatic computation possible in a computer database.

Figure 2-10 gives a database structure subset that supports both TBD/TBR and primitive requirements capture with computer sentence generation. The author has experimented with this structure in a prototype requirements tool and found that all of the normal requirements tool features are supported as well as TBD/TBR. This structure simply groups all fields together for simplicity of description of the TBD/TBR concept and is not intended to reflect a fully normalized form.

The other TYPE_REQUIREMENT cases envisioned in this database are as follows: (T) text only, (R) reference to an applicable document defined in a companion relational database, and (H) header or title only. The RELATION field values have to account for the following relationships between values and attributes: (L) less than or equal to, (G) greater than or equal to, (S) less than, (M) greater than, and F (equal without tolerance). Requirements of type Q and R would include the key information in the first sentence of the paragraph and the analyst could thereafter include the content of the TEXT field with additional explanatory information or leave the TEXT void. In text type requirements, the only paragraph content would be the TEXT field content. The user should, of course, be shielded from required knowledge of these codes via the human interface.

With these understandings, case statement code can easily be written to string together the fragments indicated in the SAMPLE DATA column interspersed with simple computer-generated connectives to create the specification paragraph included above in specification format. Code can also be provided to report and manage the TBD/TBR items toward a null condition. An added benefit from this structure is that the requirements analyst need only focus on identifying the attributes to be controlled, appropriate values and units, and the desired relationships between attributes and values. The software code concatenates these data into simple specification language based on TYPE_REQUIREMENT, RELATION, and MATURITY field values. In this case, the computer would generate a boring but clear paragraph as follows: "3.2.5.7 Engine Thrust. Item maximum thrust under standard day conditions at sea level shall be 10,000 pounds plus or minus 500 pounds (TBR-13)." The computer can be easily told to generate the underlined data.

Another benefit from the suggested structure, which carries the value as a separate numerical entity rather than burying it in a string of text, is that the numerical values can either be imported from coordinated specialty engineering models (reliability, mass properties, and so forth) or linked to them such that the desirable information system characteristic is satisfied that any one unique piece of information is stored in only one authoritative place. Margin, budget, and current capability work is simplified as well. Finally, this structure supports automated search for slack and concern cases based on a comparison between required values and current capabilities. In the Figure 2-10 example, the engine development is in trouble since capability is 800 pounds short of the required value.

In a commercial situation, there may be no formal customer with whom to interact and no contract through which TBD/TBR action will be controlled. That does not mean that one should dispense with requirements maturation controls. You will have more freedom of action in resolving problems but you will be constrained by your understanding of customer expectations and time-to-market forces.

2.5 Validation responsibility and leadership

Each program of any size must have a systems activity assigned. The leadership in some organizations, impressed by the literature on cross-functional teams, has concluded that the function called system engineering was no longer needed, that all of the cross-functional, cross-product, and cross-process work would somehow be accomplished by the several cross-functional product teams, each focused inwardly on their own small part of the overall development problem. Well, that simply will not happen. While cross-functional teams are absolutely necessary, as is their assignment based on product architecture, a system-level team is also necessary, referred to by the author as a product integration team (PIT).

The overall validation program and process should be under the leadership of the PIT. This leadership should take the form of at least the following provisions and actions:

a. Provide direction that requirements in need of validation be identified for each item undergoing development. Describe for the program the requirements subsets to be respected in the validation program. This may include the classes noted in Figure 2-3.
b. Define exactly how these requirements will be identified within databases, word processors, and/or separate media used on the program.
c. Provide the means to identify and capture the requirements validation identification and management information. This may be as simple as a set of matrices for each item with one or more requirements identified for validation or the logic of use and/or preparation for database use through schema modification.
d. Plan the validation program; review the ongoing work by product teams; provide feedback and direction to those teams on their progress, status, and nearterm work; assemble all of the reported information into a program-level status report; brief management on status periodically; and manage the overall program to yield planned results.
e. Identify, plan, and accomplish system specification–level validation actions.

We might ask ourselves who should be responsible for identifying the need for validation action. Clearly, the need for this testing must be identified by the designer or team responsible for the design because it is conducted based on their uncertainty of success and the degree of risk they feel that the program and company are subjected to as a result. Note that this offers the program two failure modes. First, the designer or team may fail to recognize the need for validation action either through oversight or unwarranted respect for their infallibility and design prowess. Second, validation action may be identified where none is needed through faulty reasoning or excessive caution. In the first case, the outcome will commonly be surprises late in the program (often during verification testing) that lead to cost and schedule hits and/or a performance shortfall that you are forced to accept because the commitment to a design cannot be undone within the unacceptable cost and schedule constraints that result at the time the problem is uncovered. In the second case, the consequences are added cost driven by unnecessary testing and analysis.

This is a common situation in solving complex problems. You have a choice that is not entirely satisfactory between two bad extremes. But, you do have a choice and it should be consciously made rather than simply evolve as a result of your lack of attention to the details. The use of concrete validation actions also separates the great systems houses from the great integration houses. The latter are very adept at solving horrible problems that arise during production, verification testing, or customer use of the product. These horrible problems simply do not appear in great systems houses because they have thought through all of the possibilities at a time in the program when problems can be solved carefully and inexpensively.

The agency or organization responsible for the overall validation process, ideally the PIT in the context of this book, must recognize these process failure modes and ensure that planned validation actions are motivated by a realistic concern for uncertainty and risk. Validation actions planned by designers and product teams should be reviewed and approved by the PIT as a means of avoiding the two validation failure modes.

2.6 Validation expectations

One way to ensure that a design will satisfy the requirements for the design is to accomplish the design work first and then define the requirements. Yes, this does sound foolish but it is not all that uncommon that drawings are released before the specification is released in some organizations. While it is necessary to define requirements before creating a design in order for the requirements work to have been at all valuable, it is not done without a certain risk. In our zeal to do a good job in requirements analysis, we may err on the opposite end of the spectrum by defining a problem that has no solution. We can spend considerable time and money working on a design solution only to find that it is impossible to satisfy the requirements as stated. We should reasonably expect that the requirements will be developed and released in a very timely fashion, that they will be complete and correct, and that they will permit the designer the greatest possible solution space within which to synthesize a solution that will play synergistically with all of the other parts of the system. So, the validation effort can be thought of as a component of the risk management strategy to uncover any problems with the requirements that would only later be caught after introducing cost and schedule problems.

2.6.1 Requirements necessity and completeness

Requirements analysis, when done well, results in a set of organized statements captured in a specification that restricts the design solution space. This set should be characterized as being necessary and sufficient — no more and no less.

If we fail to satisfy the necessary condition, it will be possible for the designer to synthesize the requirements into a design that will not function synergistically with respect to other system elements to the end that the system need may not be satisfied. If we failed to identify a requirement, for example, that a liquid oxygen valve in a cryogenic rocket propulsion system must function at a very low operating temperature, the valve could fail in flight resulting in the loss of the vehicle. You may say, "But that would have been detected in qualification testing of the valve." If the requirement were not in the specification, it is possible that it would not be exhaustively tested for that condition. One could argue incorrectly that a test program that examined valve operation at low temperature would be a case of excessive zeal on the part of the test engineer unwarranted by the requirements.

If we fail to satisfy the sufficient condition, we will have overspecified the requirements. The common result is that the cost will be greater than necessary. The increased cost is driven by the greater difficulty of satisfying the requirements. In the worst case, it is possible to achieve a null solution where it is impossible to satisfy the requirements. Thus one of the principal expectations from a validation effort is to encourage the result that we have identified only the requirements that are necessary and that we have sufficiently done so as well. Unfortunately, there is no mechanical way to ensure that the analyst satisfies this condition perfectly. The best ways to encourage this result are to effectively apply a structured analysis process, to maintain requirements traceability, and to apply an effective validation effort.

2.6.2 Requirements value credibility

All requirements should, ideally, be quantified. Otherwise, how will the designer know how to design the product. How will we test to determine if the product design is adequate for the intended application. Qualitatively stated requirements are an invitation to surprises that are measured in overruns of time and money. But, how can we establish a reasonable value for a requirement?

The most common valuation method is budgeting, flowdown, allocation, or apportionment. In this technique, we determine a value for a requirement at the system level and partition that value in accordance with a mathematical rule to determine child values. Many specialty engineers apply this technique and the following are examples:

a. The mass properties weights model;
b. The reliability and maintainability models;
c. Life cycle cost models.

It is very hard to make mistakes when applying this technique. It provides a well-disciplined environment within which to make the numerical value decisions. In the case of weight, the sum of the weights of all of the items at one branch and level in the architecture must be no more than the weight of the parent item for those items and it is very easy for the human doing this work to determine whether that is or is not the case. The error modes that are possible with this technique are (1) the system value is too high or too low leading either to an unachievable or insufficiently demanding condition and (2) the values assigned within any one branch at a particular level are not well distributed leading to local problems that can propagate downward from that point.

We may also include margins in the allocation process to provide management space in the event we run into difficulty synthesizing the requirements. A margin is simply an unassigned portion of the value at one or more levels of indenture. In the case of weight, for example, we may allocate only

150 pounds to the next tier for an item that we have previously determined is allowed 160 pounds. We have set aside 10 pounds at the parent item level that we can use during the development process to solve problems the designers encounter. If child item 2 designer finds that she cannot synthesis the requirements at the weight allocated and also stay within the cost and reliability requirements, she can appeal for an increase in weight allocation from the parent item lead engineer.

This whole process encourages value credibility because it is comprehensive and respects the simple economic principle that the sum of the parts must equal the whole and no more. It is very simple to test the veracity of the models both by the practitioner who owns the model and those who would audit it.

There are many requirements that will not yield directly to the allocation process, however. And, these are commonly the most difficult requirements to quantify. Given, for example, that we have a requirement for aircraft maximum airspeed, how shall we determine the engine thrust needed? It is more complicated than only knowing the airspeed requirement, isn't it? There are many factors involved including amount of drag entailed in the design, degree of asymmetries tolerated and the surface smoothness planned, and lift vs. gross weight considerations. The mathematical relationships between these parameters is complex and cannot accurately be resolved on the back of an envelope. Commonly, we have to apply some form of computer simulation to try various combinations of values working toward identification of a good and achievable mix of values.

In a less grand sense, we can also apply the best mix approach in pairs in what is called parametric analysis. We may inquire, for example, how reliability changes with cost and determine how much reliability we can afford. Ideally, the cost numbers should include the cost effects of poor reliability on customer loyalty, possibly in the form of two curves. One curve could simply show the effect of the cost of reliability while the other reveals the cost of unreliability on customer loyalty. Figure 2-11 illustrated this relationship.

We could include additional parameters in this parametric decision-making process, but there quickly becomes a point where the human, without special tools, cannot deal with the myriad of possibilities and complexity of the relationships leading to a need for modeling and simulation. Spreadsheet software can be used for low-end models but specialized, product-specific software will be required for more complex situations. Often this same software or derivatives thereof will be required in the development of the product design so the requirements modeling application is not necessarily money down a rat hole.

Another pathway to value credibility is through historical precedent in the form of recognized standards, professional experience on the part of the designer gained from having worked on similar applications in the past, or supervision dictation. These are appeals to authority and can produce valid requirements as well as acceptance of wildly flawed requirements. For all

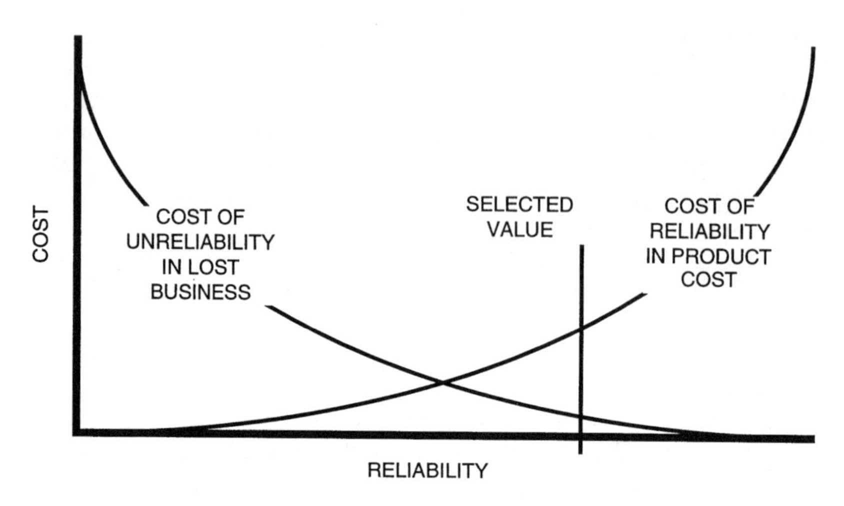

Figure 2-11 Parametric analysis of cost and reliability.

requirements that appeal to authority for their values, it is very important to identify a source and rationale and capture this data in the requirements database or in a tabular form in the notes section of the specification created with word processors. These source and rationale statements should be subjected to critical review by peers and team or group leadership.

2.6.3 *Synthesizability*

Given that we clearly understand the requirements for an item, have defined what we believe to be appropriate values for them, and validated those values through the techniques discussed above, there may still be problems in the synthesis of them. There are two possibilities with any set of requirements: (1) it is possible within the laws of science to synthesize them into a design or (2) it is not. Given that they can be synthesized based on current science and technology, the synthesis may be otherwise constrained due to cost, available time (schedule), or unavailable technologies. There are, therefore, two sources of nontechnical or programmatic synthesis constraints that we have to guard against as a prerequisite to technical validation of synthesizability — cost and schedule budgets consistent with the nature of the design problem and technology availability.

The designer or design team must review the perceived degree of difficulty in implementing a design based on the approved requirements and programmatic constraints. If the conclusion differs from previously allocated resources in time and money, they must make an appeal to management for a change in the allocations. Those changes may be forthcoming or not depending on circumstances and the strength of the case made for the change. If the additional resources cannot be acquired, the design approach or technology appeal may have to be changed for consistency with available resources.

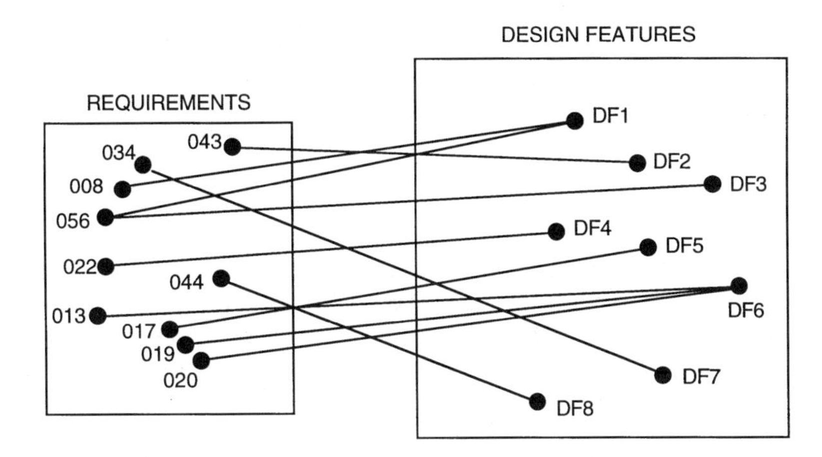

Figure 2-12 Validation traceability.

The validation process must be accomplished in the context of one or more design concepts. For this reason, many designers think of validation more in terms of design validation than requirements validation. The author has no problem with this conclusion since the validation process covers the transform between requirements and design but chose to focus the process on the requirements in this book.

One or more viable item design concepts that satisfy the item requirements without identification of any associated risks suggest that the item requirements can be synthesized, therefore, validating the requirements. In order to ensure that we have not overlooked anything, we should have some organized method for keeping track of the relationship between requirements and design features as suggested in the Venn diagram in Figure 2-12. The elements of the requirements set are identified by requirements IDs from the database in use. We could use paragraph numbers appearing in the specification but at the time the requirements are being validated this can be fairly volatile, whereas the requirements IDs should be stable since they are assigned by the computer in requirements databases and remain stable despite paragraphing and content changes. The design features appearing on engineering drawings or sketches have been arbitrarily identified by codes DF1 through DFn. We could conceivably keep track of the relationship between requirements and design features in a requirements database system, but the author knew of no database system on the market at the time this book was written that offered this capability.

Clearly, the technique suggested is a traceability technique. The difficulty is in identifying the design features in a way that they can be easily linked to the requirements. The requirements are in text form commonly in a database within which traceability can be easily maintained with respect to other text information. The design features are graphically expressed on engineering drawings and sketches and it is difficult to connect these graphical structures

ID	TITLE	DF ID	DESIGN FEATURE	DRAWING REF
043	Power Transfer	DF2	Chain drive	23-3334, ZN 5C
044	Speed-Power Control	DF8	Gear shifting mech	23-3334, ZN 8D

Figure 2-13 Synthesizability validation traceability record example.

in one system to text information in another system. The final answer to this difficulty will have to come through coordination of requirements tools and computer-aided design (CAD) systems. In the meantime, one could identify and maintain synthesis validation traceability by referencing drawing features defined in terms of a feature name and a drawing zone reference. Figure 2-13 illustrates one implementation for such a validation traceability record. The understanding here is that a database (ideally an integral part of the CAD system being used) identifies design features using what is called here a DF ID. The validation traceability database would simply make the connection between the requirements IDs in the requirements tool and the design feature IDs in CAD.

These validation traceability techniques provide a comprehensive way of ensuring completeness of the task. The argument can be made quite effectively that on most programs, however, the benefits would not match the cost of implementing. Is there a backup approach that may entail a little more risk but offer some assurance that it will be possible to synthesize the requirements? Yes, one could use the design or peer review process to achieve credible validation goals. In this approach, two or more reviewers are exposed to the principal design engineer's explanation of the design concept and they reach a conclusion about the credibility of the design solution relative to the requirements. If the conclusion is that the requirements have not been satisfied, unless it can be shown that it is not possible to synthesize the current requirements, then the designer must try again and submit to another review.

2.7 Validation methods

The purpose of validation action is to develop credible evidence of viability. The discussion on values and synthesizability imply that human mental action (which could be called analysis in the context of our current discussion) may be solely sufficient to determine if the requirements of an item are valid. Neither idle thinking nor analysis is sufficient. The methods that we might apply to accomplish this result are essentially the same as applied in verification. The principal methods involve testing as well as analysis but there are cases where inspection and demonstration can be applied effectively as well. The results of this work must be captured in some enduring way, such as in computer media or paper reports, so that they may be referred to over the life of the program.

2.7.1 Development evaluation testing

The testing accomplished in support of the preliminary design process is commonly called development or design evaluation testing (DET). Given that the designer or team determine that testing is the most cost-effective method for validating a requirement, they must determine how that testing will be accomplished and what resources will be required to accomplish the testing. A set of test requirements should be created and a plan designed to satisfy those requirements.

Testing involves an organized process of stimulating a product (or representation of the product) and measuring responses using special test equipment. The test plan commonly includes a sequence of events each with a predefined setup and an anticipated response. These test events are structured to prove that a particular design concept offers a viable solution to the item requirements. As a result of the tests, we can conclude with confidence that either we can or cannot satisfy the requirements with current design concepts and technology.

Clearly, testing most often offers the most credible evidence of validation, of the four methods discussed, because it places the design solution in the most realistic situation relative to its use and it involves a very controlled stimulus–response exercise that exposes detailed operation characteristics to clear human evaluation. It also most often requires, as might be suspected, more resources in time, money, and supporting resources than other methods. The costs include both labor and material costs. This is the way it should be, of course; the degree of credibility is directly related to cost and the amount of residual risk inversely related to cost.

Some examples of validation testing:

a. We have a requirement to provide a particular degree of guidance accuracy that is tighter than anything we have done before. We select a guidance concept, build a breadboard system or buy and integrate off-the-shelf components, and fly the result in a Lear jet with an instrumentation package and autopilot link in accordance with planned mission scenarios. We collect position error data, account for any differences in the control dynamics of the test and planned airframe responses, and compare the results to predictions derived from prediction, analysis, or simulation of system performance. As a result, we gain confidence that a final design can be created that satisfies the requirements. In the process, we also verify our simulation and gain confidence that it can be used to predict performance quickly and relatively inexpensively over a wide range of questions that will evolve during development and subsequently to analyze performance in the context of postproduction engineering changes being considered.
b. We have identified requirements for flight speed, maneuverability, and flight stability for a missile in flight to target. Captive carry space requirements dictate relatively small control surfaces that are unfolded

upon air launch. We have created a flight simulation that appears to support the current design concept but we decide to subject the complete airframe to wind tunnel test not only for general airframe aerodynamic characteristics (lift, drag, and airframe assembly asymmetry effects) but for control surface effectiveness as well. As a second thought, we wonder what would happen if the pilot had to jettison the weapons load in an unpowered state to save the airplane. Further testing reveals that a missile on one station would take the horizontal stabilizer off the aircraft. Modifications are introduced that force a missile dropped in this fashion to clear the aircraft.

c. A new sonar system has a requirement to be able passively to detect man-made underwater objects that emit a particular sound spectrum in both fresh- and sea-water over specified depth and distance ranges. To validate these requirements, we construct a test range with salinity control of the water for shorter distances and identify salt- and fresh-water test ranges in local water bodies. We intend to determine if it is possible to detect the required sound sources with laboratory equipment as a precedent to developing design concepts that can also withstand the rigors of naval shipboard use.

d. A new military jet transport must have an electronic flight control system (fly by wire) controls with a mechanical backup. An entirely new airframe, hydraulic system, and control system must be designed. As the control system component and installation designs become clear, a Flight Control System Simulator (iron bird) is developed that duplicates the configuration and operation of the aircraft so that interfaces between the flight control computers (and included software) and the hydraulic system can be tested during development under realistic conditions. Simulated flight control surfaces and hydraulic components must be in their same relative positions as they would be on the aircraft so a structural steel frame is built that duplicates the dimensions of the aircraft, with one exception. There is limited space within which to build the iron bird, so the fuselage length must be reduced to keep the simulator components within the available space. Hydraulic lines affected by the shortened simulated fuselage are looped to preserve their true length and the mechanical backup flight control cables are configured to preserve the spring rate of the full-size aircraft. All system elements are not available at the time we must start testing so we begin with laboratory equipment and preliminary models from suppliers that will be modified as the design develops. Flight surfaces driven by the system take the form of structural iron adapted to the control system operating interfaces since aerodynamics is not important in this model. As the components are developed, the revised models are installed in place of the currently installed components and tested as part of the entire system. This example illustrates several common characteristics of validation equipment and software. We have to be very careful to duplicate the necessary features of a design concept while economizing on the trivial. Good engineering judgment is needed to make these choices and if it is not present in the team many bad days will follow.

2.7.2 Analysis

Many requirements can be validated very effectively through analysis at less cost than could be done through testing. Analysis requires no apparatus replicating the product or designed to stimulate or monitor product responses. Analysis is conducted by the human mind unaided by external apparatus. The cost in analysis is, therefore, limited to labor cost only (although computing costs may also be involved depending on how the company financially accounts for computer use).

While the kinds of resources are fewer, it is possible that analysis can cost more than testing in some few cases. Where a human observation is involved, for example, it may be cheaper and quicker to build a simple representation of the product and expose people to it and evaluate their choices and results than it is to agonize through weeks of analysis over all of the possibilities. The auto industry often will outfit a test car to qualitatively demonstrate suspension or braking system performance rather than conduct a lengthy analysis of performance or computer simulation. The immediate cost is less because of the ready availability of cars and components.

There are cases where testing is impossible or so very expensive as to be prohibitive. For example, it is very difficult to test large items in a zero- or microgravity condition. Drop towers exist that can sustain the condition for seconds. There are aircraft that can fly an arc and sustain this condition for several minutes. But these environments cannot handle very large items nor for very long time intervals. The zero G performance of a whole space station cannot be tested on Earth because of its size and it must surrender to analysis as the principal method.

2.7.3 Technology demonstration

Demonstration is a process of making something evident through reasoning and exposition. It is a process of showing how some conclusion is supported through actual use of the article or process in question. Demonstration has less utility in validation than verification because you seldom have the physical resources required. During verification work, the product design is complete and one or more articles have been produced for qualification testing, so the product articles are available for use in demonstrations of their use.

Technology demonstration is most often a special case of demonstration but it can actually be accomplished through analysis or testing. We may conclude that we do not have access to a technology the designer believes we must have in order to satisfy a particular requirement. Current technology that we have access to simply will not permit a compliant design because the result will be too heavy, insufficiently reliable or durable, or characterized by poor accuracy. A new technology that we have not yet mastered (it could already be in use by our competitors) would solve the design problem and allow us to satisfy the current requirements. In this case, we cannot simply wave our arms and loudly cry during the proposal development period, "We can do it! We're the greatest!" We need some tangible evidence of a new capability not previously demonstrated by us.

By way of example, in the early 1990s there was a lot of interest among space launch vehicle producers, such as General Dynamics, McDonnell Douglas, and Martin Marietta, in finding a way to increase payload weight capability by reducing vehicle structural weight. One promising material change involved the use of a new aluminum alloy with lithium that had evolved from research and development work. The problem was how to work it in a manufacturing sense to achieve the needed strength with significant weight savings. The designers needed information on design features that could be manufactured with available or new tooling that would also satisfy demanding structural requirements. Designers were already fully aware of designs and manufacturing scenarios that had proved effective with conventional aluminum alloys and stainless steel but the knowledge base was not well developed for the aluminum–lithium alloy.

As a result, companies undertook technology demonstrations to gain the knowledge base needed to proceed with confidence in the design of lighter-weight structures providing higher payload performance. These demonstrations involved testing of samples in bending, torsion, and shear under vibration and thermal conditions at levels and frequencies anticipated in the application, as well as evolution of manufacturing methods and tools effective in working the material without degrading its structural properties.

In the interest of simplicity, both demonstration and inspection have been eliminated from Figure 2-1 but the reader can simply add them to each string of the methods included.

2.7.4 Inspection

Inspection is a formal or official viewing of an item or process in contrast or comparison with some standard of behavior or configuration. The human senses are the principal instrument of inspection but simple manually operated measuring instruments may be used such as rulers, scales, or templates. Specific features of a product or process are observed and studied by one or more humans, generally in a static situation, and features compared with a standard. An inspection of a dynamic condition might better be referred to as a demonstration.

The most common use of inspection is in acceptance of product after manufacturing. Product finish, fit, color, and feature locations and orientations are examples of things that will yield to inspection by a quality assurance inspector. Inspection does require a physical reality to be effective so it is not so commonly employed in validation as it is in requirements verification and product acceptance.

2.7.5 Combined methods

Often life is more complicated than suggested in the discussion of the four isolated methods. We may have to implement a test or demonstration that produces response data that is very complex and requires analysis to understand it. A particular validation action can involve strings of actions not all of the same method. The four methods offered provide a convenient way to describe all validation actions in a comprehensive way, but it is not all that important what methods are used. It is important that a cost-effective way is determined to validate all requirements.

Simulation offers a very common case in point for combined methods. Early in a program, we create a simulation to help develop a design concept and through multiple simulation runs we uncover a best guess at a set of values for some very troublesome performance requirements. We then develop real hardware and software based on the results of our simulation and update the simulation to replace some of the simulation code with actual product hardware and software. Finally, as the completed product becomes available, we subject it to system-level testing and use the results to verify the simulation. Subsequently, we can use the simulation to predict product performance under new conditions related to modifications or previously untested scenarios. Throughout the development and use of the simulation we cycle through sequences of test and demonstration followed by analysis leading to new rounds of testing and demonstration iteratively tuning the simulation for realism and the preferred product design for optimum performance.

2.7.6 Validation by review

The reader, especially one with a lot of commercial experience, may have reached a tentative conclusion that this validation process has been overdone and, as described, it is simply too expensive to implement for the amount of benefit derived. Those with some experience in the development of complex systems will realize that the number of requirements that will require something more than a cursory glance in the validation process is very few, perhaps on the order of 5% or less. It is commonly the requirements that offer the greatest benefit from validation that are the requirements that most powerfully drive the design solution. It cannot be denied that this thoughtful process of ensuring that the synthesis will be successful does cost money and the cost can be considerable. However, in the defense and space industry, where some very large and expensive problems have come into view too late to deal effectively with them, validation is accepted as an example of the famous Fram Oil Company oil filter commercial where the mechanic is holding an oil filter saying, "Pay me now or pay me later."

If cost is a severe program constraint, however, one could implement a partially effective validation effort by insisting that the requirements in each specification be reviewed prior to release of the document by an engineer with experience on several development programs and one who was not involved in the definition of the content of the specification. This engineer should simply look for problem areas and report them to the specification principal engineer, team leader, or program chief engineer. This simple act accomplished by someone with no ego investment in the document will repay the small cost of 3 to 4 h per specification many times over. This is essentially a peer review process.

This review process can be expanded by holding in-house design reviews attended by the development team who present their design concept to a review team, ideally selected from the functional management staff of the company. The presenters should be required to follow a presentation pattern as follows:

a. Team leader or principal design engineer demonstrates his/her understanding of the item requirements by describing how they influenced the design. A subset of the item requirements should be identified that caused the team or designer the most concern and it should be discussed how the team or designer mitigated these concerns.
b. Describe the design solution concept and link the features with the requirements previously discussed.
c. Report on the results of any formal validation efforts undertaken and the status of any actions that have not been completed.
d. Provide time for the person who did the requirements peer review to give a report.
e. Insist that the specialty engineers who reviewed and contributed to the preliminary design work offer a brief statement about the design concept from their specialized perspective. A "no comment" should not be accepted; rather a clear statement of the viability of the design from their perspective should be expected.
f. Remaining concerns and risks should be cited and the status of the work that remains to be accomplished to mitigate them discussed.

Given that the review team has been populated by people experienced in asking penetrating questions and that they have some knowledge of the product line, this review will pay big dividends. The presenters should not have to prepare special materials to satisfy the needs of this review. It should be possible for them to simply expose the review team to the results of their work using design sketches, results of analyses, and other materials prepared as a normal consequence of doing the design work. If the organization has a computer network upon which has been stored the product development work in an organized fashion, it should be possible to project that information directly upon a presentation screen in the meeting room where the review is held.

As a result of the review, the review team chairman should offer a clear response to the development team or engineer supporting the work done to date, if appropriate, or noting specific concerns that the review team still has. Action items should be assigned to accomplish any specific remaining actions to resolve those concerns. These action items should be assigned to a specific person and have a clear completion date negotiated with the person responsible for closure. If the review team is generally satisfied with the progress, they may simply require that responses to any action items be reported to the meeting chairman as closing action and, so long as the results of those action items do not expose further concerns, accept the design concept and permit the design process to proceed as scheduled. If serious problems have been uncovered in this process, the review may have to be rescheduled for a specific date agreed upon by the design team or designer.

While reviews are discussed here as an alternative to a formal validation process, the reader should be impressed with the need for a review process regardless of what is done about validation. The point is that a sound in-process review can satisfy the *minimal* needs of a validation process.

2.8 Product representations

During the verification activity we have available to us the actual product to test and inspect. At the time validation work is accomplished, we only have ideas, concepts, and models. During the verification work, we have a fully approved set of engineering drawings subjected to continuous configuration management controls. During much of the validation work in many companies, configuration management may not even be funded to support early program work because formal drawings have not yet been produced. These factors make it very difficult to control the configuration of the product representations used in validation. Chapter 3 focuses on this in detail.

An electronics item development program may employ breadboards and brassboards to evaluate design concepts. Other technologies may appeal to physical models and mock-ups that have to capture the current design concept. Test benches, mathematical models, and simulations may all be useful in validating item requirements and design concepts. As Chapter 3 points out, we must have an effective way to maintain control of the configuration of these product representations such that they remain in synchronism with the design concept. Otherwise, as the design and the test representations diverge, the results from any tests accomplished using these representations become suspect.

2.9 Whole program phases

DoD and NASA recognize a multiphased approach for development of systems to solve complex problems. One of the DoD phases is called Demonstration Validation, often shortened to DemVal. The purpose of this phase is to ensure that the system requirements are mature and clearly understood, that the technologies are mature, and that a viable concept has been defined. Sometimes, this phase can entail designing and building an actual full-scale operating product by multiple contractors with flight or ground tests to prove out the design concept. Many aircraft programs in DoD have applied this approach, most recently the F22 program. Two contractor teams each designed and built several full-scale aircraft and completed a flight test program as part of this demonstration. The customer then selected the design that most effectively satisfied the requirements. This is a pretty grand form of validation not commonly available due to cost limitations and the likelihood that something less costly will be effective in developing an adequate product.

You could look at the spiral development model, involving multiple cycles of analysis, design, build, and test, as described briefly in Chapter 1, as a kind of iterative validation process combined with a search for necessary or desirable characteristics. A principal motivation for using the spiral model is a lack of confidence that the problem is understood with sufficient clarity that a specification can be completed as a prerequisite to design work. This is a valid basis for validation for the whole set of requirements for the item/system involved and the iterative maturation process of the spiral development model satisfies the validation need.

chapter three

Product representations control

3.1 The many views of the product

For many people in engineering, the only view of the product during the development period is composed of engineering drawings and lists. Just from a documentation perspective, we would also have to accept that the following are representations of the product as well: specifications; test plans, procedures, and reports; analysis reports; and functional analysis data. It could be argued, however, that these documents are also engineering drawings since they can be identified using the same drawing numbering system used for drawings proper.

We would all argue that these documents must refer to a known configuration and commonly the engineering drawings are used as the master definition of the configuration. After we start preparing drawings on a program, their numbers can be used in combination with dash numbers and revision letters to refer very specifically to items under development. All of the companion documents can be similarly referred to and mapped to their corresponding drawing numbers. Prior to drawings being prepared, architecture IDs can be used as this master configuration identification reference. As drawing numbers are assigned to the architecture items they may be allowed to take precedence as the primary configuration identification. Figure 3-1 illustrates an architecture diagram with architecture ID numbers in the lower right corner of the items represented by blocks. This model uses a base 60 system making decimal delimiting unnecessary. Alternatively, decimal points can be used to separate levels from one another.

Specific configurations can be defined in terms of sets of documents in particular configurations forming what are called baselines. The control of the evolution of these sets of data is the principal activity of a configuration management organization. The methods for accomplishing configuration management are well developed and most often this job is done very well

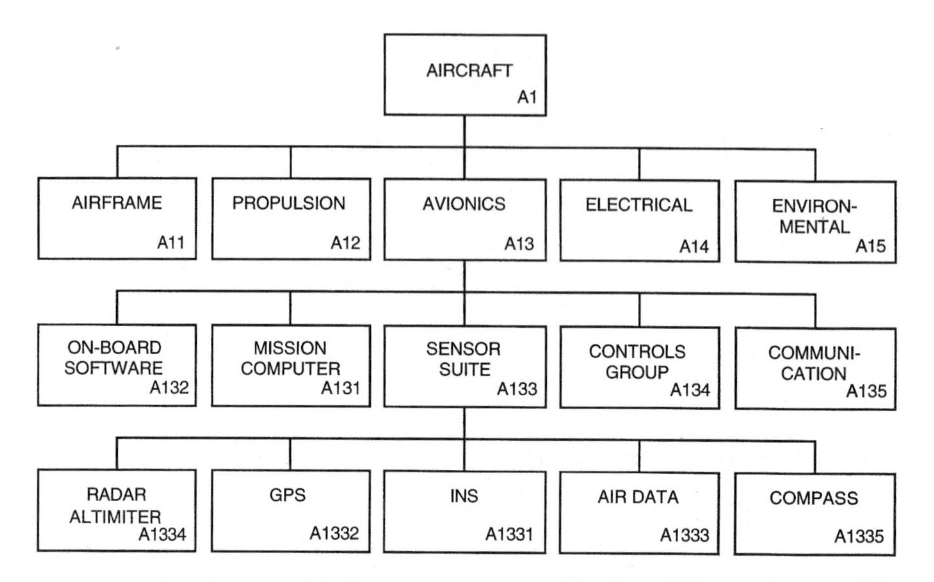

Figure 3-1 Typical architecture block diagram.

in engineering organizations. As important as these documentation baselines are, however, they are not the only representations of the product. The configuration of the other representations is seldom managed well. What might these other representations be?

Throughout the early development of a system, special test articles are required to support the validation of certain requirements in development evaluation testing. In the development of an aircraft system, it may be necessary to build a hydraulic control system test bench to evaluate design concepts and their suitability for satisfying required control surface shaft torque under flight conditions. This test bench must comply with the current configuration as defined by the corresponding product item documentation set. If this test bench configuration includes actuators with different static and dynamic characteristics than those currently identified in the design baseline documentation, the results from any tests involving the test bench are brought into question.

If differences between product representations and the current product design go unnoticed and tests or analyses are completed based on those representations, the resultant data may be used to make critical decisions that are flawed which may lead to other flawed decisions in a pattern that could be very difficult to unravel when we inevitably find that a problem exists.

In addition to special development test articles, other examples of representations include: physical models and mock-ups; mathematical models (finite element analysis, reliability model, weights model, etc.); and simulation software and data sets.

Table 3-1 Sample Representations (Rep) Identification Matrix

Item ID	Item name	Rep. number	Product number	Principle engineer	User data
001	Avionics Test Bench	34E1212B	34E4242C	E. Jones	
002	Guidance Simulation	34A4511	34E1521	A. Clancy	
003	Reliability Mathematical Model	34A1321	34E1000	P. Gordon	
004	Weights Model	34A5444	34S2400	M. McSwain	

3.2 *Representation identification*

The first step in the process of controlling the configuration of product representations is to simply list them all. Table 3-1 offers a sample identification matrix showing that we have numbered and named each item. Following that, we have identified the configuration of the representation with a unique drawing number and the corresponding product drawing number. A principal engineer is named who is responsible for knowing the configuration of the item at all times and maintaining it in a specific configuration. Finally, there may be application-specific user data of interest that one could add to the matrix. We may wish, for example, to have a column for current status.

Table 3-1 only ensures that we know, at any one time, the configuration of each representation provided that we have some way of interpreting the cryptic Rep. number. Each line item should be backed up with some viewgraph engineering, one or more drawings, or a more formal set of documentation all coordinated by the Rep. number. This may be as simple as a sketch of the item with features noted with arrows so representations identification and control need not be a costly process.

The process of identifying product representations should begin in the proposal period or commercial equivalent. People from the different functions can be polled or interviewed by a system engineer to uncover representations people working on the proposal think will be needed during program implementation. Some of these will be identified in proposal write-ups, required material lists prepared in support of the material cost estimate, or be obvious from a description of work required in the statement of work. In some cases, the responsible proposal person may not have overtly identified a representation that will be needed, and it may require some discussion to bring that need to the surface so that it can be listed and included in cost estimates. In other cases, we will find that items that are listed cannot be justified by the person who added it to the list. It may be possible to achieve the objectives of the representation by other means using resources that are required for other purposes. So, this list or matrix can be established as a part of the proposal work that should be accomplished anyway to coordinate needed product representations with planned work and quantification of the proposal material estimate.

3.3 Representation management

Given that we have a means to record the identity of each representation and clearly define its current configuration, we should also make provisions for capturing the history of each item. We must know the representation configuration at all times, not just when the program begins. So, we must recognize that the matrix illustrated in Table 3-1 will change over time. As we become better informed about the best product design solution, the design configuration will change. We may also have to modify some representations into different configurations or create multiple versions of them as part of the process of evaluating two or more alternative design solutions.

Ideally, we would immediately think, all active representations should be maintained in a configuration that reflects the current product design status, but that may be both unnecessary and counterproductive as well as costly without corresponding benefit. The first requirement is that we know the representative configuration and be capable of determining its relationship with respect to the current product design. Some representations will have to be applied fairly continuously throughout the development period and these will have to be maintained in the current configuration. An example of this is the weights model.

There may be long periods of inactivity for some representations following an intense application. During this period of inactivity, the related product design may change several times. The question is, should we accept the cost burden to upgrade such a representation each time the product design changes or only when and if we have to apply it again. This would be a good factor to include in our identification matrix (as part of the user data). Program management could review the list and place the items in one of several categories as follows: (1) representations which must be maintained current at all times, (2) representations for which the cost–benefit, for each change, should be evaluated and the change accomplished, if appropriate, and (3) representations which should not be updated unless directed by program management.

Audits should be accomplished from time to time of the configuration of active representations comparing them with the listed configuration. This need not be a costly or elaborate activity. A system engineer can be assigned to audit three items during a 2-week period as a supplemental duty. This process can be repeated at other times for other items. Where discrepancies are found, we may wish to modify one or more representations to a specific configuration or only update the matrix to show the current configuration.

Some of the representations will be useful in early program work and others in later program work. Program management should require that the matrix reflect only the active items or that it be annotated with a status field indicating items as active or inactive. When a particular representation is no longer required with some certainty, we may wish to dispose of it or place it in storage in the event that it may have an application on later product modifications.

Some time after a representation has served its initially conceived purpose and been put in storage, it may develop that it is needed again to perform development testing work. The current configuration of the representation and the product configuration of interest must be carefully studied and ways explored to cause equivalence to the extent necessary. It may even be necessary from time to time to deconfigure a representation to explore an product configuration of continuing interest. It is important to recognize that a particular representation need not reflect every feature of a product item it represents. While it would be possible to fabricate a general test article that had all of the features of the planned product, this course of action would probably be more costly and lead to less flexibility than using specialized representations with only the features of interest.

For example, we are developing an air-launched missile and must, at some point in the program, verify that the missile can be safely jettisoned from the aircraft for safety-of-flight reasons. We could use a complete missile, possibly in a range mode with recovery parachute, or we could use a deadweight with the same size, aerodynamics shape, mass, and center of gravity of the real article but without any of its other features. This problem gets more complicated when the missile must be driven into a particular control fin mode for jettison, and in such a case it might be more effective to use an actual missile with onboard computer to verify safe jettison.

3.4 Representations documentation

In most companies and programs there is no recognized way to document these representations uniformly. Let us propose one. During the earliest work on a program someone should be identified to collect information on these items as discussed above. This information should be captured in a product representations log or report in the form of an identification and status list or matrix and a few descriptive pages for each line item in the list that describes the configuration of the item. The list must also include identification of the engineer responsible for the item and the person to be held accountable for developing and maintaining the descriptive information as well as maintaining the actual configuration of the item.

The person who creates and maintains the overall document could be selected from a systems or configuration management function. The problem with the latter is that commonly there is no budget allocated to this function early in a program when this work must be initiated.

The documentation should be deposited on a computer network server and made available to all in read-only mode. The document need not be assembled into a paper product for publication. It can be assembled from pieces which are assigned to the several principal engineers who have write privileges for their portion of the document. The principal engineer for an item should maintain some form of simple configuration identification documentation whether this information is collected at program level or not. This whole matter should not add significantly to program cost because the principal engineers should keep track of the configuration of their representations. It will take some system engineering discipline and force to encourage all of these principal engineers to write down the configuration and share it with the program in general.

Someone in systems or configuration management should be aware of all of the representations the program is depending upon and ensure that they are in a configuration appropriate to their use. With the cost distributed as noted above, the cost of the oversight function also will not be significant. What will be significant is the positive results that occur on the program by avoiding failures driven by inappropriate configuration of product representations. This is another one of those situations where system engineering value goes undetected when it is done well and provides glaring evidence of failure when it is not.

chapter four

Verification requirements

4.1 Verification documentation

Requirements are defined for items as a prerequisite to design and are written for the specific purpose of constraining the solution space such that it is assured that the item designed in accordance with them will function in a synergistic way with respect to other items to achieve the system function, purpose, or need. The requirements for an item are captured in a specification. The structure of a specification may follow any one of several formats but a very popular format, recognized by DoD, entails six sections. The requirements for an item in this format are documented in Section 3 titled "Requirements." A specification in this format also includes a Section 4 titled "Quality Assurance Provisions" or "Verification." Section 4 defines verification requirements for all of the item requirements defined primarily in Section 3.

Verification requirements define precisely how it will be determined whether or not the item requirements have been fully complied with in the design. For each requirement in the requirements section (Section 3 of a DoD specification) we should find one or more verification requirements in Section 4 (DoD). It is good practice to also include in Section 4 a verification traceability matrix that establishes the relationship between Sections 3 and 4 content and tells the methods that will be used to accomplish the indicated verification actions. If Section 3 includes a weight requirement, such as

> 3.7.2 Weight. Item weight shall be less than or equal to 134 pounds

Section 4 could contain a verification requirement, such as

> 4.3.5 Weight. Item weight shall be determined by a scale, the calibration for which is current, with an accuracy of plus or minus 6 ounces. The item shall be placed on the scale located on a level, stable surface and a reading taken. Measured weight shall be less than 134 pounds and 11 ounces.

MIL-STD-490A, the military standard for program-peculiar specifications for several decades until 1995, named this Section 4 "Quality Assurance Provisions." The replacement for 490A, Appendix A of MIL-STD-961D, refers to this section as "Verification" reinforcing the author's selection of the meaning of the V words included in Chapter 1 and assumed throughout this book.

This book accepts the six-section DoD format as follows:

1. Scope
2. Applicable Documents
3. Requirements
4. Verification
5. Preparation for Delivery
6. Notes

The majority of requirements that should be verified will appear in Section 3. But, it is possible that Section 5 could include requirements that should be verified. A specification may include appended data that may also require verification. An example of this is where the specification contains classified material which is grouped in an appendix that can be treated as a separate document under the rules for handling classified material. The majority of the specification, being unclassified, can then be handled free of these constraints.

Section 2 includes references to other documents that are identified in Sections 3 and 5 either requiring compliance or cited as guides for the design. This is a essentially a bibliography for the specification and should not include any references not appearing in requirements paragraphs elsewhere in the specification. Therefore, it is inappropriate to verify the content of Section 2 directly. Rather, the requirements that contain these references in Sections 3 and 5 must be verified in accordance with these documents where compliance with the documents is required. The way that the document is called in the body of the specification must be considered in determining verification needs because it may limit the content to method 2 or schedule 5, for example, or otherwise limit the applicability. Section 2 may include tailoring that must be considered as well. Generally, this tailoring will make the applicable document content less restricting. Sections 1 and 6, in this structure, should contain no requirements.

It should be said that not all organizations capture the verification requirements in the specification covering the item they concern. The author encourages that the verification requirements be captured in the corresponding specification because it brings all of the item requirements into one package through which they can be managed cleanly. As we will see in this chapter, there are other places to capture them including not at all. Failure to identify how it will be verified that the product satisfies its requirements will often lead to cost, schedule, and performance risks that we gain insight into only late in the program. It is a fact proven on many programs that the later in a program where faults are exposed, the more it costs in dollars and time to correct them. Alternatively, these failures may require so much cost to fix that it is simply not feasible to correct them, forcing acceptance of lower than required performance. Timely application of the verification process will help to unearth program problems as early as possible and encourage their resolution while it is still possible to do so within the context of reasonable cost constraints.

An additional reason to coordinate the preparation of Sections 3 and 4 is that the content of Section 3 will be better than if the two sections are crafted independently. It is very difficult to write a verification requirement if the corresponding Section 3 requirement is poorly written. This will lead to an improvement in the Section 3 requirement to make it easier to write the Section 4 requirement. If one person is responsible for writing the Section 3 requirements and someone else must then craft Section 4, or the equivalent content included elsewhere, it removes a significant motivation for quality work from those responsible for Section 3.

4.2 *Item planning fundamentals*

A program involving some complexity will likely require verification for several, possibly many, items. The verification work for each item can follow essentially the same process within the context of an aggregate program that manages the complete effort at the system level. This chapter and Chapters 5 through 8 focus on item qualification verification. Each item that must be qualified may be treated as discussed here. The developer should organize the complete process for all of the items into a system verification program as discussed in Chapter 9.

4.2.1 *Traceability matrix*

The verification planning process begins with the verification traceability matrix that should appear in each specification. All of the verification traceability matrices in all of the program specifications provide the principal input to a planning process that must integrate these many requirements into a least-cost process composed of verification strings of the several kinds defined in paragraph 4.2.2. Table 4-1 offers a fragment from a verification traceability matrix for an integrated guidance set. The matrix coordinates the requirements (generally in Section 3) with verification methods to be applied, the level at which the verification action will be accomplished, a reference to the paragraph in Section 4 (in the context described above), and a verification string number (VSN).

Some engineers insist on assigning the same paragraph numbers in Section 4 that are used in Section 3, with the exception of the leading 4 being added. This may require a leading subparagraph as well, such as 4.2. While this does make it easy to understand the correlation between them, it places

Table 4-1 Sample Verification Traceability Matrix

Section 3 Paragraph	Title	Methods T	A	E	D	Level	Section 4 Paragraph	VSN
3.2.1.5	Guidance Accuracy	X				Item	4.2.37	0032
			X			Item	4.2.38	0033
3.2.1.6	Clock Accuracy	X				System	4.2.41	0102

an added and unnecessary burden on those who must maintain the document. Specification paragraph numbering will change from time to time and it will ripple through Section 4 when this approach is used. Granted, much of this can be avoided by retaining the paragraph numbers of deleted requirements with a note "Deleted," but this is not a total solution. As the reader can see from Table 4-1, this matrix provides clear traceability between the two sections so it matters not how you number Section 4 paragraphs so long as the numbers are unique.

In Table 4-1, we have also added a VSN field. A VSN is a combination of an item requirement (complete paragraph or parsed fragment), a verification method, and its corresponding verification requirement (complete paragraph or parsed fragment). Some people prefer to call these verification event numbers, but the author concluded that the word event should be used in its normal planning data context to represent an instant in time corresponding to some condition. Also, the word string has an immediate appeal in a physical sense in that we will attach these strings to verification tasks.

Those familiar with database design will recognize this as a possibly redundant field since each line (record) in the table appears to be unique without it. The problem, suggested by the parenthetical reference to paragraph fragments, is that all specifications will not be perfectly prepared. Even if our organization is staffed by perfect analysts, our customers and our suppliers may foist upon us imperfect specifications containing page-long paragraphs stuffed with multiple requirements. The VSN allows us to identify atomic requirements verification components uniquely.

Even if all specifications were prepared with the greatest respect for the one-requirement–one-paragraph rule, the VSN is also useful in joining a requirements database with separate test and evaluation and analysis databases for performance purposes as well as to provide a clear set of hooks between the databases that may be developed at different times. Figure 4-1 illustrates, in a Venn diagram format, an example of a fairly complex paragraph in a specification Section 3. The paragraph consists of three fragments each of which must be verified through an application of a different combination of verification methods. Each combination of a fragment, a method, and the corresponding verification requirement in Section 4 corresponds to a unique VSN. In the case where a paragraph was properly written and contains only a single fragment, and that requirement will be verified by a

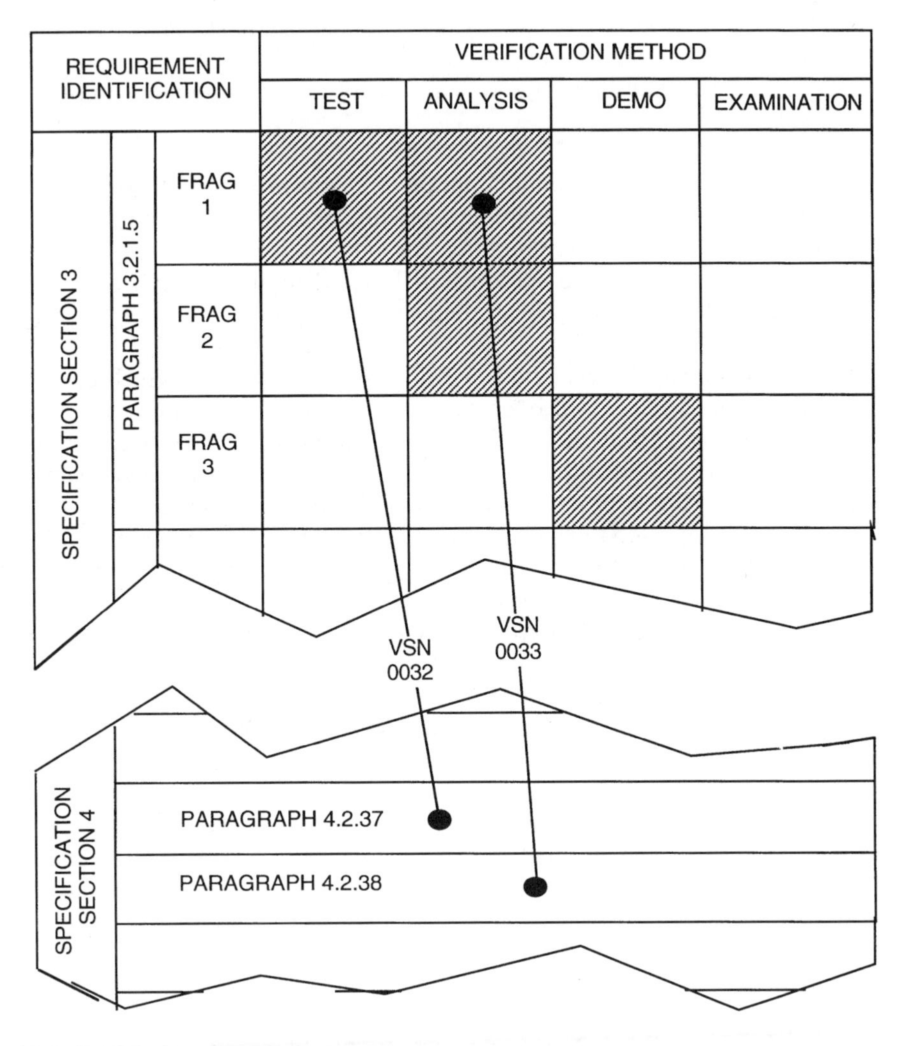

Figure 4-1 Verification string numbers (VSN).

single method, this identification approach will collapse into a single VSN assigned for the paragraph.

4.2.2 Verification methods

Table 4-1 also identifies the methods to be used in verifying the listed requirements. The four principal methods employed are test (T), analysis (A), examination (E), and demonstration (D) described below. The examination method was referred to as inspection in MIL-STD-490A but this often resulted in confusion in that all verification methods were also referred to under the general heading of quality inspections. MIL-STD-961D uses the term inspection as a generic method of which the other four are examples and uses the word examination for one of the four specific methods in place of inspection.

a. *Analysis.* An element of verification or inspection that utilizes established technical or mathematical models or simulations, algorithms, charts, graphs, circuit diagrams, or other scientific principles and procedures to provide evidence that stated requirements were met. Product item features are studied to determine if they comply with required characteristics.
b. *Demonstration.* An element of verification or inspection which generally denotes the actual operation, adjustment, or reconfiguration of items to provide evidence that the designed functions were accomplished under specific scenarios. The items may be instrumented and qualitative limits of performance monitored. The product item to be verified is operated in some fashion (operation, adjustment, or reconfiguration) so as to perform its designated functions under specific scenarios. Observations made by engineers are compared with predetermined responses based on item requirements. The intent is to step an item through a predetermined process and observe that it satisfies required operating characteristics. The items may be instrumented and quantitative limits of performance monitored and recorded.
c. *Examination.* An element of verification or inspection consisting of investigation, without the use of special laboratory appliances or procedures, of items to determine conformance to those specified requirements which can be determined by investigations. Examination is generally nondestructive and typically includes the use of sight, hearing, smell, touch, and taste; simple physical manipulation; mechanical and electrical gauging and measurement; and other forms of investigation. A test engineer makes observations in a static situation. The observations are normally of a direct visual nature unsupported by anything other than simple instruments like clocks, rulers, and other devices that are easily monitored by the examining engineer. The examination may include review of descriptive documentation and comparison of item features and characteristics with predetermined standards to determine conformance to requirements without the use of special laboratory equipment or procedures.
d. *Test.* An element of verification or inspection which generally denotes the determination, by technical means, of the properties or elements of items, including functional operation, and involves the application of established scientific principles and procedures. The product item is subjected to a systematic series of planned stimulations often using special test equipment. Performance is quantitatively measured either during or after the controlled application of real or simulated functional

or environmental stimuli. The analysis of data derived from a test is an integral part of the test and may involve automated data reduction to produce the necessary results.

4.2.3 Product and verification levels

Table 4-1 includes a column titled Level. This column identifies the architectural level of hardware or software at which that requirement should be verified. Generally, we will conclude that a requirement should be verified at the item level, that is, at the level corresponding to the specification within which the traceability matrix appears. This means that we would run a test, examination, demonstration, or analysis focused on the specification item. Other possibilities are that we could do the verification work at a higher or lower assembly level. If the item is a component, we could do the verification work at the subsystem, end-item, or system level. Similarly, if the item a subsystem, we could do the verification work at the component level. Most often, when the level is other than the item level, it will be a higher level.

Suppose that we were dealing with the requirements for a liquid propellant rocket engine. We may require the engine supplier to verify the procurement specification liquid oxygen and hydrogen flow rates corresponding to full thrust at sea level in a test prior to shipment and also require that these levels be verified after installation in a launchpad test. We would mark the requirement in the engine specification for verification at the item level (engine) and at the system level (launch vehicle on the pad). Note that in this case we are dealing with acceptance requirements verification rather than development requirements verification (qualification).

The "V" model discussed in Chapter 1 and illustrated in Figure 1-7 offers a very illuminating view of the verification levels applied in product development. At each stage of the decomposition process we are obligated to define appropriate requirements constraining the design solution space, and, for each requirement, we have an obligation to define precisely how we will verify that the design solution will satisfy the requirement. This verification action should normally be undertaken at the item level, but it may make good sense for particular item requirements to be verified at a higher level such that several items may be verified in a coordinated series of tests and analyses.

Also as noted above, there is a third level of verification not covered in qualification and acceptance verification activity commonly. That third level is system verification. This would normally only be accomplished in association with the qualification process and only done once. Following the completion of the qualification process for configuration or end items, a program should have arrived at a point of some confidence that the system will function properly. It is still possible, of course, that when the verified items are combined into a system that they will not perform as specified. When dealing with very complex systems, like aircraft, for example, this final form of qualification verification must be accomplished within a very special space and under very controlled conditions to minimize the chance of damage to property and threat to the safety and health of persons. This is especially true if the system in question involves military characteristics.

In the case of aircraft, we call this activity flight test. The aircraft passes through a series of ground tests including a taxi test where all systems can be proved to the extent possible without actually leaving the ground. This will be followed by actual takeoff and landing exercises and flights of increasing duration, increasingly demanding maneuvering trials, and gradually expanding airspeed–altitude envelope excursions and increasingly demanding loading conditions. Similar graduated testing work is accomplished on land vehicles, ships, and space vehicles.

It is possible that an ongoing acceptance kind of verification may also be applied on a program. If the product enjoys a long production run with multiple blocks of product, as in a missile program, each block may have to be tested in a flight test program to ensure a condition of continuing repeatability of the production process. Even where the production run is constant, individual product items could be selected at random to undergo operational testing to accumulate statistical data on production quality, continuing tooling trueness, product reliability, and storage provisions adequacy.

Most often, system-level testing is accomplished by the builder sometimes under the supervision of the customer. In DoD this kind of testing is called development test and evaluation (DT&E). Customers may conduct their own form of these tests once the product has completed DT&E called operational test and evaluation (OT&E). The later may be broken into interim (IOT&E) and follow-on operational test and evaluation (FOT&E) activities. Operational testing is commonly accomplished with actual production items while development testing is accomplished using special production units that will generally have special instrumentation equipment installed.

4.2.4 Verification phases

There are at least two phases of verification, and some would say three. First, we should verify that our product design solution satisfies the requirements in the item development or Part I, specification. Subsequent to the DoD move to performance specifications in the mid 1990s, these specifications would be called performance specifications in DoD. Some engineers refer to this as verifying the "design-to requirements." This process is also referred to as qualification of the item for the application. The associated verification work is commonly only accomplished once in the life of the item on the first production-representative article. The program manufacturing capability may not have matured sufficiently to produce the articles required in the qualification process but these low-production-rate units should be as representative as possible to ensure that the verification results can be extrapolated to full-production units. Chapters 6 through 9 deal with item qualification verification and the related functional configuration audits. Chapter 10 covers system verification.

The second verification phase involves acceptance of each product article as it becomes available from manufacturing or other sources to determine if it is acceptable for sale or delivery. In this case, we seek to verify that the individual item complies with "build-to requirements" sometimes captured in a product, or Part II, specification. These specifications were renamed detail specifications in MIL-STD-961D, Appendix A. This activity could also be driven by engineering drawings as well as quality and manufacturing planning documentation without reference to product specifications. This process is referred to as product acceptance and is either accomplished on each product article delivered or within some sampling regime. Chapters 11 through 13 deal with acceptance verification and the related physical configuration audits.

Some organizations insert a first article inspection between these two phases that may call for special inspections that will not be required in acceptance of each article. The author has collapsed this into the acceptance phase in this book but the reader may wish to separate these and differently characterize them. An example of a case where this might be useful is where the first article is being used as a trailblazer to check out the process as well as to manufacture or process the product.

4.2.5 *Items subject to qualification and acceptance*

Qualification and acceptance verification is commonly accomplished on items variously referred to as configuration or end items. These are items through which a whole development program is managed, items that require new development and have a physical or organizational integrity. They are designated in early development planning, and major program design reviews and audits are organized around them. Ideally, every branch in the system architecture can be traced up through at least one of these items such that verification work accomplished on them will ensure that the complete system has been subjected to verification. Very complex systems may include more than one level of these items.

It is very important to list specifically the items that will be subjected to formal verification of requirements. It is not, in every situation, necessary to verify everything at every level but it is possible to make outrageous mistakes where we do not consciously make this list and publish it exposing it to critical review. Unfortunately, these mistakes are not always immediately obvious and are only discovered when the product enters service. Verification of the requirements for components does not necessarily guarantee the next-higher-level item will satisfy its requirements. The Hubble Space Telescope will forever offer an example of this failure. In computer software, most people would be rightly reluctant to omit testing of any entity in the hierarchy. At the same time, there are cases where it should not be necessary to verify at all levels. It requires careful analysis and good engineering judgment built through experience to make these calls.

4.2.6 Verification directionality

Things in a system are most often verified from the bottom up, where, as each item becomes available, we subject it to a planned verification sequence. The results of the verification work for all of the items at that level and in that architecture branch are used as a basis for verifying the parent item. And this pattern is followed up to the system level. Other directions are possible. The top-down approach is not often applied in hardware but can be used in software in combination with a top-down build cycle. Initially, we create the code corresponding to the system level complete with the logic for calls to lower-tier entities that can be simulated in system-level tests. As the lower-tier entities become available they can be installed in place of the simulating software and testing continued.

4.2.7 Product verification layering

A complex product will consist of many items organized into a deep family tree. When following a bottom-up directionality, it may be necessary to verify component-level items followed by verification applied to subsystems formed of components followed by verification of systems composed of subsystems within end items. Generally, the qualification process will subject the product to stresses that preclude its being used reliably in an operational sense and may argue against use of components in subsequent parent-item qualification testing. The part of the hardware qualification testing that has this effect is primarily environmental requirements verification.

Where we apply bottom-up testing with component-level environmental testing, we may have to produce more than one article for qualification testing. After component qualification testing these units may have to be disposed of or used in continuing functional testing in an integration laboratory or in maintenance demonstrations. The next article will have to be brought into the subsystem testing. If that level must also be subjected to environmental testing, yet another set of components may be required for system-level qualification. In this case, the system is a collection of cooperating subsystems to be installed in an end item rather than the complete system. It might be possible to avoid multiple qualification units if the environmental testing can be delayed until after functionally oriented system integration testing. We will return to this problem under system verification in Chapter 9.

4.2.8 Verification requirements definition timing

The right time to define verification requirements is at the same time that the item requirements are defined. The reason for this is that it encourages the writing of better item requirements. It is very difficult to write a verification requirement when the item requirement is not quantified because you cannot clearly define an acceptable pass–fail criteria. For example, how would you verify a requirement stated as, "Item weight shall not be excessive." On the other hand, if the requirement were stated as "Item weight shall be less than or equal to 158 pounds;" it can be very easily verified by weighing the item using a scale with a particular accuracy and determining if the design complies.

Ideally, all requirements analysis work should be accomplished by a team of specialists supporting one item specification principal engineer. The principal, or person assigned by the principal, should accomplish the structured decomposition process for the item and assign responsibility for detailed requirements work, such as value definition. For example, the principal would depend on the reliability engineer to provide the failure rate allocated to the item using the system reliability math model. Whoever accomplishes the detailed analysis work should be made responsible for determining how the requirement will be verified and for writing one or more verification requirements. Given that the verification method is test, a test engineer should be consulted in writing the verification requirement. If any other method is selected, there probably will be no organized functional department that focuses on the method, so the engineer responsible for the specific requirement must write the verification requirement(s) without the support of a methodological specialist.

4.3 Verification requirements analysis

The traceability matrix should be fashioned as the requirements are identified. Where you are using a modern requirements database, this is done automatically. Otherwise, you will have to create a separate table in a word processor, spreadsheet, or database and suffer the potential problems of divergence between the content of the specification and the verification traceability matrix. The verification requirements must be developed in combination with the item requirements identification. Ideally, the person who writes an item requirement should immediately write the corresponding verification requirement or requirements and complete the other matrix entries as noted earlier.

4.3.1 Selecting the method

The verification requirement must be written in the context of the planned verification method. So, for each requirement in the requirements sections of a specification, we have an obligation to define how we shall prove that the design solution is compliant with that requirement. How shall we proceed to define that verification requirement in the context of a method? There are two routes that we might consider.

a. *Full Evaluation.* Determine how we can verify the requirement using each of the four methods and select the one method and its corresponding verification requirement based on a predetermined criteria oriented toward cost and effectiveness.
b. *Preselect the Method.* Based on experience, select the method you think would be most effective and least costly and develop the verification requirement based on that method.

The full evaluation method appears to be a very sound system engineering approach involving essentially a trade study for each requirement. Before we eagerly adopt this alternative, however, we should reflect on the cost and the degree of improvement we should expect over the preselect method. Given engineers with experience, the preselect method can be very successful at far less cost than the full-evaluation approach. The experience factor can be accelerated by applying written criteria.

The first element in that criteria should be to pick the method that provides the best combination of good effectiveness and low cost. A good rule is to pick the method that is the most effective unless it is also the most costly. In that string, determine the value of the effectiveness as a means of making the final selection. In the case of software, select test unless there is some overwhelming reason not to do so.

4.3.2 *Writing responsibility and support*

It is a rare person who can do all of the requirements analysis and specification writing work without assistance from specialists in several fields. This is especially true for verification requirements. There should be assigned a principal engineer for each specification and that person should ensure that all of the verification requirements are prepared together with the item requirements by those responsible. If the test method is selected, someone from test and evaluation should assist the engineer responsible for that requirement in writing the corresponding verification requirement as noted in Figure 4-2. Other methods will require support from the responsible specialty engineering discipline. For example, if the requirement was crafted by a reliability engineer, that reliability engineer should provide a requirement for verifying that requirement, perhaps through analysis involving prediction of the reliability from a piece parts analysis or via reliability testing.

The responsibility for writing the Section 3 requirement should flow through the diagram to the responsibility for writing the verification requirement. In the case of specialty engineering requirements, the specialty engineer can be depended upon to provide self-help in the areas of analysis, demonstration, and examination. This same engineer may need help from a test engineer in the case of test verification.

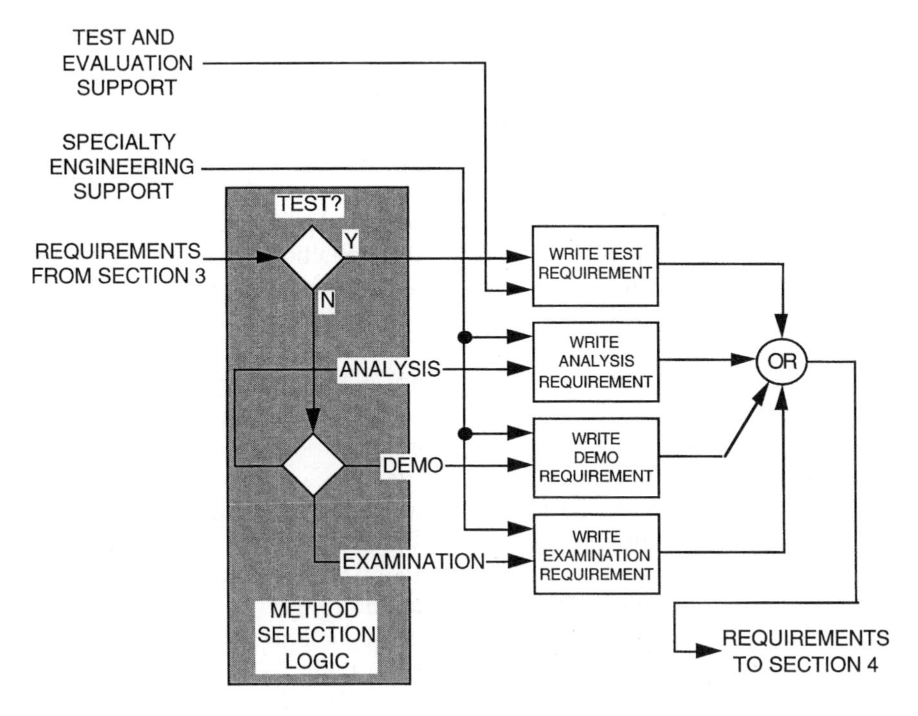

Figure 4-2 Method selection criteria.

4.3.3 Writing the verification paragraph

Given a particular method has been determined or is being considered for a particular item requirement, we must conceive what results will satisfy us that the product design does satisfy the item requirement. We wish to define the requirements for the verification process not the procedure, which should unfold from the requirements. The verification requirements should constrain the verification procedure design as the item requirements constrain the item design. Some examples are offered below. Paragraph numbering follows the MIL-STD-961D pattern.

a. *Example of Qualification Inspection For Compliance.* Note that we have accounted for the possibility that when the scale reads 133 pounds and 13 ounces (134 pounds less half of the worst-case plus or minus error figure); it is actually responding to 134 pounds because of inaccuracy in the scale.

"3.1.19.3 *Weight.* Item weight shall be less than or equal to 134 pounds.

4.3.5 *Weight.* Item weight shall be determined by a scale, the calibration for which is current, with an accuracy of plus or minus 6 ounces. The item shall be placed on the scale located on a level, stable surface and a reading taken. Measured weight shall be less than 133 pounds and 13 ounces."

b. *Example of Qualification Analysis for Compliance.* Note that we have allowed for an analysis error margin of 100 hours.

"3.1.5 *Reliability.* The item shall have a mean time between failure greater than or equal to 5000 hours.

4.3.32 *Reliability.* An analysis of the final design shall be performed to determine piece part failure rate predictions and the aggregate effects accumulated to provide an item mean time between failure figure. Predicted mean time between failure must be greater than 5100 hours."

c. *Example of Qualification Test for Compliance.* Note that design faults could pass through our verification process. Faults could exist that result in this mode being entered from an improper mode. We have tested for improper exit but not for improper entry.

"3.1.3.5 *North–South Go Mode.* The item shall be capable of entering, residing in, and exiting a mode signaling North–South auto passage as appropriate as defined in Figure 3-5 for source and destination states and Table 3-3 for transition logic.

4.3.15 *North–South Go Mode.* The item shall be connected to a system simulator and stimulated sequentially with all input conditions that should trigger mode entry and mode exit and the desired mode shall be entered or exited as appropriate. While in this mode, all possible input stimuli other than those which should cause exit shall be executed for periods between 100 and 1000 milliseconds resulting in no mode change. Refer to Figure 3-5 and Table 3-3 for appropriate entry and exit conditions."

d. *Example of Acceptance Inspection for Compliance.* Note that a qualification test could also be run to demonstrate durability of the markings under particular conditions.

"3.1.13 *Nameplate or Product Markings.* The item shall be identified with the item name, design organization CAGE code, and part number using a metal nameplate in a fashion to survive normal wear and tear during a lifetime of 5 years.

4.3.42 *Nameplates or Product Markings.* Inspect the item visually for installation of the nameplate and assure that it includes the correct item name, commercial and government entity (CAGE) code, and part number."

e. *Example of Qualification Demonstration for Compliance.* In this verification requirement, one could criticize the way the demonstrator is to be selected, it being possible that a tall person could have small hands.

"3.1.17 *Human Factors Engineering*

3.1.17.1 *Internal Item Access.* All internal items shall be accessible for removal, installation, and adjustment.

4.3.38 *Internal Item Access.* A man shall be selected who is at or above the upper extreme of the 95th percentile in stature (73 inches) and that person shall demonstrate removal and installation of all parts identified through logistics analysis for removal and replacement and unpowered simulation of any adjustments in the item."

4.4 *Verification planning, data capture, and documentation*

Many system engineers have acquired their experience while working on military programs using various military standards as the basis for their work. Many of these people used the specification model contained in MIL-STD-490A, Specifications Practices, for years. In response to the DoD interest in performance specifications and a changeover to commercial standards wherever possible, DoD canceled MIL-STD-490A in 1995 except for existing programs. The program-peculiar specification content material was moved to MIL-STD-981D, Appendix A. The specification formats were simplified and the name of Section 4 changed from "Quality Assurance Provisions" to "Verification."

This section of the specification, by whatever title and section number, provides a place to capture the verification requirements that tell how to prove that the design solution does satisfy the requirements contained in the "Requirements" section, which is Section 3 in the DoD model. Development organizations have used two principal alternative ways to document verification requirements, test and analysis procedures, and test and analysis reports. Some of those alternatives are

a. Specification Section 4 provides a definition of verification responsibilities and a verification traceability matrix like the one shown in Figure 4-1 but no detailed verification requirements. A verification requirements document or test requirements document defines the detailed verification requirements.
b. Specification Section 4 provides a definition of verification responsibilities, verification traceability matrix, and the detailed verification requirements.

Alternative "b" is encouraged in this book because it keeps all of the item requirements (development and verification) together. Often when alternative "a" is employed, different organizations develop the specification and the verification document. By hooking the two together tightly, we encourage simultaneous definition of both, leading to identification of better requirements. It is difficult to write poor requirements when you must also write the corresponding verification requirements. The difficulty in writing the latter based on a poorly defined requirement will force a discipline in

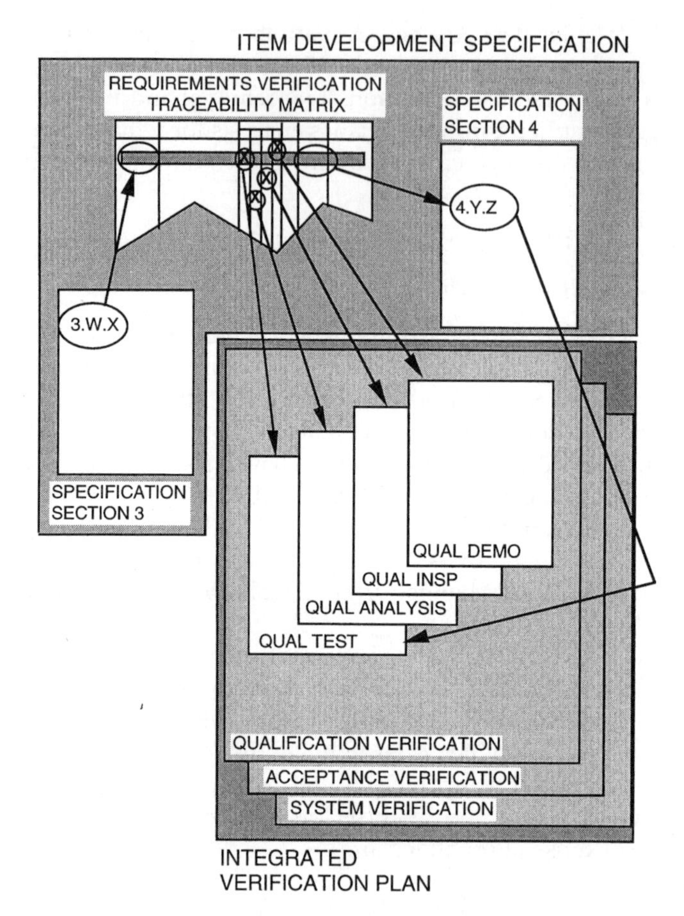

Figure 4-3 Verification planning documentation suite.

requirements writing leading to many subtle improvements in the development process as well as an effective verification process.

Figure 4-3 illustrates the recommended approach for capture of verification requirements and planning information documentation. The item requirements are determined (and in this example loaded into specification database as paragraphs in Section 3) and listed in the requirements verification traceability matrix, assumed to be included in specification Section 4 (Quality Assurance Provisions or Verification depending on the standard followed). For each requirement listed in the matrix (such as 3.W.X), the person who authored the requirement or the item principal engineer enters an "X" in the appropriate methods column or columns. The requirement author must then write one or more verification requirements for each verification method for inclusion in the specification (Section 4 paragraph 4.Y.Z in this example) and section cross-references are entered into the traceability matrix as well.

Table 4-2 Sample Verification Compliance Matrix Fragment

Section 3 Paragraph	Title	Methods T	A	E	D	Level	Section 4 Paragraph	VSN	VTN
3.2.1.5	Guidance Accuracy	X				Item	4.2.37	0032	F412314
			X			Parent	4.2.38	0033	F412822
3.2.1.6	Clock Accuracy	X				Item	4.2.41	0102	F412314

The next step is conditioned by how the organization has ordered and assigned the verification planning work, so the approach discussed here should be understood to be one approach that will work. Let us assume that this is a development specification with which we are dealing and that we have previously agreed upon the four verification methods noted in Figure 4-3. An engineer has been assigned the responsibility for creating and maintaining an integrated verification plan with sections for item qualification, item acceptance, and system verification. Since we are dealing with a development specification, our interest is drawn to the item qualification section in this case. Within that section we have organized the planning data under the four possible methods of verification (test, analysis, examination, and demo). The aggregate input to the test subsection of the qualification section is every verification requirement marked for the test method in every development specification on the program. Clearly, there are many other ways we could organize this data.

Every one of these verification requirements need not be copied into the indicated verification plan section verbatim. Our test-planning challenge is to create the least-cost way to achieve the intent of the aggregate of all of the verification requirements within available budget and schedule. If we simply respond to each verification requirement in a mindless copycat fashion, it is likely that duplication of effort will drive the cost to a level higher than necessary.

Staying with the test method for the moment, we must determine a list of test tasks that will accomplish the work defined in all of the verification requirements. Within each test task, we must then connect up a set of verification requirements (with a test method indicated) that can be accomplished in that task. The union of all test tasks should cover all of the verification requirements identified with a test method and comprise the qualification (in this instance) test program. The result should also produce the initial input to an expansion of the verification traceability matrix into a verification compliance matrix, an example of a fragment of which is shown in Table 4-2. The principal change from the verification traceability matrix is the addition of the verification task number (VTN). In this example, two of the listed requirements map to the same verification task. A task is a collection of related work that can easily be planned, managed, and performed as a single entity, a single test or analysis. These tasks should be correlated with items in the system architecture. A suggested VTN identification numbering technique will be offered in Chapter 5.

The verification traceability matrix is oriented toward a single document but the verification compliance matrix is oriented toward the system. Therefore, our matrix would have to include some additional columns to the left of the Section 3 paragraph column to uniquely differentiate the specification of interest from all of those on the program. We could assure uniqueness for these specifications by including a specification number, part (1 or 2), and a revision letter. These are omitted here in the interest of space. Note that the VSN appears redundant because our example does not parse particular item or verification requirements into fragments. Our table does not include fragment identifiers for Section 3 paragraph or Section 4 paragraph in the interest of space. In order for these identifiers to be effective, of course, a definition would have to reside somewhere identifying the extent of these fragments. Where the organization is using a requirements database tool that can capture parsed paragraph information, this is assured. Where an organization is still operating with typewriter thinking in a computer world, this will be very difficult to accomplish. Traceability and verification are the big drivers in movement to computer databases for requirements capture.

The numbering of the verification strings should be assigned for uniqueness within each specification, made unique by the addition of the specification identification. The verification task numbers, on the other hand, should be unique systemwide. One can use smart numbering systems where each place value has special meaning (once you learn that meaning), but more often than not these methods blow up eventually, leading to a lot of unproductive busywork to sustain a bad idea.

The verification compliance matrix, depicted in Table 4-2, must be supplemented, in the integrated verification planning data, by a companion table (database), called a verification task matrix, that defines the verification tasks in terms of procedure reference, responsible engineer, planned and actual procedure release dates, planned task start and complete date, and current status. The final verification compliance matrix is then a join of these two database tables plus the specification tree table as suggested in Figure 4-4.

The verification traceability matrix in each of the specifications should remain fairly stable throughout the development process. The verification compliance and task matrices will obviously change as the program progresses and status changes, so it may be preferable to prepare them as separate documents and reference them in the integrated verification plan. This is especially useful where the customer is a DoD component or NASA center and contractual requirements include periodic delivery of these data items. This avoids bumping the integrated verification plan revision letter each time the matrix changes.

The next verification planning step is completed by preparing verification task plans and procedures for each verification task. The work up to this point has been focused on the verification strings hooked to item requirements. Now the focus must shift to verification tasks formed of work

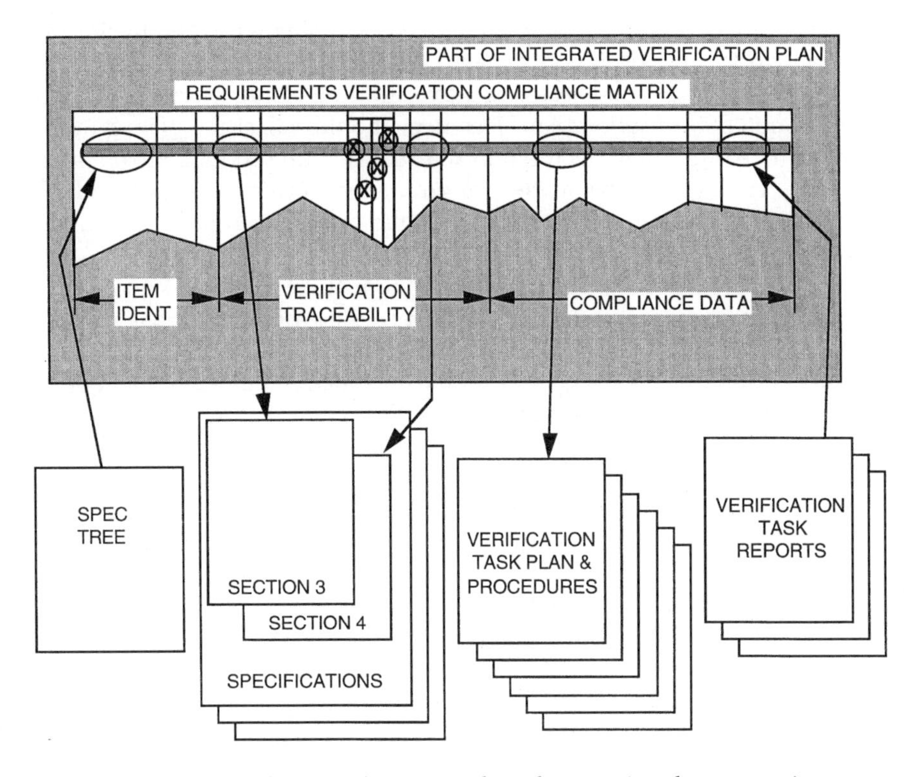

Figure 4-4 Verification planning and implementation documentation.

required to satisfy a particular collection of strings generally oriented toward the items to be qualified.

The verification compliance matrix identifies all of the requirements that must be covered in the task plan and procedure for a given VTN, the engineer responsible for crafting the task plan and procedure, and other management data. The responsible engineer must prepare a plan and procedure that accomplishes all of the verification strings (defined by VSNs) mapped to the task. A series of task steps are defined involving test stimuli, responses, and decision criteria; test resources (test equipment and apparatus some of which may have to be created especially for this program) are identified; and cost and schedule constraints are identified and balanced with available resources.

At this point, the planning work has been completed for the single string we have traced. There will, of course, be hundreds or thousands of such planning strings to be accommodated. The integrated verification plan satisfies a central organizing function but need not be a single large document. It essentially consists of the verification compliance and task matrices (included or referenced) and all of the verification task procedures, each of which could be separately released and referenced in the plan.

Intervening between the completion of the planning work and the implementation work is the design and manufacturing process that produces one or more items that will be subjected to the verification planning data contained in the integrated verification plan. This item should reflect the planned production item features manufactured in accordance with production planning.

At the appointed time in the program, the verification task will be accomplished by, or under the direction of, the principal engineer identified in the matrix and results of the task captured in a verification task report. The reported evidence should clearly tell the results of the task making it possible for an engineer or manager to determine whether the task proved that the requirements were satisfied by the design or were not (qualification case still in mind). Based on a review of the evidence contained in the report, a decision must be made along these lines, "Were the requirements verified or not?" If so, the report results may be filed away, a reference to the report included in the matrix (not shown in Table 4-2), and the status changed in the matrix to complete the verification implementation action for that task.

If problems are identified in the report, another course of action must be undertaken. It must be determined by the technical community and management just where the problem lies. For example, is it based on: a failure of the design to satisfy requirements, verification stresses in excess of the requirements, human failure during the verification task, a flawed test article (defective or wrongly built), or some other phenomena? Depending on the cause, we will have to take action to complete the verification action. This may require a design change, a change to the requirements, or an agreement with our customer on a waiver that permits us to deliver product that does not meet requirements in a particular way for a specified period while we correct the problem. After corrective action is completed, we will likely have to rerun the verification task, possibly with changed requirements, design, and procedures and publish a report revision.

Figure 4-4 illustrates this whole cycle in terms of the documentation discussed above. Clearly, the integrated verification plan plays a critical role in pulling the many pieces of this puzzle together into a unified verification activity that stretches across much of the program time span from definition of item requirements through the completion of verification audits that certify customer acceptance that the verification evidence contained in the verification reports actually proves that the product design does satisfy the previously reviewed and approved requirements.

One of the principal services provided by the plan is the reference to the report documentation that must be reviewed in order to reach a conclusion about requirements compliance. On a program where a government customer requires delivery of data, the compliance matrix should be one of the deliverable documents in order to keep the customer apprised of the availability of report data that must be reviewed by the customer in preparation for program audits. This matter will be picked up again in Chapter 8.

Many engineering organizations prefer to prepare a very simple Section 4 for specifications containing only definitions and a verification traceability matrix. The verification requirements may be captured in a separate document called a test requirements document, the integrated test plan, or the appropriate detailed test procedure. This solution can be made to work for requirements that will be verified through test, but there is commonly no comparable solution for requirements that are to be verified by other means. The approach described above ensures completeness of the verification effort.

4.5 Section 4 structure

4.5.1 MIL-STD-961D structure

Figure 4-5 gives a suggested verification section structure based on MIL-STD-961D. The specification under this standard is intended to progress from a performance specification that would be used as the basis for qualification verification into a detailed specification through the addition of content to paragraph 3.3, Design and Construction, and additional changes. Therefore, the verification section may contain the verification requirements for the qualification process and later have the acceptance verification requirements added as the specification transitions from a performance specification to a detailed specification.

The author hastens to mention that Figure 4-5 includes the formatting encouraged by MIL-STD-961D. The author's preferred style involves no title underlining, capitalization of all words in the title, and no periods at the end of titles with the title and paragraph number on their own line. These are small things and you are encouraged to use whatever format your customer base prefers unless there are conflicts across all of the members of your customer base. In this case, you would do well to find out how to serve all of your customers using a common formatting.

The term "verification traceability matrix" is used in paragraphs 4.1 and 4.4 of Figure 4-5 consistently with its use throughout this book. Table 4-1 of this book offers an example of this matrix. MIL-STD-961D refers to this matrix as the cross-reference matrix. That standard does not offer a preferred physical location for this matrix in a specification and does not provide a home reference for it in any paragraph in Section 4 as Figure 4-5 does in paragraph 4.4. Reference to this matrix as 4-1 and its location at the back of Section 4 presumes that there is only one table in Section 4, this matrix. If there is more than one table in Section 4, the reference in paragraph 4-1 would, of course, have to be incremented. With the table at the end of the section it can be easily used independently from the specification as a working paper if desired.

Paragraph 4.2 of Figure 4-5 offers a place to capture information about the several classes of verification actions. The subparagraphs should identify the sections of the verification traceability matrix that apply. Where the specification covers more than one form of verification, it may be useful to partition the matrix into design (or qualification), first article, and acceptance

4. SECTION 4 — VERIFICATION.

4.1 Methods of verification. Methods utilized to accomplish verification are defined below. The letters preceding each method appear in Table 4-1, verification traceability matrix, to designate the methods.

A. Analysis. An element of verification that utilizes established technical or mathematical models or simulations, algorithms, charts, graphics, circuit diagrams, or other scientific principles and procedures to provide evidence that stated requirements were met.

D. Demonstration. An element of verification which generally denotes the actual operation, adjustment, or reconfiguration of items to provide evidence that the designed functions were accomplished under specific scenarios. The items may be instrumented and qualitative limits of performance monitored.

E. Examination. An element of verification and inspection consisting of investigation, without the use of special laboratory appliances or procedures, of items to determine conformance to those specified requirements which can be determined by investigations. Examination is generally nondestructive and typically includes the use of sight, hearing, smell, touch, and taste; simple physical manipulation; mechanical and electrical gauging and measurement; and other forms of investigation.

T. Test. An element of verification and inspection which generally denotes the determination, by technical means, of the properties or elements of items, including functional operation, and involves the application of established scientific principles and procedures.

4.2 Classes of verification. Three classes of verification work are identified and described in subordinate paragraphs. Design verification provides for the qualification of new items not yet proven in their application. First article and acceptance inspections provide for verification work oriented toward specific product articles during or subsequent to manufacture.

4.2.1 Design verification
4.2.2 First article inspection
4.2.3 Acceptance (conformance) inspection
4.2.3.1 Sampling inspection
4.2.3.1.1 Inspection lot
4.2.3.1.2 Classification of defects
4.3 Inspections
4.3.1 General inspections
4.3.1.1 Inspection conditions
4.3.1.2 Inspection equipment
4.3.1.3 Toxicological product formulations
4.3.2 Detailed inspection requirements
4.3.2.X Detailed inspection element X
4.3.2.X.1 Methods of inspection
4.3.2.X.2 Special inspection conditions
4.3.2.X.3 Special inspection equipment
4.4 Verification Traceability Matrix

Figure 4-5 Specification verification section structure.

sections so that the appropriate parts can be easily referenced. If the verification requirements become mixed in the matrix with respect to the classes, this referencing may become fairly complex. One approach to the partitioning arrangement would be to create the following paragraph structure:

4.3.2	Detailed inspection requirements
4.3.2.1	Development inspection requirements
4.3.2.1.1	Verification requirement 1
4.3.2.1.2	Verification requirement 2
4.3.2.1.N	Verification requirement N
4.3.2.2	First article inspection requirements
4.3.2.2.1	Verification requirement 1
4.3.2.2.2	Verification requirement 2
4.3.2.2.N	Verification requirement N
4.3.2.3	Acceptance inspection requirements
4.3.2.3.1	Verification requirement 1
4.3.2.3.2	Verification requirement 2
4.3.2.3.N	Verification requirement N

If it is necessary to partition each verification requirement further into methods, equipment, and conditions, then this will add complexity to the paragraphing structure but the clarity of the mapping to classes will be worth the added complexity.

Paragraph 4.3.2 provides the place to capture all of the verification requirements numbered in any order you choose. As noted earlier, some engineers like to number these by adding 4.3.2 to the front of the corresponding Section 3 paragraph number. Unless the inspection action is very complex and requires a lot of special equipment and special control of conditions during the verification task, subparagraphs under paragraph 4.3.2.X are not required. If the same equipment and conditions apply across the complete verification process, this content can be included under general inspections (4.3.1).

4.5.2 A simple structure

The structure of Section 4 can be much more simply structured where the specification only has to refer to a single verification class such as qualification. This would be the common case in a Part I, development, or performance specification. Similarly, a Part II, product, or detail specification, can focus on the acceptance requirements. MIL-STD-961D and Figure 4-5 cover a structure appropriate for a specification that covers all item requirements, performance, and detail. This is very common for in-house procurement specifications but uncommon for DoD specifications.

Figure 4-6 offers a simple structure for a performance specification. There is only one class, qualification, or design verification so paragraph 4.2 focuses only on that class. Paragraph 4.3 is partitioned into the four methods, which

4.1	Verification Methods
4.2	Design Verification
4.3	Detailed inspection requirements
4.3.1	Test Verification Requirements
4.3.1.1	
:	Test verification requirements 1 through X
4.3.1.X	
4.3.2	Analysis Verification Requirements
4.3.3	Demonstration Verification Requirements
4.3.4	Examination Verification Requirements
4.4	Verification Traceability Matrix

Figure 4-6 Simple verification section structure.

is not actually needed. An even simpler structure would simply list all of the verification requirements under paragraph 4.3 and 4.3.1 through 4.3.X. These are the paragraph numbers that will appear in the traceability matrix and the methods entries will tell which method applies. The other information (test resources and equipment, for example) in the MIL-STD-961D structure would be found in the verification planning and procedures data.

4.5.3 External verification requirements documentation

The verification requirements could be placed in the verification plans and procedures documentation or in a separate test requirements document used as a transform between the specifications and test plans. Both of these solutions leave a lot to be desired for the reasons already covered. If separate documents are used, then the verification traceability matrix will have to identify the document or documents the verification requirements appear in as well as the paragraph numbers corresponding to specification Section 3 requirements.

4.6 Verification computer databases

The utility of relational databases for verification management should be clear from the discussion up to this point. Figure 4-7 illustrates the structure of such a database comprising requirements, test, and analysis (combined with demonstration and examination) components. This diagram was assembled as a means to communicate possible data relationships to the system analyst responsible for designing a particular verification database.

Given a requirement R1 in a requirements database, it may have to be verified using methods M1 (test), M2 (analysis), and M3 (demonstration). The verification analysis requirements may include two paragraphs for inclusion in Section 4 both of which trace to a particular test task. This path can be traced, finally, to test procedure and test report content. Requirement R3 was apparently crafted with more than one requirement in a paragraph so

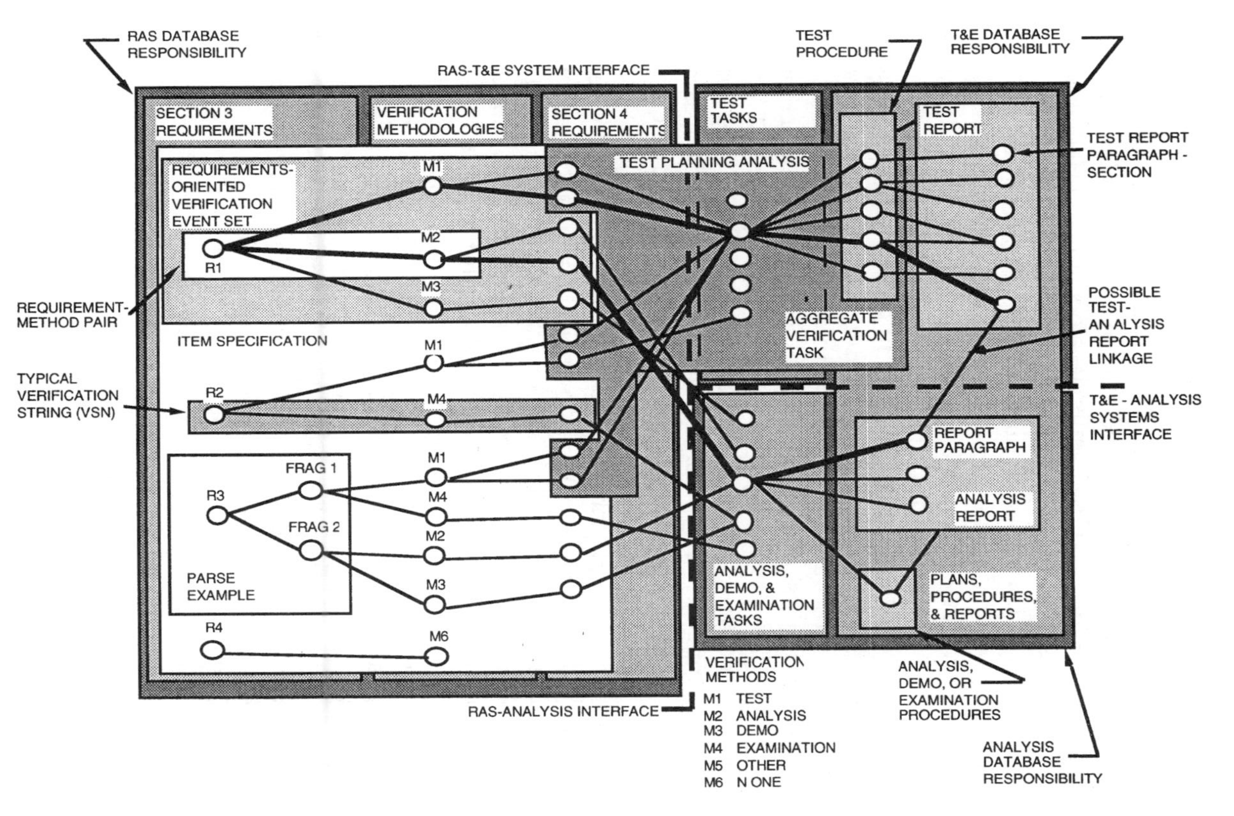

Figure 4-7 Verification data structures.

it has been parsed into two fragments, each of which is traced to multiple verification tasks.

The database system in question here was being created in three parts due to the budget profile. Money was available immediately for the development and use of what is called here a requirements analysis system (RAS) followed at a later date by interface with a test and evaluation (T&E) database to be developed, and finally, with a database that would provide similar coverage for analysis, demonstration, and examination tasks. Hooks had to be planted in the RAS for the subsequently developed relational databases.

But, a simpler view of the utility of databases is available to us. The requirements analyst is interested in a list of requirements for the item in paragraph number sequence. The person doing verification requirements work may be interested in the Section 4 sequence or only those requirements to be verified by the test method. If this data is captured in a relational database structure, the tables can be called up in any fashion desired with only a single source for any unique piece of information. Most of the popular requirements tools on the market include a verification capability, but they may not extend into the test and analysis planning and reporting activities. One can build his or her own system but should first recognize that a great deal of money will become tied up in maintaining the system once it does start functioning properly. We are easily drawn into self-help databases by the high initial cost of commercially available systems only to find in the long run that we have committed much more money to our home-grown tool. One should study the available commercial products carefully before embarking on an in-house design. In most organizations that try this route, management eventually concludes that the continuing cost is insupportable and that, "... we make airplanes (or computers, or rockets, or ...) not computer tools." Several good tools have evolved through this path like Slate that was originally developed in-house by Texas Instruments and later spun off to TD Technologies.

The one advantage in building one's own database is that you will become a better system engineer in the process. The relationships between the many pieces of information involved are sometimes very subtle and complex. You cannot help but become more proficient in verification work as a result of database analysis work even if you do the work with typewriter technology thereafter.

chapter five

Top-down verification planning

5.1 A matter of scale

Up to this point we have been viewing the verification process from the microperspective. Before proceeding to amplify on that perspective in relation to specific methods and techniques, let us step back and consider the verification process from a macro/perspective. There are planning activities appropriately done from the bottom up and those best done from the top down. In Chapter 1 we implied that we should plan the whole program applying a top-down functional decomposition process. Figure 1-14 provided a life cycle functional flow diagram and the intent was to decompose each block in that diagram to encourage concurrent development and optimization of product and process together. Function F4 of that diagram is titled System Verification and in this chapter methods for decomposing this function from the top down will be offered. In subsequent chapters, we will fill in the details from the bottom up.

5.2 Expansion of function F4

There are three major elements of the verification process as discussed earlier and illustrated in Figure 5-1. They include the item qualification process (F41), system test process (F42), and item acceptance process (F43). Task F44, development evaluation test (DET) program, may be included in the verification activity or located under validation, which is where the author would prefer to place it along with all of the analyses that are accomplished to evaluate the relationship between item requirements and design alternatives for the purpsoe of reducing program risk.

The item qualification process is driven by the item development requirements as discussed in Chapter 4. Item qualification test work must be accomplished on test articles representative of the final design fabricated in a fashion as close to the final manufacturing process as can be arranged.

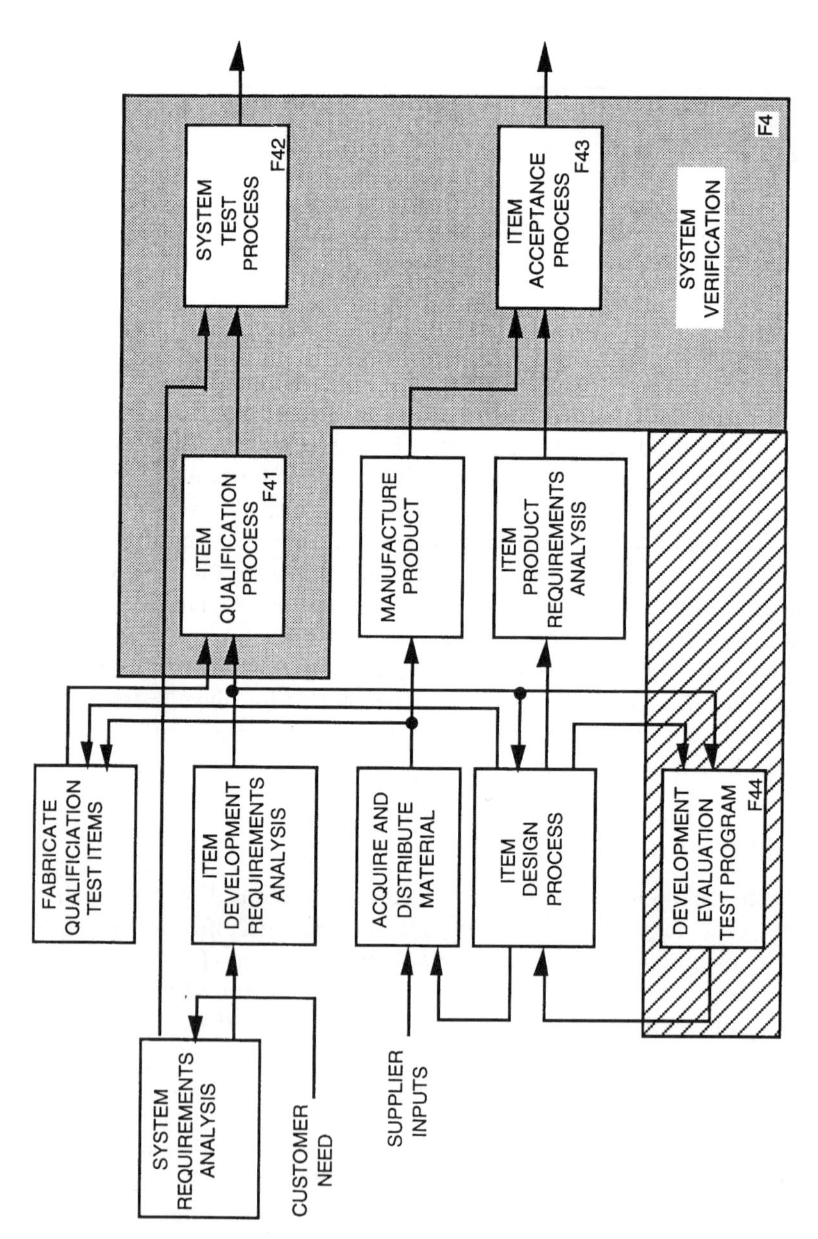

Figure 5-1 System verification context.

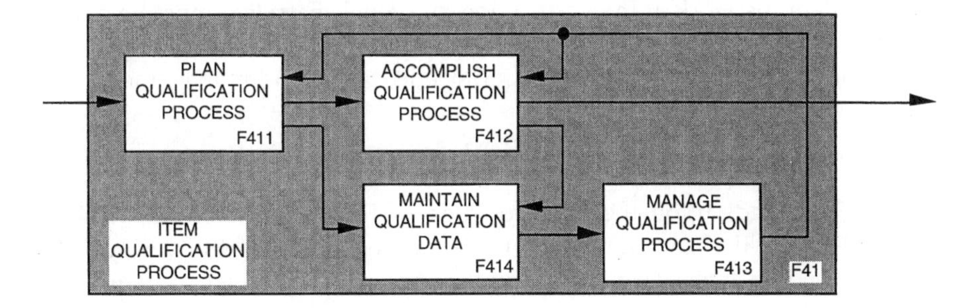

Figure 5-2 Qualification process flow.

All of the item qualification work should be complete as a prerequisite to system testing where we seek to provide evidence that the whole system satisfies the system specification content. On a DoD program, the results of all of the item qualification tasks will be audited at a functional configuration audit (FCA) that may be accomplished in a single audit or a series of item-oriented audits capped by a system FCA.

The item acceptance process is driven by the item product requirement and accomplished on product intended for delivery to the customer. The first article that flows through the acceptance process will produce evidence of the degree of compliance of the design and manufacturing process with the product requirements. The results of the acceptance process on the first article will be audited at a physical configuration audit (PCA) on a DoD program.

Every program that an enterprise undertakes will likely fit into the generic process illustrated in Figure 5-1. Therefore, this work may be planned generically for this level of detail. The procedures for accomplishing this work may be captured in generic practices at this level and referenced in program-peculiar planning.

5.3 *Item qualification process*

Figure 5-2 represents a decomposition of the item qualification process identifying four significant subtasks. We must first plan the work extending our understanding of the verification requirements into a definition of tasks that collect all of the verification requirements and correlating these tasks with system items in the form of item verification task groups. Each of these tasks is then planned in detail and associated with persons or teams responsible for accomplishing the work, thereby identifying verification task responsibility groups. The result is an integrated verification plan (IVP).

The plan defines data that must be tracked over the whole process best accomplished in the context of a relational database as discussed in the last chapter. The two principal manafestations of this work are the verification compliance matrix and the verification task matrix. The former lists every requirement in all of the program specifications and correlates the details of the verification process. The latter is the principal process management tool defining all of the work at the verification task level. This data must be maintained throughout the verification process as planned work is accomplished to provide management the information needed for sound management.

5.4 Qualification process implementation

Figure 5-3 expands upon the F412 function of Figure 5-2. Each task F412L represents a verification item task group focused on all of the verification tasks mapped to item L. A program will require some number of item verification processes based on how many items we organized all of the work within. In an extreme and unlikely case, one could accomplish all qualification work at the system level making for a single process. Each of the item-oriented processes F412L will include some combination of verification tasks applying one or more of the four methods noted. This may include three test tasks, four analyses, one demonstation, and two examinations making a total of ten verification tasks that have to be completed to accomplish all of the verification work for this item. Each of these tasks should be included in the IVP with appropriate planning and procedures content, and each of these tasks should produce a formal written report that forms a part of the program verification documentation base in the integrated verificatin report (IVR) or independent reports as selected for the program. These written reports need not be voluminous. Lest this terrify some readers with a vision of mountains of data, some of these reports may be a simple as a single page in the IVR.

It should be noted that one or more of the item qualification processes may be associated with system-level work. For example, we may accomplish all reliability, availability, and maintainability (RAM) analysis work within the context of a system RAM analysis reported upon in a single analysis report. Other work may be accomplished at the end-item level as well as subsystem and component levels. The point being that the term *item* should be understood to have a broad association with respect to the product architecture.

Figure 5-3 shows the item qualification tasks in a simple parallel construct. On an actual program, this arrangement will commonly be a series-parellel construct both within the task F412L and at the next-higher level shown on the figure. We must determine what all of the tasks are and how best to structure them relative to each other on a particular program. Some of these tasks will properly be prerequisites to others so there is a sequence relationship not reflected in Figure 5-3 that we need to respect in specific program planning work. On a particular program and for a particular item, there may not be any required verification tasks for one or more of the four methods noted and these simply become voids.

A functional configuration audit may be accomplished for each item as suggested by Figure 5-3. On a specific program, we may hold individual FCAs or group them in various ways depending on the needs of the program. Where there is more than one FCA, there should be a system FCA to evaluate the results of all of the individual FCAs and assure that all issues have been addressed.

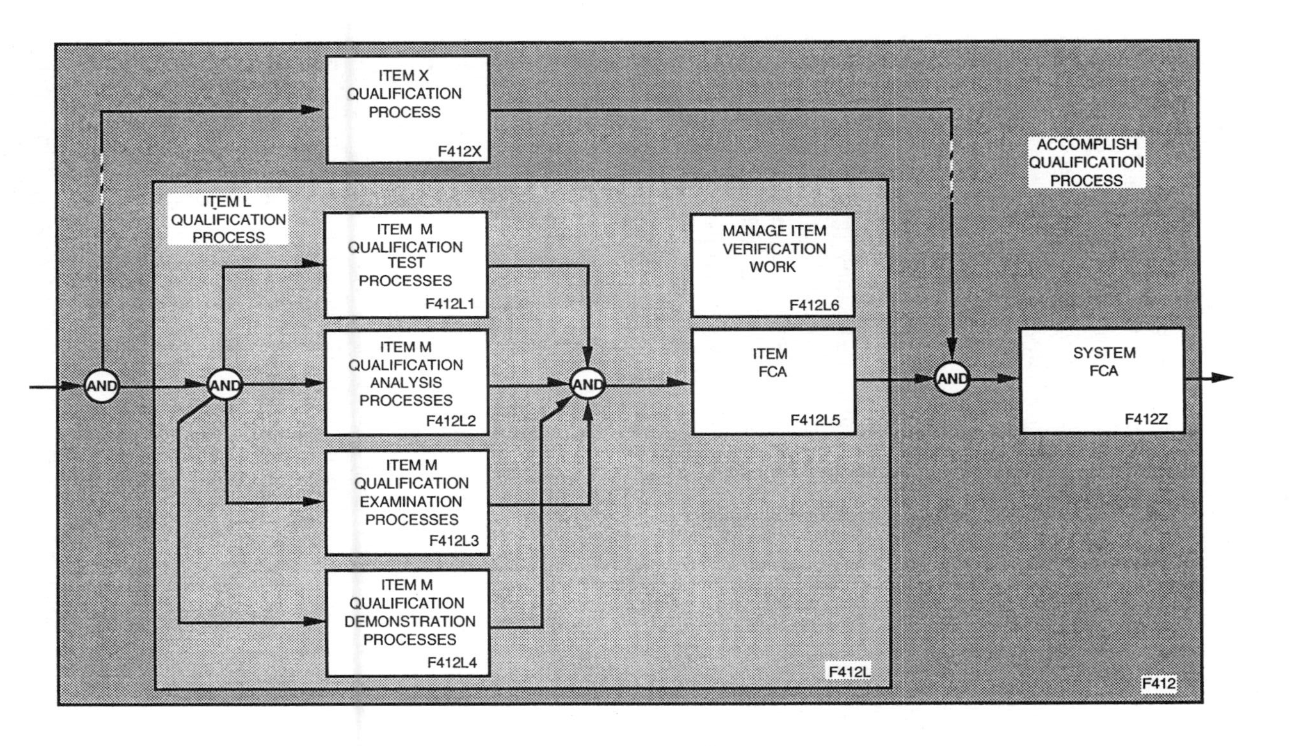

Figure 5-3 Accomplish qualification process.

Each of the item verification processes must be managed by an assigned authority, a team or an individual within the context of overall verification process management. If this process corresponds to a team organizational responsibility, then the team should be the agent and the team may appoint someone to act as its verification lead engineer. Given that there is more than a single verification task in the item verification process, that lead engineer should coordinate all of those tasks and ensure that the goals of each are satisfied in the work accomplished. This team verification lead engineer may have to accept the responsibility for more than one item task group leading to a multiple item responsibility group. Otherwise, the item task and item responsibility groups merge for a single item.

Figure 5-4 illustrates the internal complexity of a single item qualification verification task that may be composed of many verification tasks applying the four methods. As noted above, some of these methods blocks may be voids for a particular item. The figure shows all of these tasks in parallel once again, but on a particular program probably will have to follow some series parallel network of tasks. Analysis tasks may link to particular test tasks the output of which is needed in a subordiante analysis task.

The reader will note that the author chose to organize all of these tasks within the order LMN for item, method, task. This has the effect of dispersing all test tasks rather than pooling them into one contiguous group. People in a test and evaluation function will not prefer this method of organizing the tasks for this very reason. The author will choose this same structure for the IVP outline in Chapter 6. The rationale for these choices is that we should make every effort to align the management data in accordance with the team structure on a program to reduce the intensity of the relationships that cut across this structure and strengthen the links that bind together the management strings in accordance with the team structure. In that the author has encouraged product-oriented cross-functional teams, it makes sense to follow through with team, or item, high on the verification data organization structure priority.

The result will be that item teams can focus on a contiguous section of the IVP as their responsibility. Other methods of organizing this information will scatter the team responsibility in planning data. If all of this information is captured and retained in relational databases, this will not be an impediment to anyone wishing to focus on only the test planning data as the database can be structured to apply that filter.

The generic F412LMN tasks correspond to the VTNs referred to in Chapter 4 where L is an item indicator and M is an element of the set {1, 2, 3, 4} for the four methods of verification. If there are eight test tasks that have been identified for item L, they would be identified as tasks F412L11 through F412L18. Once again, these tasks will likely not appear in the detailed planning as a simple parallel string; rather they will be arranged in some series parallel structure crafted creatively by the verification work planner.

The author would choose to use a base 60 system for numbering these items and individual tasks using all of the numerals, all of the uppercase English letters, except O, and all of the lowercase English alphabet, except l.

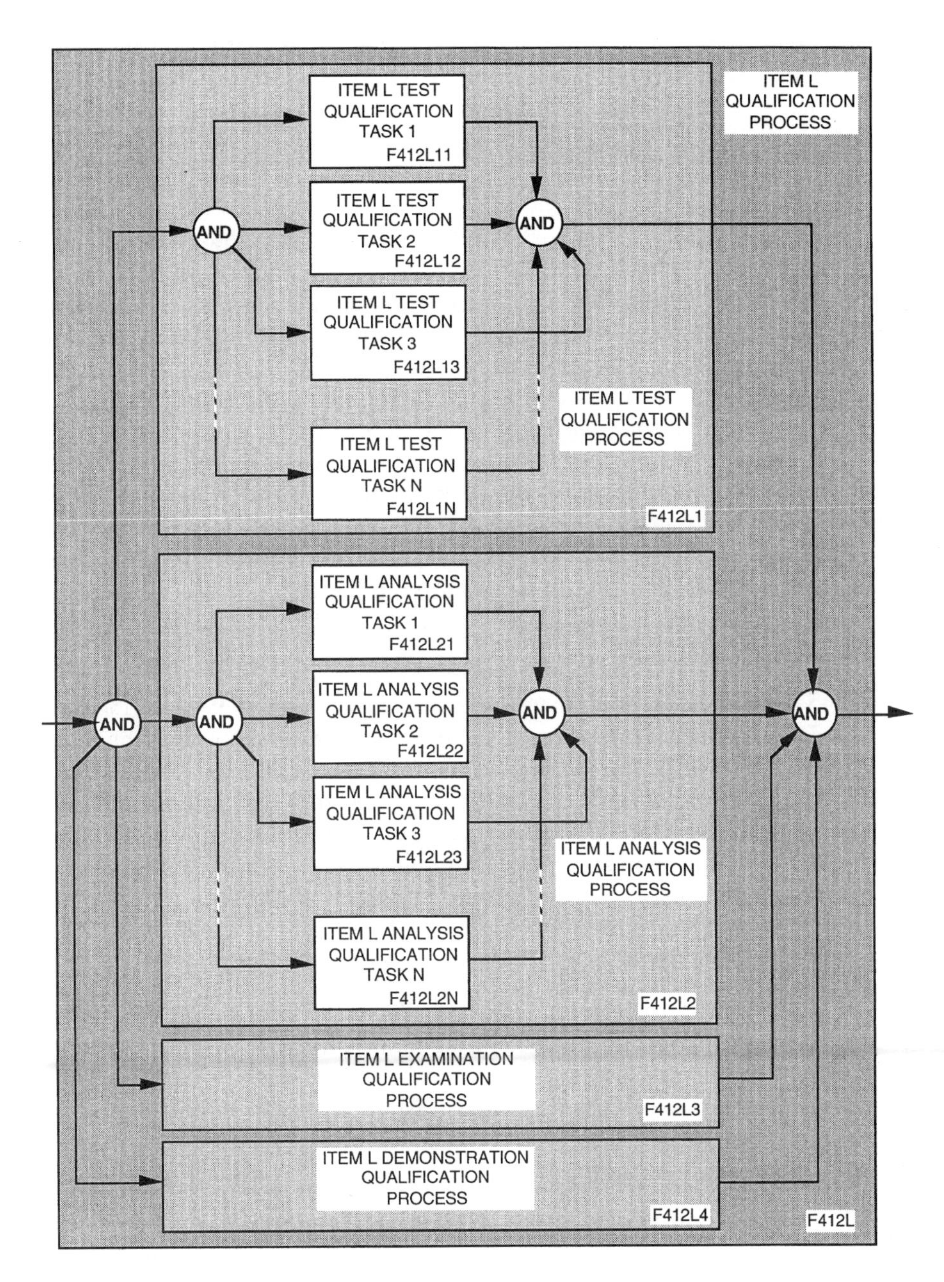

Figure 5-4 Item L verification process.

If the program is so large and complex that there will be more than 60 tasks of verification of the same method applied to one item, then it may be necessary to apply a different task-numbering scheme using decimal points, dashes, or let N take on a dual character length. In the latter case the tasks noted above would be numbered F412L101 through F412L108. If you applied the base 60 system here it would permit numbering as many as 3600 tasks which should be more than enough for the most complex program.

5.5 Specific task processes

The verification tasks that form the atomic level of the verification work illustrated as task F412LMN in Figure 5-4 can be further refined diagrammatically in our generic process description. These diagrams are useful in describing these processes but the numbering of the tasks in our generic process diagram can also be applied on a given program. Every program will have an activity F412 composed of many item qualification processes F412L. Each of these item processes can be further subdivided into the four verification methods: F412L1, test processes; F412L2, analysis processes; F412L3, examination processes; and F412L4, demonstration processes. These tasks are defined by the verification strings (VSN) that are mapped to them, the items with which they are associated (L), and the methods (M) that are selected. Subordinate paragraphs expand each method generically and these generic structures offer guidance for the planning data that should be included in the IVP.

5.5.1 Generic test task

Figure 5-5 illustrates the generic test task process which will be described in Chapter 7. The principal steps are prepare for the test, accomplish the test, and report the results. In some cases it may be necessary to refurbish resources and dispose of residual material in accordance with special instructions in the plan because they are classified or hazardous. If the results do not support the conclusion that the requirements were satisifed in the design, then it may be necessary to make adjustments and retest.

5.5.2 Generic analysis task

Figure 5-6 illustrates a generic analysis task. The same basic string of tasks is included here as in test, prepare, do, and report. If the evidence does not support the compliant conclusion, we may have to adjust our analytical approach or change the design and subject the new design to another analysis.

5.5.3 Generic examination task

Figure 5-7 offers a generic examination process. It is similar to the analysis process in that there is no special-purpose instrumentation involved. This is primarily a matter of one or more humans making observations essentially with their unaided senses.

5.5.4 Generic demonstration task

Figure 5-8 provides a generic demonstration process. It is similar to the test task described above. It is possible to damage or overstress the product in a

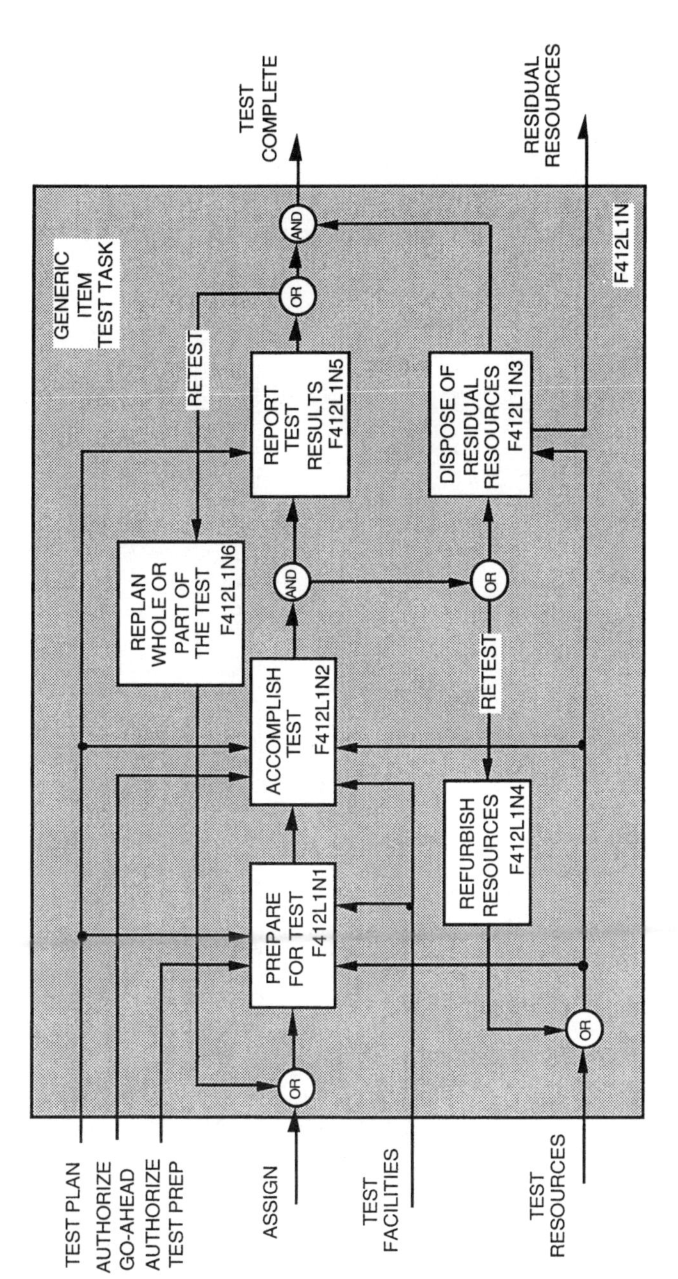

Figure 5-5 Generic item test task flow.

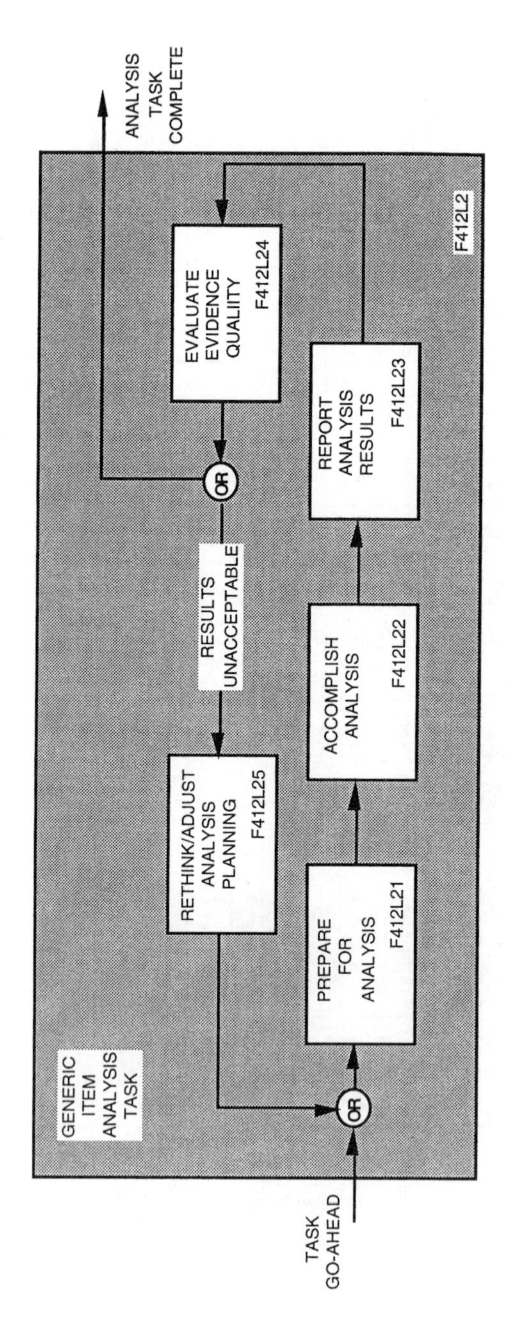

Figure 5-6 Generic item analysis task flow.

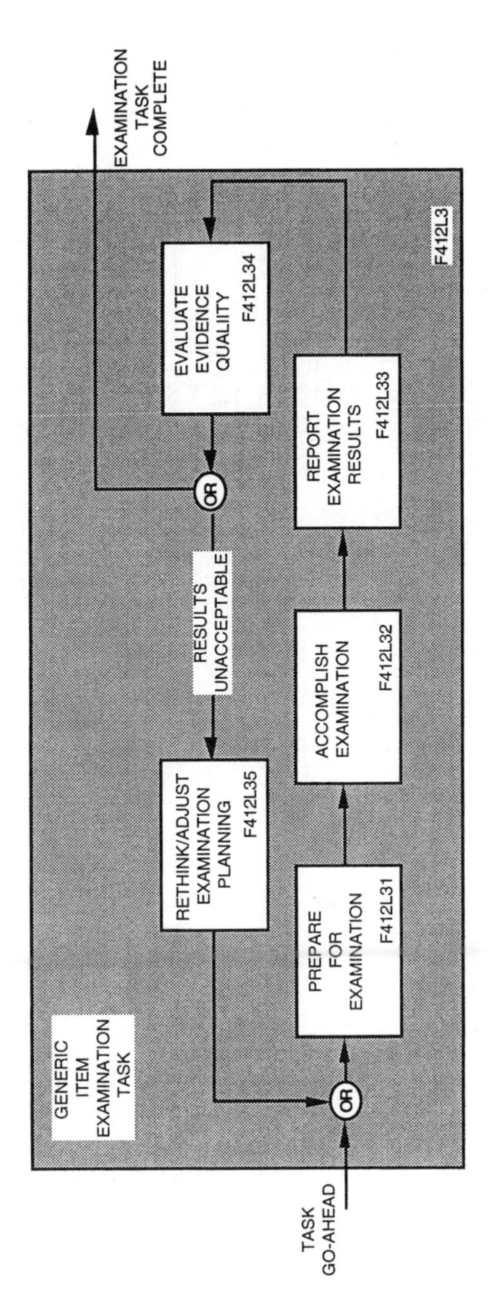

Figure 5-7 Generic item examination task flow.

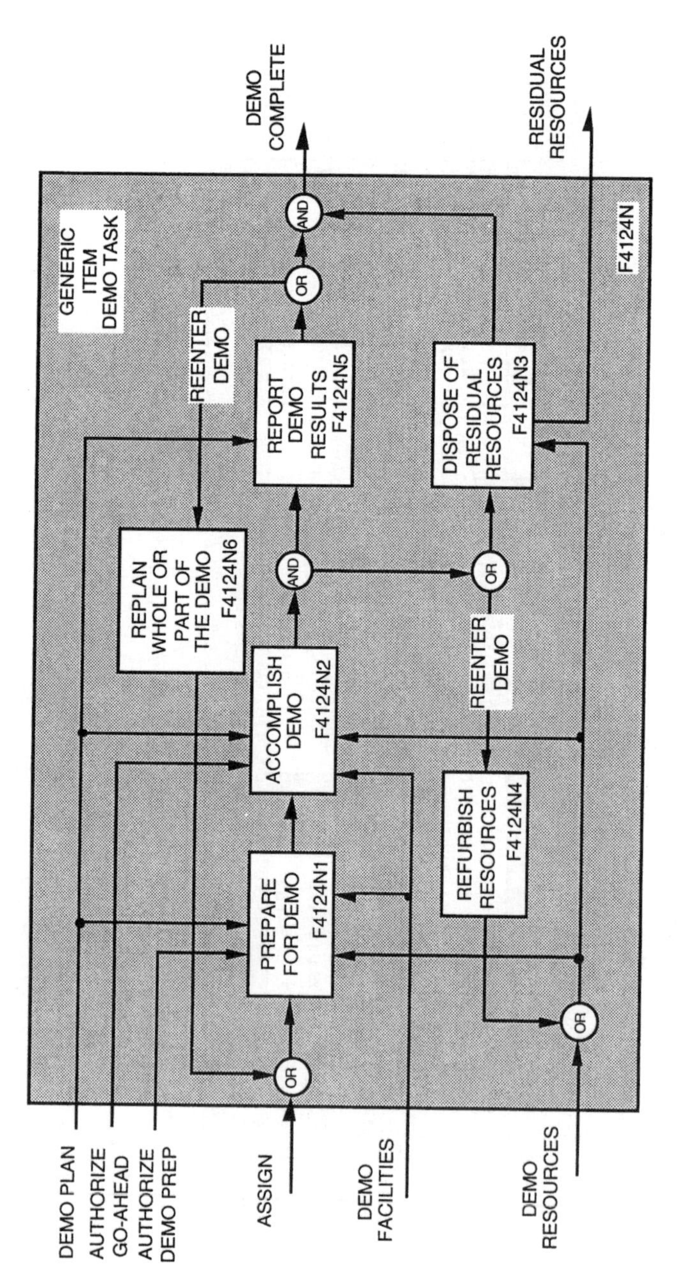

Figure 5-8 Generic item demonstration task flow.

demonstration so provisions are included for refurbishment. Disposal of residual materials may also be necessary subsequent to or during a demonstration.

5.6 *Program task structuring*

The diagrams in this chapter offer a generic view of a verification process for qualification. They can be described in enterprise practices documents in generic terms. In Chapters 6, 7, and 8 we will discuss the process of identifying all of the tasks in each method category and these tasks will fall right into the framework we have constructed in this chapter.

The specific plans and procedures related to each of these tasks will drop right into an IVP paragraphing structure introduced in Chapter 6.

chapter six

Item qualification test planning analysis

6.1 Overview

As noted earlier, there are two major steps in the verification process. First, we prove that the designed and manufactured product, described on the engineering drawings, meets the predefined development, performance, or Part 1 requirements. That is called development, design, or qualification verification or, simply, qualification. Second, we prove that the product, manufactured using the final methods and tooling, will comply with the engineering drawings and that the production process will produce repeatable results. This is called product verification or acceptance and is intended to prove that the product complies with the requirements in its detail, product, or Part 2 specification. In this and Chapters 7 through 9, we will take up qualification. Chapters 6 and 7 focus on test as the verification method and Chapter 8 on other methods. Chapter 9 is concerned with management of the qualification series leading to a functional configuration audit (FCA) and Chapter 10 extends the conversation to the system level. Acceptance will be covered in Chapters 11 through 13.

Qualification test planning is a process of transforming the detailed verification requirements captured in the program specifications, and mapped to a test method, into a comprehensive series of testing tasks that collectively and least expensively expose the relationship between item capabilities and the development requirements for all of the items in the system.

Throughout this chapter, we will try to focus on the test method so we must reserve a place in our mind for the greater picture including the other methods and corresponding work associated with planning, accomplishing, and reporting upon the verification work associated with them. The intent is to break the whole job into a series of fragments for discussion purposes, but the reader should not lose sight of the fact that when performing the work, we must be working the whole job including work related to verification by analysis, demonstration, and examination.

The approach encouraged in this chapter is based on having done a credible job of verification requirements analysis, as discussed in Chapter 4, as a part of the system and item specifications development. This work provides us with the detailed information needed to fashion a comprehensive test plan. We also need knowledge of the test resources available from our own company in the way of test laboratories and equipment, their capabilities and availability. Add to this the test resources available to the customer and other parties who may be called upon to cooperate. Finally, we also need adequate time and money resources to execute a good verification test plan properly. All of these factors, time, money, test facilities, and the verification plan, must be mutually consistent or failure in some measure is guaranteed.

We will first briefly take up the overall qualification test planning environment and then focus on the detailed planning work associated with the test planning analysis for a generic item within the context of that overall plan. In Chapter 9, we will expand upon the management process introduced here.

6.2 *Qualification test program planning*

We will consider several parts of the planning process independently but they must all be accomplished together. For example, scheduling should be influenced by what items will be tested and how long it will take to accomplish that testing, and vice versa.

6.2.1 *Test philosophy*

The testing can be accomplished in one of several sequences but it should be in some organized sequence where prior testing accomplished can be used as a basis for future testing to be done. It would make little sense to select randomly the things to be tested in time.

While a system is best defined in terms of its requirements and concept from the top down as discussed in Chapter 1, systems are commonly designed and built, or implemented, from the bottom up. After we know what all of the needed elements of a system are and their requirements, these elements, at some lower level of indenture, will surrender to design and manufacture or coding. This sequence matches the specialization paradigm perfectly since the lowest-tier items will require fewer specialized knowledge disciplines than higher-order elements. The designs for these lower-level items can therefore be developed more efficiently than those at higher levels.

When applying this method to software coordinated with bottom-up testing, it will be necessary to create some special software to enable the testing process. These modules are commonly called drivers and they simulate the higher-tier software not yet developed. It is very difficult to design hardware-dominated systems from the top down and even more difficult to manufacture them in this sequence. They consist of physical things such that it makes little sense to think about wholes as prerequisites to parts. On the other hand, computer software being an intellectual entity, it is possible to design and implement (code) software elements of systems in a top-down fashion where the operating system and the top-level structure of the system are first coded. As the analysis unfolds, the lower-level modules of the software are designed and coded, and so on down through the whole system architecture development activity. Special software-testing modules are required in this case as well, and they are often called stubs which simulate the lower-tier software not yet coded.

You could elect to design the system based on an outside-in sequence. With software, this process begins at the system terminators shown on the context diagram, the external system interfaces. The software required to interact with these interfaces is first developed followed by the internal processing elements of the system. Testing can reflect this arrangement with initial testing of the peripheral software connected to internal simulators followed by design and coding movement toward the center. This approach could conceivably be used in a hardware system being modified.

The inside-out approach could also be applied for a software system where we start with the interior elements of the system, that is, those farthest from the external interfaces. We create the core processing software and move outward toward the external interfaces testing from the inside out as well. This method could also be applied to hardware-dominated system modification.

Generally, we should align the test philosophy with the design and manufacture/coding plan. Whichever direction we choose as a basis for design or coding, we should select that same direction for design. Imagine the problems if we elected to design and code the software from the bottom up but to test it from the top down. By the time we were ready to test the software, we would have lost months of opportunity to test the lower-tier software that had already been created while we awaited topping out the software design and coding. You can imagine other inefficient coding and testing combinations. In the case of software it is best also to include an aggressive testing process interactively within the design and coding process following the spiral development sequence discussed in Chapter 1.

6.2.2 *Top-down and bottom-up planning*

Actually, test planning should be accomplished in both a top-down and a bottom-up direction simultaneously. There are macroconsiderations such as the amount of time and money available within which the testing must be accomplished. Figure 1-14 offers a top-down view of the qualification process within the context of a life cycle functional flow diagram. F4 in Figure 1-14 includes the qualification process.

But, the details can best be developed from the bottom up.

6.2.3 *Qualification test scheduling*

The overall time frame available for accomplishing the item qualification test work is constrained by the time necessary to develop the physical product, including its design and limited manufacturing of sufficient articles to support the test program, on the program front end and on the back end by the planned or required delivery of the product. On a program applying the grand development environment, this work should not be moved too far back into the program because it will interfere with doing the work required to support a sound design. The result will likely be that we will find many flaws in the design during testing that will require redesign and retest, referred to as regression testing in software. If the qualification testing is delayed too long, there is a danger of adversely effecting the system-level testing schedule or a required delivery or market-dominated ship date. Commonly, one cannot begin system testing before the items have been qualified. This is especially true for aircraft and other systems entailing operating safety concerns. As the reader can well understand, it is difficult to recruit a test pilot with good sense to fly a new aircraft until all of the items that compose it have been qualified for the application.

This model is different where a rapid prototyping environment is applied in the context of the spiral development sequence. In this case, we seek to get into testing as quickly as possible and translate the lessons learned into the next round of analysis and design leading to a future testing cycle. The same fundamentals apply as discussed here except that we are repeating the process cyclically with changing test requirements driven by increasingly detailed knowledge of the evolving product.

Much of the item qualification testing can be accomplished in parallel, especially where it is accomplished by suppliers but there are limitations on your own company's human resources and test laboratories. Most often this peak problem is mitigated by the natural probabilistic dispersion in the different development times for the items in the system. Some things will require relatively little time to develop while others will push the limit of available development time. You can take advantage of this dispersion to broaden and depress the qualification test peak so as to avoid a personnel peak that might force the company through a hiring-and-layoff cycle to support program personnel needs and avoid adversely influencing the other tasks in the development process in terms of budget or schedule allocations.

In addition to the overall test time span and its location within the context of the program schedule, there may be some particular sequence requirements that must or should be recognized in laying out the overall program. A little later we will take up the top-down vs. the bottom-up or inside-out test sequences used in software testing, but, in general, the testing process should follow the bottom-up sequence from parts and materials through components to subsystems, systems, end items, and finally the systems. This is especially useful where the lower-tier elements have to be operated as part of higher-tier tests. By ensuring that the lower-tier items function satisfactorily, they become part of the test setup or environment of the higher-tier items. In both hardware and software it is possible to test all items independently by simulating the input–output relationships but at some point all of these items must be played together.

Program schedule constraints and testing activity needs will mutually dictate the block of time available to the qualification test activity, thus defining the schedule from a macroperspective. Within this time block we need to organize all of the verification strings (identified by VSN) listed in all of the specification verification traceability matrices, that are mapped to a test method, into a series of verification test tasks (identified by VTN). These tasks have to be stitched together into a context respecting necessary sequences of tests driven by a sound test strategy and coordinated with the available time and money, thus providing a microinfluence on the test schedule.

The schedule can be constructed based on deterministically stated time or probablistically stated time. In the latter case we estimate a mean and a variance for the time where the variance is a measure of the risk in accomplishing the test on time. Planning and scheduling tools are available and inexpensive for building network diagrams in the CPM and PERT formats that provide not only a pictorial view of the plan but aid in identifying and allocating the resources and identification of the critical path as well as giving visibility into the possible effects of schedule timing variations.

In summary, from a macroperspective we need to bound the total time frame for qualification testing based on larger program concerns and on our experience with our product line. As we identify the items to be tested, we should fit the corresponding test tasks into the overall time frame recognizing we may have to make microadjustments in the program test time frame. Finally, as we accomplish the detailed test planning for each verification task, these time allocations may have to be adjusted again and a final schedule published.

Figure 1-14 illustrated a life cycle functional flow diagram as the first expansion of the customer need. One of those blocks was F4, System Verification, expanded upon in Chapter 5. We now wish to fill in some details needed to plan the verification work.

6.2.4 *Qualification test item selection*

During the period when we are identifying verification methods, developing verification requirements, and aggregating all of the verification traceability matrices into the compliance matrix, the program must finalize the list of product items that will be subjected to qualification testing and begin the test planning process for those items. On a DoD or NASA program, the item list would be predetermined in the form of the configuration-item or end-item list, respectively. That is, the items selected through which the customer will manage the development program will determine the items that must be tested because of contractual and management patterns. Nongovernment customers may also predetermine the items that will be subjected to qualification testing in accordance with some selection criteria. In all of these cases, the development team should be made aware by the customer of the selection criteria or the prescribed list of items as early in the program as possible and that list coordinated with the specifications that must be developed.

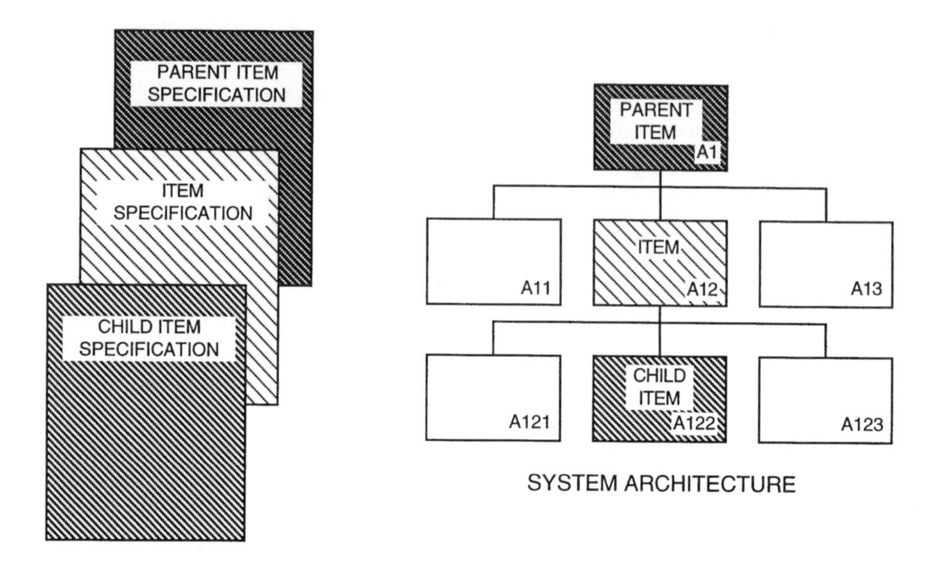

Figure 6-1 The relationship between architecture and specifications.

Where the customer does not enter into the decision-making process with respect to the items that will be tested, the program team will have to do this themselves. On a largely unprecedented system development, this process should evolve as one input to the architecture synthesis applied to the stream of allocated functionality. The architecture, an example of which is shown in Figure 6-1, should be assembled based on physical things that result from an agreement on the functional allocations and aggregation of that functionality into things, among manufacturing, test and evaluation, procurement, engineering, finance, and logistics participants. The specification tree should be overlaid upon this evolving architecture. It is generally the items which will have specifications prepared that will populate the list. If it is other than this list, we should be concerned because it is the specifications that should drive the qualification process. It is true that some organizations believe they can put together an acceptable test plan without reference to the specifications but the author does not accept this claim.

Section 4 in a DoD specification, when properly prepared, will identify most or all of the item-related verification requirements (there could be carryover from parent and child specifications). As the specifications quickly mature, the contained verification traceability matrices will define the requirements that must be verified by test and the level at which the individual verification actions shall be accomplished. This data will have been assembled from the work of many persons making verification decisions in building the many verification traceability matrices. In a simple case, where all of the line items in every item specification verification traceability matrix are marked for level ITEM, then it would be very simple to make the item list from the list of specifications. Since every matrix entry in a specification that is marked for a higher level will correspond to some other item covered by a specification, the higher-level marking for verification level does not alter the logic for verification item selection. This list can be expanded when a lowest-level specification identifies a requirement for verification at a lower level than the item. This could conceivably add hardware or software modules or other entities to the list that do not have a specification themselves. As noted above, this is a troublesome eventuality but it could occur as in the case of a subsystem that includes an item of customer-furnished equipment or software that has no surviving specification and is being used in a different application from that for which it was developed.

The list of items thus formed should then be reviewed for ways to reduce verification cost through pooling testing tasks together at higher levels. We may conclude, for example, that the testing called for in five item specifications at the item level can be achieved at the next-higher level just as effectively but at less cost. We will still need all of the items as units under test, but we may be able to gain some time or cost advantages from aggregating items under test not possible through individual item testing.

The final list, however derived, defines the items about which we must build the test tasks and task procedures. Generally, this is simply going to be a list of the new items for which specifications have been or will be prepared or any items from previous applications that are going to be modified or used in a significantly different way. While it is possible to apply the top-down item definition approach, the details of the test program should be based on the content of the aggregate of all of the specification verification traceability matrices. This chapter proceeds based on the assumption that the detailed test program will be developed based on that content.

The reader should see at this point that the planning of the verification work, like many other things, must proceed from two perspectives simultaneously, the macroperspective or system perspective and the microperspective or task perspective. We will discuss responsibilities for these planning focuses shortly. As the plans are formulated and conflicts are identified, we will find cases where the macroperspective should prevail and others where the microperspective must prevail. So, there should be a lot of give-and-take between the people developing the verification planning rather than a rigorous top-down hierarchical direction style applied. At least, the former will commonly be more effective.

6.2.5 Item-to-task transformation

The verification requirements analysis work should be accomplished in association with the item view while the verification planning work should be accomplished in association with the verification task view. That is, the planning work that we must do must be collected about the verification work to be done which we have characterized by VTNs assigned within the context of a dumb numbering system respecting uniqueness from a system perspective. If, as noted earlier, all requirements are verified at the item level, then the item list and list of VTNs may be identical. But, even in that case,

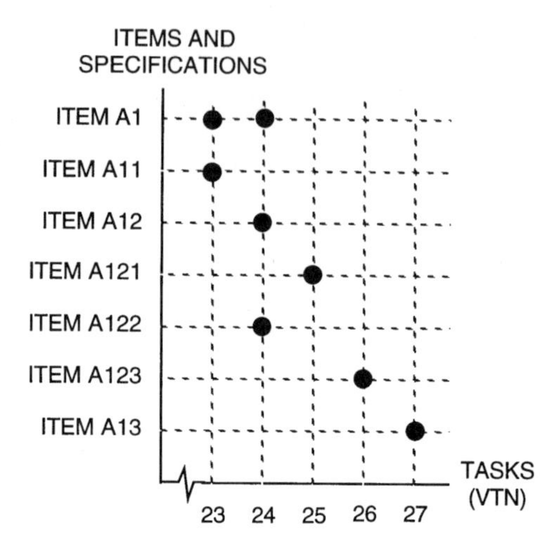

Figure 6-2 Item VTN map.

we may have to run more than one verification task on one item, there may be specifications that include the requirements for the item and its children, and some items will require the application of methods other than test and they should be identified by a different VTN. Figure 6-2 suggests a way to correlate items and tasks in a simple matrix. On one axis are the VTN numbers, described or defined in a separate listing, and on the other axis are the architecture IDs, referred to in Figure 6-1.

Note that some tasks will verify requirements for an item and its parent or child item and that one item may have verification strings in more than one task. The rationale for these apparent anomalies was noted above. If there is a one-to-one relationship between items and tasks then it is not worth the effort to create and maintain an item VTN map, but otherwise it offers visibility into the verification planning effort to ensure that all of the items fall into at least one VTN. This matrix could be located in the IVP (under paragraph 1.2 or 2 in Figure 6-4).

There is much to be learned about your development process from within the verification planning work. Verification planning and traceability work drove early system engineers to mainframe computer systems and continue to influence organizations to search for better methods than they currently apply. Specifications that are late in release, that still contain values to be determined, that contain requirements qualitatively stated without values, that contain paragraphs two pages long sprinkled with multiple requirements are important messages that there is something wrong with your process or the skills of the people doing the work that the verification process must use as its input. Verification is a rich source of lessons learned or metrics that have a broader influence than simply how well you are doing verification work.

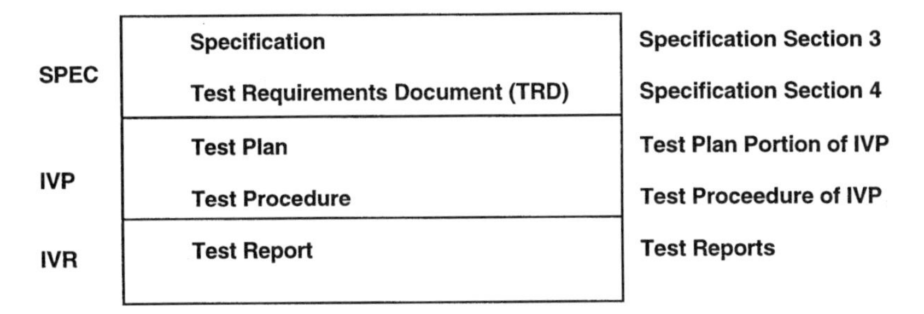

Figure 6-3 Verification documentation structure.

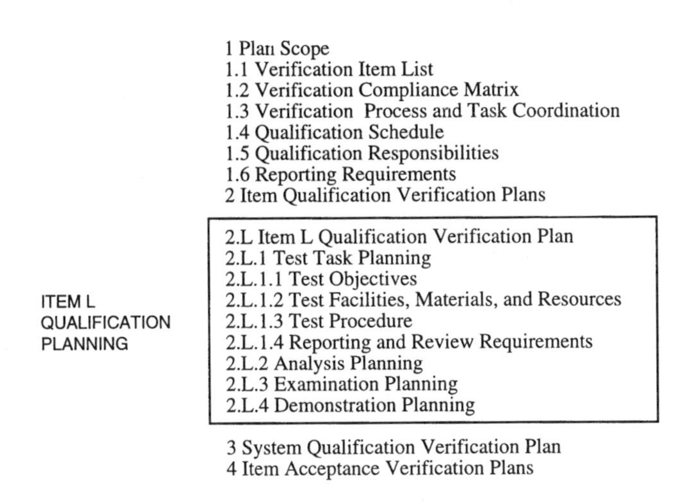

Figure 6-4 Integrated verification plan outline.

6.2.6 *Verification information organization*

The stream of test verification documentation includes those identified in Figure 6-3. These may be separately documented or integrated into some fewer number of documents. It is recommended that the verification requirements for each item be captured in the same specification wherein one finds the corresponding item requirements. The individual test plans and procedures should be captured in an IVP or at least an integrated test plan (ITP). These plans all fall into paragraphs 2.L.1.1 and 2.L.1.2 of Figure 6-4 while the procedures drop into paragraphs 2.L.1.3. Similar paragraphing exists for nontest verification methods. The reporting of test results can be done using separate reports for each test task or they could be grouped into an IVR mirroring the structure of the IVP. It would be possible, as well, to include the reporting in the IVP but it is probably better to separate these two streams.

For each item that requires qualification (development verification) by means of testing, we must perform a transform between the aggregate of all of the test VSNs assigned to that item to a verification task plan, based on the methods to be applied (test, analysis, demonstration, or examination) and finally to a verification task procedure, giving step-by-step instructions on test performance, that will produce good evidence of compliance. The plan should identify goals, resources, schedule, and the general process for achieving the planned goals. The planning data could be captured in a separate document for each item or all of the planning data could be collected in an integrated program verification plan as suggested in Figure 6-3.

Figure 6-4 offers an outline for the integrated plan. Sections 1 and 2 focus on the organization of the overall plan and item qualification, respectively. Paragraph 1.1 provides the item list discussed earlier. These are the items for which we must provide verification planning and conduct verification tasks. Paragraph 1.2 of the plan refers to the existence of the compliance matrix published separately. We do not want to include this in the test plan because it will drive an endless revision cycle of the whole document. Paragraph 1.3 of the plan provides an overall task process (graphically or outline structured) and links together any verification tasks that must be coordinated. Paragraph 1.4 refers to a separately maintained overall schedule coordinated with the compliance matrix. Paragraph 1.5 identifies the overall pattern of responsibilities for verification. Detailed item verification responsibilities are coordinated in the matrix. Paragraph 1.6 of the plan gives reporting requirements including who must be told of the results, how quickly or through what means they must be told (you may require an urgent summary or flash report followed by a final report), and in what format the report will appear.

Section 2 of the plan provides all item qualification planning organized into one subsection for each item L listed in paragraph 1.1. Each of these sections may include subsections for (1) testing, (2) analysis, (3) examination, and/or (4) demonstration. This chapter covers testing so only the plan structure related to item test (2.L.1) has been expanded in Figure 6-4. The plan outline also assumes that an item X has only a single verification task associated with it for a particular method (test). It is possible that a given item will require two or more test tasks, in which case the 2.L.1 section may have to be further divided into 2.L.1.1 through N corresponding to test tasks 1 through N and each subsection would then be further divided as shown in Figure 6-4.

The structure depicted in Figure 6-4 does not include development evaluation testing (DET) because that is a validation activity and in the author's view it would be captured in the validation planning stream. The reader may conclude that he or she would rather partition the whole information problem differently such that all test-related planning data are together including DET. Alternatively, the program could prepare an integrated validation and verification plan, a very grand goal requiring the application of tremendous management skills that encourage good program discipline that those on the program readily adapt to and even come to appreciate because they understand the benefits. The integrated verification plan will be quite difficult enough to create on a program and an organization that succeeds in crafting and implementing one with good results should be very proud of its achievement.

As noted, one could partition the information to focus first on method rather than item such that there was contained in the whole plan an integrated test plan, an integrated analysis plan, and so forth. In an organization with a strong functionally based and collocated test and evaluation department, T&E department management will likely support this structure so that they may focus their energy on their own section of the document rather than supply fragments of a whole. The author believes that the organizational strength must be in the program organization in the form of cross-functional program teams participated in by T&E people as well as everyone else. In this context, it is far better to make the first document structural cut based on the items in the system about which the teams will collect. The author has chosen to focus the IVP content on the items in Section 2 rather than verification tasks (VTN), once again, because the work relates much more directly to the teams organized about the items. The VTNs will end up mapped to, or consistent with, the items on our verification items list so perhaps this is only a semantic difference.

Another reason that the IVP may be difficult to create is that the customer may require separate delivery for review and approval of the plans and procedures. In this situation, it may be possible to convince the customer that their needs can be satisfied by the IVP through a two-step release, the first issue including the planning data and the second adding the procedures.

6.3 *Item planning cycle*

Given that we have assured ourselves that the item verification requirements analysis job has been done well, we are ready to prepare a detailed plan. This process includes four steps that will be taken up in turn:

a. *Verification Matrix Test Data Fusion.* Collect the verification strings about the item to be tested and identify the test as a verification task with assigned VTN.
b. *Qualification Process Design.* Define test objectives and prepare an item test process flow diagram. Map resources and procedural needs to the blocks of the diagram. Identify test resources for each step and as a whole.
c. *Prepare the Procedure.* Create a step-by-step procedure for employing the test resources for each block of the test process diagram and fashion forms or computer data entry regimes for capturing test results.
d. *Validate the Plan.* Expose the item plan and procedure to critical peer and/or management review and correct any discrepancies.

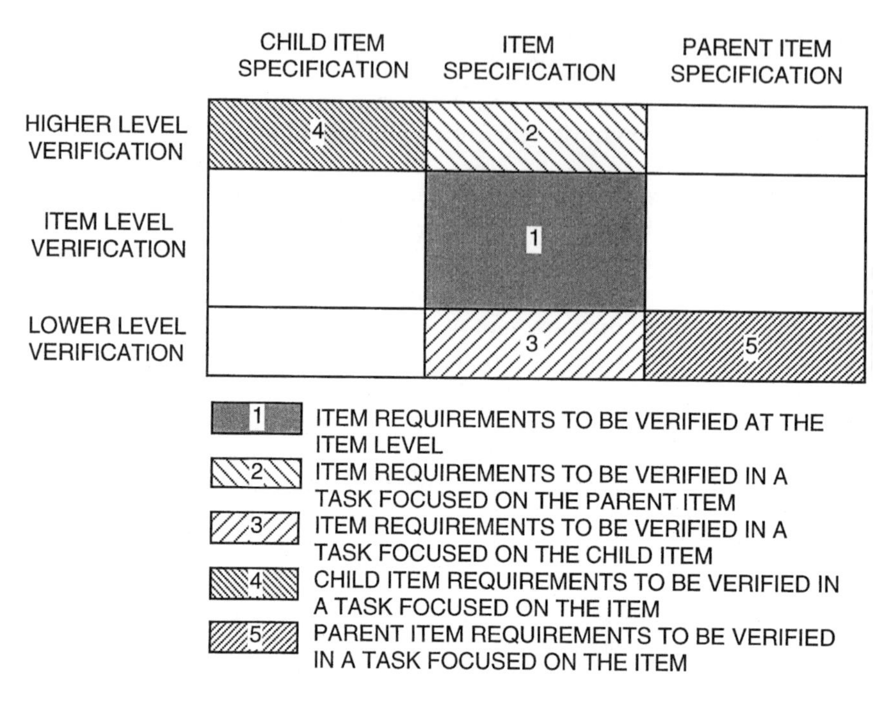

Figure 6-5 Item verification string fusion.

6.3.1 Verification matrix test data fusion

The verification requirements that were mapped to the test method in the item verification traceability matrix marked for item level have to be combined with the test verification requirements identified in lower-tier specifications for a higher level of verification and the test verification requirements in the higher-tier specification that were identified for lower-tier verification. Figure 6-5 illustrates this item string fusion. Zones 1, 2, and 3 cover all of the requirements in the item specification. Zones 1, 4, and 5 correspond to all of the VSNs that map to a common verification task focused on the item in question. Zones 2 and 3 relate to item requirements that will be verified in tasks focused at lower or higher levels, respectively. This same pattern can apply to every item specification on the program, some subset thereof, or none of them depending on the density of the requirements that are verified at other than the item level throughout the program specifications.

In the simplest case where every verification requirement is marked for verification at the item level, the item specification space (Zone 1) fills with all of the item requirements and the other two item spaces (Zones 2 and 3) shown in Figure 6-5 retract to a void. In this case, the verification test program for the item becomes very simple, being composed of a one-to-one matching of the items to be tested and the verification task plans.

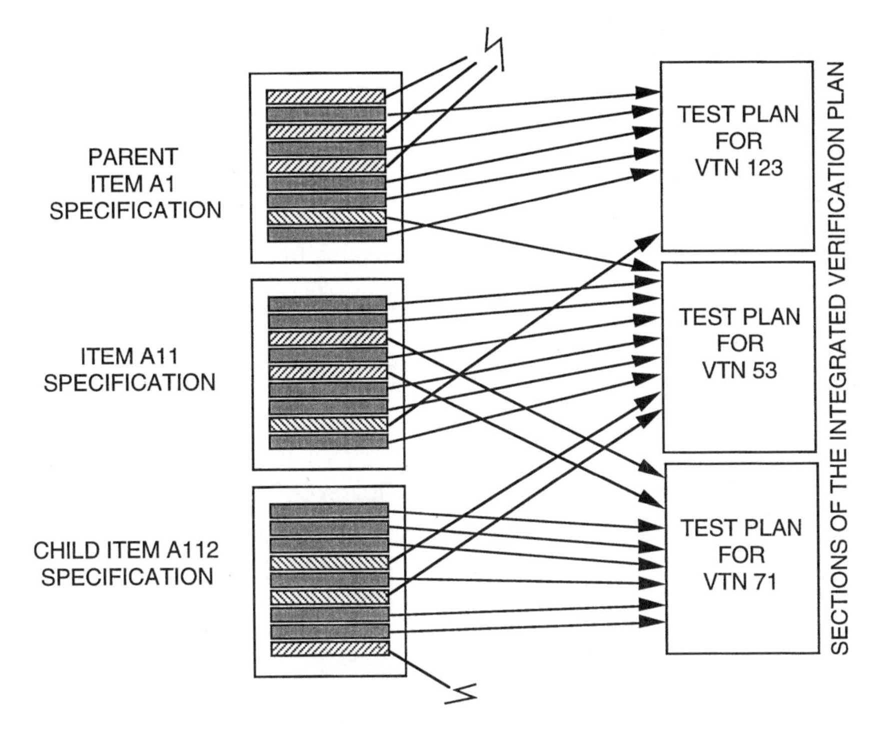

Figure 6-6 VSN-to-VTN transform.

The union of all of the VSNs mapped to a VTN focused on a particular item yields the verification requirements that should be the basis for the development of the test task or tasks for an item. This same process must go on for all other items on the verification items list. Figure 6-5 is focused on only a trio of specifications, but the real situation is more complex. Any one item specification will most often have more than one child item, and the requirements in the item specification that are to be verified at a lower level could be verified in a task oriented toward one, some subset of, or all of those child items.

Figure 6-6 offers an alternative view of this process of coordinating the many VSNs from three specifications (and we may have a hundred specifications on a program) into three test tasks (and there are also analysis, examination, and demonstration tasks as well) focused on the items covered by these three specifications. Each specification consists of some number of requirements, and they have been mapped in their own verification traceability matrix to a method and identified as a specific VSN. Those requirements that will be verified at a higher level are hooked into the VTN for the parent item, those to be verified at a lower level are hooked into a VTN for a child item, and those to be verified at the item level are hooked into a VTN oriented toward the item. Once again, as the number of requirements on the program that are verified at other than the item level decrease, the non-item

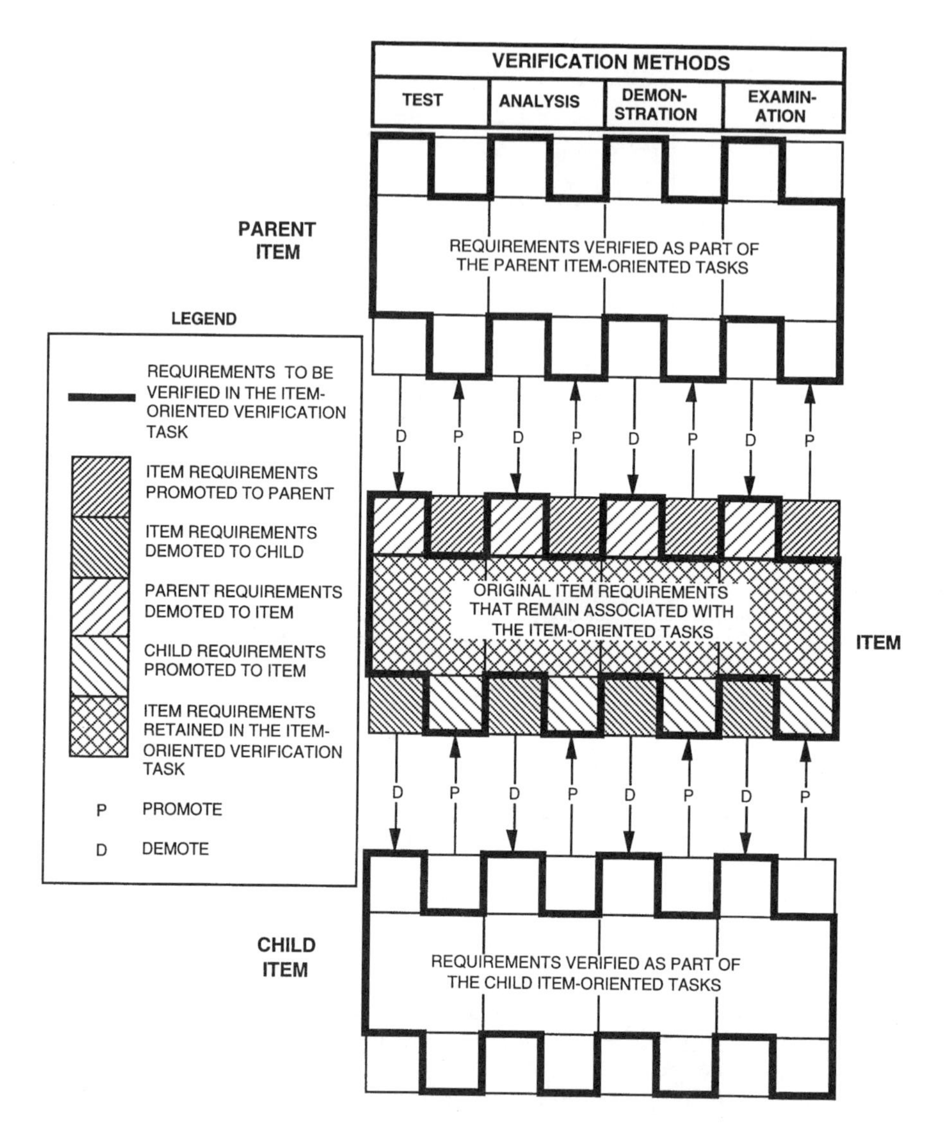

Figure 6-7 Promoting and demoting VSN.

links in this diagram would all move toward a one-to-one relationship between specifications and test plan data identified by a VTN.

A third view of this data fusion process is exposed in Figure 6-7. Most of the requirements (possibly all) that become part of the VTN associated with item test verification come from the item specification. Some of these requirements are promoted to the parent item specification test VTN and other are demoted to a child item specification test VTN. There may be parent item requirements demoted for inclusion in the item VTN and promoted from child item VTN to the item level. The left-hand end of the item diagram captures all of the item, child, and parent item requirements that must be connected together for inclusion in the test VTN focused on this item. In a more complex situation, there may be more than one test involved but all of the corresponding requirements for all of those tests are illustrated in this subset.

Where a program is organized into cross-functional product development teams, these teams should accomplish the data fusion work for their item or items and a system team should integrate and optimize the planning data that crosses item boundaries beyond the team responsibilities boundary. Where a program is served by a functional test and evaluation group, this group should accomplish the test string data fusion and test integration work. In this latter case, we have a remaining problem, however, with who should have the responsibility for the nontest verification actions. One solution to this problem is to apply the broadest definition for the word test and accept that all four methods are subsets of test resulting in the T&E organization being responsible for management of all verification work.

6.3.2 Qualification process design

6.3.2.1 Test task goals

The general goal for each verification task is to establish unequivocally whether or not the item satisfies the requirements for the item. The part of this total job that we will focus on in this chapter is composed of the test tasks. So, a general goal that could be applied to most tasks is, "Show by test means to what extent that the item satisfies each of the requirements mapped to this task." If the task is fairly complex, it may be useful to define more than one goal focused more finely on specific objectives. The problem now remains to expand this goal (or these goals) into supporting details that tell how they shall be satisfied in terms of work and resources.

6.3.2.2 Process analysis

The recommended approach to expanding upon task goals is process diagramming that encourages synthesis of the goals into a process that achieves the goals in sufficient detail to permit preparation of procedures for each of the blocks on the process diagram. If the test task is sufficiently simple, one block on our process diagram may be sufficient, but generally we can characterize every test task by a generic string of activities as illustrated in Figure 6-8 in IDEF0 format. The differences from test to test will correspond to the ACCOMPLISH TEST block. We may have to decompose that block further if it represents a complex undertaking. The decision on whether or not to decompose should be based on whether the block is sufficiently detailed that a procedure can be visualized for it or not. If not, then further decompose the activity until that is the case.

In the IDEF0 format the general flow through the process is indicated by the arrow-headed lines entering and departing the blocks on the left and right ends. The lines entering the top correspond to controlling influences

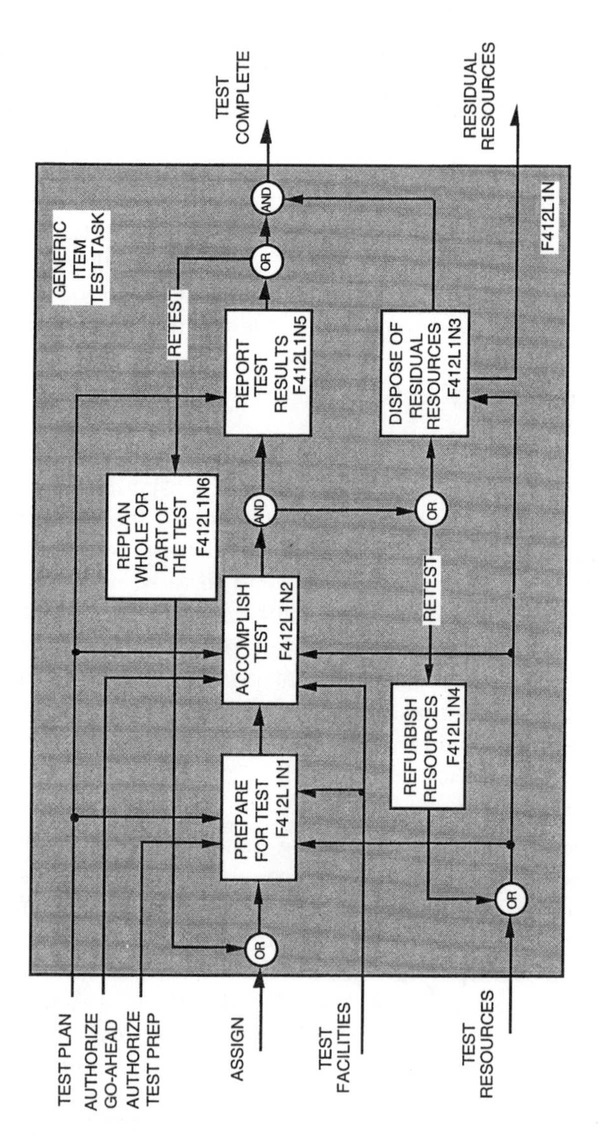

Figure 6-8 Generic test task.

such as procedures and rules. The lines entering into the bottom surface relate to mechanisms or resources needed to support the work related to the block. If the task runs according to plan, there will be only one passage through the task, but, if as sometimes occurs the test fails to produce evidence of compliance, then it may be necessary to rerun the test with possible changes in the test procedure before or after design changes are accomplished and introduced into the item under test.

The tasks are numbered in Figure 6-8 based on the master life cycle diagram in Figure 1-14 which includes a function F4 for system verification. In Chapter 5 we expanded this function to identify a generic test task repeated here in Figure 6-8. The L is the identification of the specific item being tested or the VTN corresponding to it, the 1 following the L is for the test category and N corresponds to the specific test task for item L as described in Chapter 5. The blocks on the diagram are then simply numbered in some order starting at 1. This diagram uses the base 60 system to avoid decimal delimiting, but, on a large program, it may be necessary to use decimal delimiting in order to identify all of the very large number of verification tasks uniquely. It is not essential to create a complete flow diagram, of course, if Figure 6-8 fairly represents all of the tasks generically. One could simply refer to the generic process diagram numbering the tasks as noted here for use as a key field in computer databases. Where a particular test is very complex it may be necessary to fashion a specific process diagram showing how the individual steps will have to hook together. If the only block that has to be expanded is F412L1N2, then one could simply craft an expansion of this one task to show more detail and leave the others in generic reference.

In this chapter we are primarily interested in a single generic verification task. We are implying that all verification tasks are independent and would appear as many parallel channels in the overall verification process diagram. In Chapter 10, we will consider the more complex and realistic situation where the tasks are related in series parallel arrangements and these connections must be foreseen and documented as part of our overall plan.

6.3.2.3 Resource identification

The planning of a test is much like logistics support analysis. We have to determine all of the resources required to achieve the goals of the activities identified for the block on a process diagram. It is only the work that will have to be accomplished in F412L1N2, and F412L1N1 to a lesser degree, that will drive the need for resources. Granted, you will commonly have to have these resources on hand to accomplish F412L1N1, but here we are interested in our process diagram as a model to determine what resources will be necessary. One of the resources is, of course, one or more articles of the product to be tested. Generally, this should be as much like the production articles as possible, but in some cases it may be necessary to instrument the article under test to acquire the needed test data. If this is the case, then the test analysis work should identify the specific requirements for the differences that must be introduced into the product for the purpose of testing, and it should be determined to what extent, if any, that these changes will have on the performance of a normal production article. These test-related changes should have no effect on the validity of the test data collected in the planned test applying to the question of whether or not the item design satisfies its requirements.

If the item under test includes software, either that software or special or modified software developed specifically for the test will be needed if the test requires operation of the item. Some items which include computers must be qualified in a series of tests that examine parts of the complete operating capability of the item. This may include tests to determine correct operation of the computer program in solving problems presented as different sets of input data as well as environmental tests that examine the ability of the item to survive while impacted by various combinations of temperature and vibration stimuli. Another test may include the item in a larger subsystem test where all of the subsystem items interact to accomplish a higher-order function.

In addition to the item to be tested, we will also need some special test apparatus. This may include things that simulate particular environmental stimuli (temperature, vibration, shock, sunlight, wind, etc.); simulate operational conditions [turn rate table, guidance model running on a mainframe computer, or a global positioning system (GPS) signal generator]; or provide a system infrastructure interface permitting operator input and status monitoring either automated or manually implemented. Special test equipment may be required to stimulate and monitor performance. Signal generators may create input signals with observation of outputs on an oscilloscope, for example, in a manually implemented test. In the case of an automated test, we may have to build or buy a special computer circuit board for a personal computer (PC) to interface the item under test with the controller provided by the PC. This may also require some special software crafted in a particular language.

There is also a need for more mundane things like workbenches, chairs, pencils, special forms upon which to record test results, special instrumentation and recorders, paper, and hand tools. In some cases we may have to develop and build special test stands as in the case where we are testing a new rocket motor. In the way of facilities, we may have to book time on an automotive test track, an environmental test chamber, or other laboratory.

Every task is going to require personnel to perform the test. Our plan must identify personnel requirements in terms of how many people over what time frame and the qualifications of those personnel. The personnel identified at the microlevel for each task may or may not combine at the overall qualification program level so as to be compatible with the planned program-staffing profile. As this information accumulates, the system agent for the qualification program must pile up the personnel requirements in time and adjust the schedule, if necessary, to avoid exceeding planned staffing. Alternatively, we must gain acceptance of a higher personnel peak and

2.L.1.1	Test Objectives	
2.L.1.2	Test Facilities, Materials, and Resources	
2.L.1.3	Test Procedure	
2.L.1.3.1	Prepare For Test	F412L1N1
2.L.1.3.2	Accomplish Test	F412L1N2
2.L.1.3.3	Dispose of Residual Resources	F412L1N3
2.L.1.3.4	Refurbish Resources (if required)	F412L1N4
2.L.1.4	Reporting and Reviewing Requirements	F412L1N5

Figure 6-9 Item L test task plan fragment.

possibly higher aggregate cost or extended schedule time frame selected to reduce the personnel peak.

The way to determine these material and personnel needs is to determine how each step in the process will be accomplished and to write down all of the resources required to achieve the intended results mapped to the task in which it is needed. We can then sum up the items needed for the test task we are planning and enter that in the appropriate paragraph of our IVP. As in the case of personnel, a system agent needs to also accumulate the aggregate resources across the qualification program in several other categories, such as facilities, items under test, test equipment, and consumables.

6.3.3 *Test procedure development*

In addition to physical resources, each step in the process description must be described in terms of how the task will be accomplished by the test personnel, that is, a test procedure. Figure 6-9 suggests an expansion of the IVP shown in Figure 6-4 for the generic item L test plan. This outline can be expanded in paragraph 2.L.1.3.2 for a very complex test involving many steps on a process diagram. As a minimum we should have one paragraph for each block on our process diagram. These paragraphs can be constructed based on a description of how we would accomplish each detailed step. If it isn't obvious how a step can be performed, then that step should probably be expanded in process diagram form until it is possible to describe how the lowest-tier tasks can be performed.

The content of the plan coordinates with the generic flow diagram in Figure 6-8 as noted by the process block references in Figure 6-9. It would be hoped that it is not necessary to replan any part of the test because it will be concluded as planned to good result. Therefore, any replanning instructions can be covered more generally in the IVP and should not be repeated for each test task. Generic content can also be included about disposal and refurbishment but any specific tasks should be included in the task plan.

Figure 6-10 offers an excerpt from the specification and IVP showing one whole verification planning string. This starts with a very simple weight requirement in Section 3 of the item specification and is reflected in a

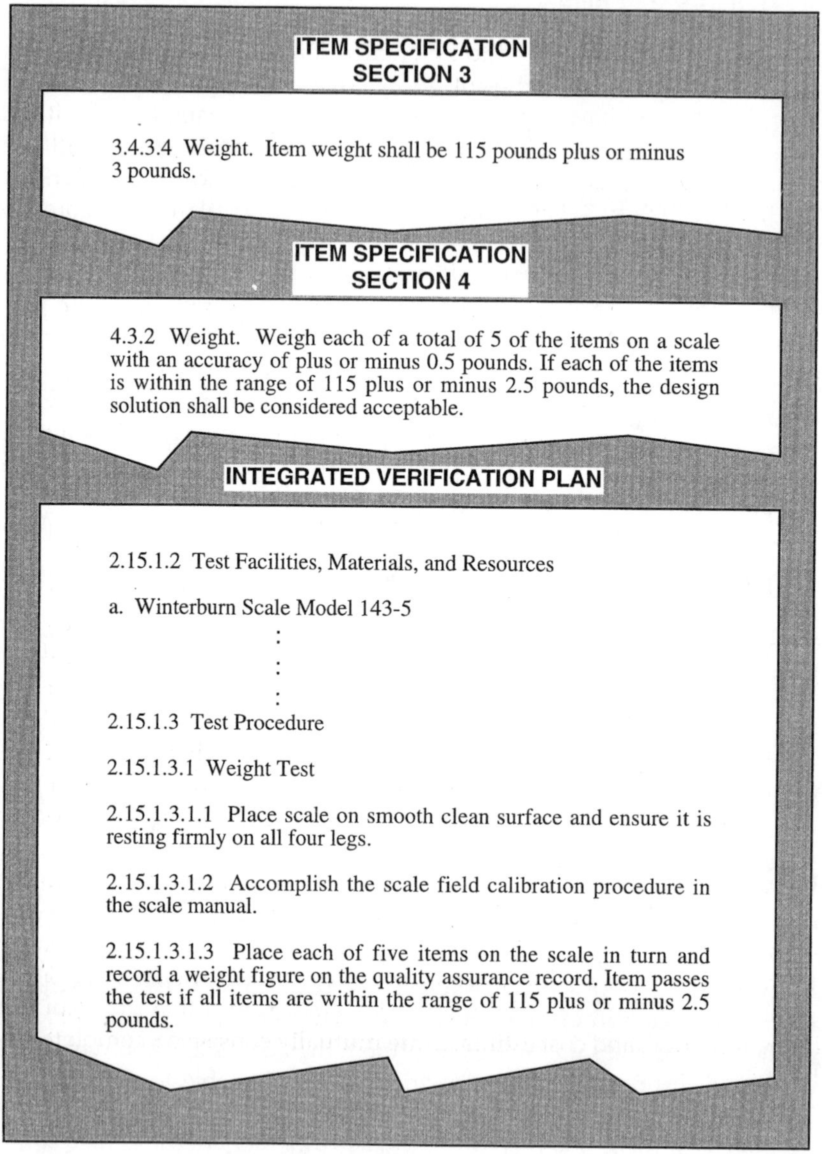

ITEM SPECIFICATION
SECTION 3

3.4.3.4 Weight. Item weight shall be 115 pounds plus or minus 3 pounds.

ITEM SPECIFICATION
SECTION 4

4.3.2 Weight. Weigh each of a total of 5 of the items on a scale with an accuracy of plus or minus 0.5 pounds. If each of the items is within the range of 115 plus or minus 2.5 pounds, the design solution shall be considered acceptable.

INTEGRATED VERIFICATION PLAN

2.15.1.2 Test Facilities, Materials, and Resources

a. Winterburn Scale Model 143-5

:
:
:

2.15.1.3 Test Procedure

2.15.1.3.1 Weight Test

2.15.1.3.1.1 Place scale on smooth clean surface and ensure it is resting firmly on all four legs.

2.15.1.3.1.2 Accomplish the scale field calibration procedure in the scale manual.

2.15.1.3.1.3 Place each of five items on the scale in turn and record a weight figure on the quality assurance record. Item passes the test if all items are within the range of 115 plus or minus 2.5 pounds.

Figure 6-10 One verification planning string.

verification requirement in Section 4. These would be coordinated in the item verification traceability matrix which would also show a test method for verification at the item level. This, and other item verification requirements (VSN), are collected into an item 15 verification plan and related procedure. Finally, the results are reported upon in the item verification report (not shown).

6.3.4 *Validate the plan*

Our verification plan will have to be created long before we implement it so it could suffer from insufficient knowledge and require update as the program knowledge base grows and testing activity nears. Clearly, it will have to contain a great deal of information prepared by many different people and could suffer from poor integration across these many perspectives. There are many other sources for error in our plan. Therefore, we should resort to some form of validation of our plan as early as possible before it is implemented. This validation should proceed on two levels: review and approve each IVP item section and the overall plan. The most cost-effective approach in validating either the individual item section or the complete IVP is through a formal review and approval process.

On a simple program it may be possible to merge all the detailed reviews into one overall review. If the planning documentation is quite voluminous, as it can easily become on a large program, the content can be broken into pieces with reviewer subteams. It is best to break the whole into parts along the lines of the items to be verified rather than by verification methods because of the better item synergism that results.

It is not a good idea to permit product development teams, where the teams are responsible for implementing the plan, to review their own portions. One approach that can be effectively used in this case is to ask the teams to cross review the plan parts. This has the advantage of improving overall team knowledge of the verification work yet to be performed and takes advantage of some healthy native competitive spirit between the teams. When reviewer subteams are used, there should still be a system-level review that evaluates the results of the subreviews and looks at the plan from an overall perspective.

Each plan section should be reviewed by a decision maker or team that has the responsibility for the design of the item covered by the plan section supplemented by system-level people. The plan should contain content for each of the basic sections noted earlier, and that content should be reviewed to ensure that all assigned VSN map to the content and that the resources, facilities, schedule, and cost estimates are mutually consistent, complete, and believable.

The presenters for the aggregate plan review should be those responsible for implementing the plan, and the reviewers should be recruited from program and functional management people familiar with test and analysis work. The review of the aggregate plan should touch on the following:

a. *Coverage* — Is everything in the system covered?
b. *Completeness* — Will the plan comprehensively prove that the system and item requirements have been satisfied?
c. *Depth and Detail* — Is the testing and analysis work conducted at sufficient depth to ensure good evidence?

d. *Least Cost* — Are there any ways remaining that the required resources could be reduced or rescheduled for lower cost and shorter schedule without reducing the effectiveness of the verification work?
e. *Believable* — Is the content of the planning data believable in terms of appropriate resources, time, cost, and expected results.

6.4 *Integrated verification plan test integration*

On even a moderately sized program, it will not be possible for one person to do all of the test planning work. One person should be tasked with the responsibility for the overall plan but that person cannot do all of the work. For one thing, there will be so much work to do that it must be done in parallel, and even if one person could physically do it all, no one can possibly understand and apply in an economic manner the breadth and depth of knowledge required to do this work for it exceeds anyone's capacity for knowledge.

Since the aggregate work will be contributed to by many persons, each trying to do his or her best in some relatively narrow portion of the whole, it will be necessary for a system agent (the person with overall responsibility or an integration engineer assigned to the team) to integrate and optimize the planning at the system level. The good news is that each item or test principal will properly try to optimize about their own sphere of interest and responsibility. The bad news is that their best efforts may not be in the best interests of the overall program. The process of recognizing the difference between individual excellence for the parts of a whole and aggregate adequacy and doing something about those differences to enhance the aggregate results is called optimization.

Generally, we cannot depend upon those responsible for the parts of the whole to do this work well. On a small program the integrating agent can be formed by a single engineer at the overall plan or test level who has established an integration team composed of representatives from each item team. On a large program, it may require a dedicated integration team that is also responsible for the overall IVP.

The first test integration tasks to tackle are the overall test schedule, identification of the items that will be the focal points for the test program, organization of the test planning data in the IVP (or alternative documentation structure), and assignment of responsibility for all of the parts of the whole test-planning activity. Just this work is the result, of course, of a very difficult integration job that must balance many competing and conflicting constraints. Some of the technical and programmatic integration work that must be done to get this far includes

a. Identify the major facilities or laboratories needed for each verification task or item and assure that any internal program conflicts with respect to these resources are identified and either resolved or carried as a program risk. It will not be possible to do the integration work for all of the resources required for each test at this point because you are depending on the people who will plan each test to identify these. But, the major facilities can be planned from the top down and they must be identified as early as possible so that their availability can be assured. This is like developing the long lead items list for a design to assure that parts that are hard to get will be available when needed. There is some risk in this, of course, because early guesses may not be good ones, but planning work often must proceed based on our best information as we seek to refine that information. This is often a matter of balancing a schedule risk against a cost risk. These major facilities may include one or more of the following examples: wind tunnel or test tank, electromagnetic test range or chamber, automotive or tracked vehicle test track, integration laboratory, structural test laboratory, propellant test stand, environmental test chamber or laboratory, or engine development test cell.

b. Assembled preliminary schedule derived from identifying overall program needs and time allocations to tasks.
c. Review the relationships between tasks identified for test and other methods for a given item in search of missing links and redundant work. It often occurs that a test must be preceded by an analysis to establish the right test parameters and followed by one to understand the data produced, for example.
d. Ensure that reasonable budgets have been identified for all planning tasks and that there are no program constraints in the way of authorizing its use at the right time (as scheduled).
e. Ask if all of the test tasks make sense. Are there alternative ways to achieve qualification that might be less expensive and less time-consuming? Can any tests be combined or better coordinated? Are there any items that can be legitimately converted to analysis because they are being used in an application similar to their previous use (qualification based on similarity)?

Given that all of the work is identified, scheduled, budgeted, and assigned, subject to revision, of course, the administrative or programmatic system agent (verification manager by whatever name) should authorize that planning work begin as scheduled and the technical system agent (possibly one and the same person) should make ready to monitor that work.

As the work described in paragraph 6.3 of this book begins and carries forward, it will produce a stream of documentation defining needed resources and procedural details that the technical system agent(s) need to be familiar with. There was a time when this work had to be done with typewriter technology making it very difficult for anyone but the item principal to be fully informed about its content. Today, with most large organizations performing this kind of work having effective networks of desktop computers and engineering workstations, there is no excuse for failing to produce this stream of data in a fashion that encourages rapid development and timely technical oversight. The ideal environment within which to capture this planning information is a relational database interfaced with the requirements database as suggested in Chapter 4. All of the planning data have been fashioned into a paragraphing structure just like the specifications. True, there will be tables and figures as well as text, but they can be included in graphic fields in the database in the records corresponding to the paragraphs that call them. Indeed, today you can also include sounds and video strips in databases if needed.

So, given that the item principals collect their test planning data in an integrated on-line database (respecting network access rules assigned to the participants to control write access), the technical system agent can have ongoing access to that data as they are assembled. The system agent, ideally, should have the ability to comment about the evolving data in the context of that data without changing the source data. Alternatively, the system agent can simply call or e-mail the right person to tell them about his or her concerns. There should also be periodic meetings at some level of indenture specifically for integration purposes. The team responsible for a particular test (if it is large enough to require more than one person) should meet periodically, and the system agent should attend especially when there appear to be integration problems. On a relatively small program, possibly on a large one as well, the technical system agent should have a periodic meeting of all of the lead people involved at the time in verification planning work. At these meetings, they can use computer projection from the database to discuss and even correct in real time any integration concerns that prove to be problems.

Computer database systems or tools are sometimes so intricate and difficult to understand that people who do not use them every day cannot maintain the skills to use them efficiently. Like some other companies, McDonnell Douglas on their FA-18 E/F program has solved this problem and, in the process, probably created a workable programwide common database. At the 1996 International Council on Systems Engineering (INCOSE) Symposium in Boston, Ken Kupchar from McDonnell Douglas demonstrated how they had linked their requirements database captured in Ascent Logic's RDD-100 through Interleaf to their intranet home page. With this arrangement, where the connectivity extended to the test database (which could be retained in RDD-100 as well or any other well-engineered tool), a system agent could look at any portion of the expanding test plan within the context of a word processor view of that content by calling up the program home page and selecting down to the test plan content of interest.

Networked computer projection capability within the work spaces or meeting rooms will permit this data to be reviewed and changed in real time based on review comments. This can place explosively powerful capabilities in the hands of the development team. In this application, it may require one machine in the review space to change the source data and another to project the data via intranet web access if it is desired to view the data in word processor format for the meeting.

At this point the reader should have come to realize that if the test planning data is organized as outlined in this chapter, there is nothing to keep us from capturing the content in a database rather than a word processor (the modern typewriter). Once it is captured in a database, we can use that same data in many other ways. We can establish traceability among the source requirements in the specifications, the verification requirements, hopefully, but not necessarily, captured in the specifications (rather than in separate test requirements documents), and the test procedures. We may even extend that traceability to the test reports since they are also created in the same format (paragraph numbers, paragraph titles, and text with referenced graphics). Systems can be devised that would not only capture the data but could tremendously support the integration of the data. Systems so constituted should be able to provide management statistics on the evolving test planning process.

In addition to the data-related aspects of the integration job, there remain the pure technical aspects that cannot be done through any conceivable machine. But, the people who do this work need the information to feed their mental analysis and synthesis processes. To the extent that the organization does not have the data-related capabilities discussed above, the technical integration job must be more difficult because of the increased difficulty in obtaining the information needed to stimulate the integration process. This mental process seeks out inconsistencies and conflicts, nulls, redundancies, and mismatches.

When we identify problems in the evolving test planning data, we must have the machinery in place to hear it and act upon it if appropriate. This is where the periodic test planning meetings at some level of indenture come into play. Wherever possible, we should allow each team to identify and resolve problems in their test planning work. Where problems are identified externally to a team and that problem is perceived to impact on others, the problem should be reviewed at a higher level if it cannot be resolved between the two agents most intimately involved. On very large programs it may be necessary to create a special form to identify test planning problems, much like an action item form. This form follows the problem until it is resolved.

6.5 Special problems in verification through test

6.5.1 Life cycle requirements

Let us assume that our product is to have a mean life expectancy of 25 years. That is, the customer should be able to plan on the product being useful under the conditions defined in the specification for a period of 25 years. They will make replacement financial projections based on that period and develop other aspects of their business around the stability of that product's availability to them. The customer may wish to gain some assurance that the product will really last that long under the conditions described in the specification and may require the contractor to prove that it has a 25-year life expectancy.

One easy solution to this is to delay sale or deployment for 25 years during which the product is used in a fashion defined in the specification. If it passes that test, then it can be delivered to the customer. Obviously, this is not going to work. The customer wants the product now *and* they want it to have a 25-year life expectancy upon delivery. After all, these are not unreasonable demands.

There are two techniques that can be applied to satisfy this difficult requirement. First, the product can be designed with margins on key life cycle requirements to encourage longevity even if the engineers are a little wrong in their analyses. Examples of this are

a. The weight and strength of a structural member can be selected with an adequate material fatigue margin with weight minimized to reduce the wear and tear on associated moving parts that are in its load-bearing path.
b. The components used on an electronic circuit board can be selected for a thermal rating significantly above the expected operating temperature.
c. The capacity of an environmental control system can be designed with both high and low temperature margins and sized to loaf through periods of heavy demand.
d. Although small-automobile-engine design has come a long way in the last decade, the big thundering V8 engine that goes 100,000 miles before its first tune-up is contrasted with the screaming little 4 that needs radical work at 60,000 miles.

This first step gives the customer and the design team confidence in the product, but it will still be necessary to produce some credible test results that support the long life conclusion. The common way of doing this is through accelerated life testing. The item is overstressed in one or more parameters for a reduced life period. The theory is that there is a predictable relationship between stress and life. Clearly, we could apply sufficient static stress to any new steel I beam that would cause it to fail in a few seconds. This is an extreme case that does not fit the intended meaning of accelerated life testing. The item should satisfy all of its requirements subsequent to the life test. The other extreme is to test the item for its planned life expectancy. Neither of these extremes is satisfactory.

In many systems, it is not the continuous stress (electrical, thermal, physical, and so forth) applied that affects the life cycle, rather the cycling of parameter values linked in some way with the use of the product. Some examples of this cycling are

a. A Boeing 737 jet flying in the Hawaii area lost a sizable portion of its structural skin over the forward portion of the passenger cabin while in flight. Luckily, the remaining fuselage had sufficient integrity that the aircraft was miraculously able to land and the problem fairly quickly understood. Each time a jetliner takes off it endures a change in outside pressure between airport level and 20,000 to 30,000 feet. The inside pressure and oxygen content must be maintained at a level encouraging continued human life. The result is a considerable pressure differential applied across the skin cyclically throughout its life. Add to this, the temperature differential between sea level and high altitude and the normal stresses of flight. The repeated flexing of materials over the life of this airliner resulted in small cracks in the skin which eventually connected up sufficiently to cause explosive cabin decompression in flight. The SR-71 and the space shuttle offer extreme examples of this phenomenon with tremendous thermal and pressure differences observed.

A credible accelerated life-testing plan can most often be built on a cycle rate higher than that normally realized in use and to more extreme values in key parameters during each cycle. In the case of an airliner, it may experience three takeoff and landing events per day (more for a commuter airliner, perhaps), that is, three use cycles per day. We might accelerate the life cycle in this case by subjecting the test article (a whole fuselage or a representative sample) to 30 cycles per day, depending on the test resources available, and use 10% higher temperature and pressure differentials than expected in the worst case. We could conclude that the 30 cycles per day would allow us to complete the testing in one tenth the time. If our aircraft had to exhibit a 25-year life expectancy, we could conceivably complete the testing in two and one half years just based on the testing frequency. By developing a relationship between the parameter value extremes and lifetime, we could further reduce the test time span. The result would be an acceptable test time span consistent with the duration of other testing that can be accomplished in parallel.

b. An automobile company decides that it will build a car that will have a nominal life of 10 years. They require that the engine and all major supplier items satisfy that requirement. An automobile engine endures a lot of stress in starting because of initial lubrication problems that are not exercised by simply running the engine in some continuous life test. One may conclude from studies of product use, that average drivers start their car three times per day. Over a 10-year life, this involves $365 \times 3 \times 10$ or 10,950 engine starts. We might add a 10% margin for a total of 12,045 operations. An analysis shows that during a 24-hour period, we could run one start every 10 minutes without exceeding starter and test battery simulator temperature limits. Therefore, we conclude that we can complete start tests over a period of $10{,}950/24 \times 6 \times 30 = 2.788$ months, or about 3 months.

We will have to combine the results of these tests with results from endurance tests based perhaps on an accelerated model crafted with a randomly changed rpm characterized by 80% at highway speed over a continuous period of 30 days. The logic for the duration could be based on 5 hour/day normal engine run $\times$ 365 days/year $\times$

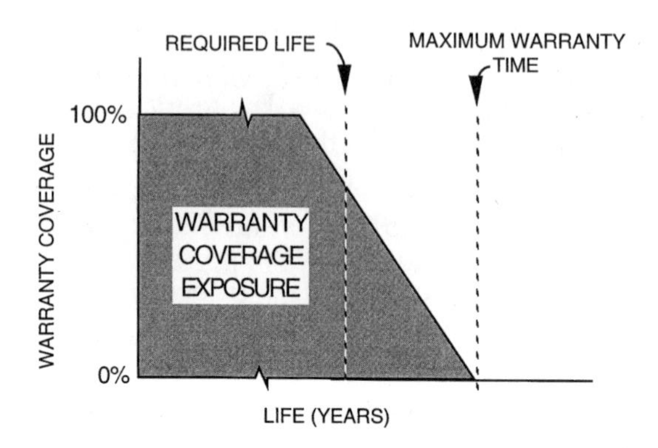

Figure 6-11 Life warranty utility curve.

> 10 years = 18,250 hours. If we run 24 hours per day, the 18,250 hours are achieved in 760 days. Previous testing and field failure data may have shown that this kind of testing stresses the engine at a level of 1.2 relative to real life so we can reduce the number of days to 760/1.2 = 633 days. It may require more than a single test in each of these cases in order to achieve statistically valid results, but they may be run in parallel if the supporting resources are available.
>
> Many components in automobiles operate on different schedules like doors, windows, arm rest, radio, and air conditioner/heater. So the complete test program will have to be pieced together from the related realities.

Now, the logic for these cases may or may not be perfectly valid so it is and should be clearly defined, open to critical review, and comparison with real-world conditions.

A system that does not present a human safety hazard could be life tested in use by agreeing upon a lifetime warranty adjustment as illustrated in Figure 6-11. Prior to the completion of the required life cycle, the contractor would shoulder the warranty cost for failures, but after the life requirement has passed the customer would pay the contractor for repairs based on some mutually agreed upon utility curve. This curve could begin prior to the absolute end of the life cycle period to respect a reality that maintenance does normally swing up as wear-out is approached on the right end of the reliability bathtub curve. The main problem with this arrangement is the questionable contractor and customer longevity over the life of a system. Some portion (all or part) of the warranty funds might have to be put into escrow to mitigate this risk from the customer's perspective. Also, from the customer's perspective, this agreement should be transferable to preserve the product value in the event of transfer of ownership of the product or operating organization. These are cost and risk questions that have to be balanced as part of the contract negotiation.

6.5.2 Rare environmental conditions

One of the most difficult circumstances to account for in testing is a condition that very seldom materializes naturally. If you could wait until the condition occurs, it would cost very little to execute the test because the conditions would simply occur and you would have to have come up with ways to observe behavior in some fashion. Let us say, by way of example, that our project must build a protective wall that will prevent flooding of a new mall area created along a riverbank that has been known to flood every few years in the past. We might wonder how high is high enough and how strong is strong enough.

One accepted approach to this problem is to appeal to historical realities. By researching records we could discover the water depth for floods that occurred over the past 100 years and pick the maximum as the design goal (the 100-year flood). The thinking is that 100 years is a long enough period of time to consider in defining the extent of a natural periodical phenomenon for the purposes of identifying the worst-case peak value. It is unlikely that a worse condition will be achieved in an economically significant period of time in the future. As time progresses and the period of time over which good records are maintained is extended, we may have to deal with the 500-year flood in such exercises.

It would not be reasonable to now wait around until we observed this flood condition as a way of verifying our design solution on the actual construction. This condition may not occur again for 70 years and when it does it may take out our whole project. We could simply conduct an analysis of alternative structures impressed with the forces derived from our 100-year flood. We could also build laboratory models and extrapolate from the results of tests conducted on them. It would not be sufficient in this case to simply base our conclusions on the depth of water in that floods bring with them dynamic conditions as well as static loading. We should be interested in the relationship between the foundations of our wall structure and the rate of flow of passing water to ensure that it is not possible for erosion to wear holes in our protection. We may have to conduct special tests of erosion over predetermined time periods with particular soil conditions.

In cases involving a smaller geographical space than a flooding river, such as the temperature extremes influencing a new bulldozer design, we may be able to run tests of our product in a special thermal test facility. The bulldozer has been built for use in arctic conditions operating down to –40°F and we want to test it to that level. We could test, at the item level, the engine, drivetrain, and other functional systems by simply chilling the items in a small environmentally controlled test cell. But, the whole bulldozer will require some maneuvering room. There are test facilities adequate for such tests, but they are often scheduled far into the future and the cost of operating them is considerable. So, assuming that we do not wish to have to build a special facility for a one-time test, we will have to project our need for this facility into the future from early in the program and in the process identify schedule risk to be a concern for a considerable period of time.

Test planning conducted well at the right time is part of the systems approach for avoiding future problems, a part of the risk identification and mitigation process that assures us that we (customer or contractor) will not be caught short later with a mess on our hands.

6.5.3 An infinite multiplicity of combinations

When testing for some conditions it is impossible to account for every minute possibility. For example, we wish to examine the performance of a new business jet in different airspeed–altitude combinations. There are clearly an infinite number of possibilities over the altitude and speed range of the aircraft. One of the conditions of interest is the stall characteristics within the planned flight envelope. While we cannot test for every possible speed–altitude combination, we can pick several points that exercise the aircraft at airspeed–altitude envelope extremes and interpolate results between these extremes.

6.6 Product-specific test planning

The discussion up to this point generally applies to verification planning for any item, but there are unique problems and solutions related to testing a product as a function of the technology to which it appeals.

6.6.1 Hardware

Hardware comes in many forms focused on an appeal to some combination of technologies such as mechanics, electronics, hydraulics, optics, pneumatics, and structural entities that tie hardware elements into a functional whole. Most of the characteristics of these items are observable through features of their physical existence. All four verification methods can be applied to these items since they have a physical existence and their features and operations can be observed through the human senses either directly or through indications on special test equipment.

Procedures for hardware verification using the test method tend to be dominated by human or machine actions to stimulate test sequences by manipulation of controls and inputs possibly provided by special test equipment and observations of product responses. Physical characteristics can be measured and checked against required values. Compliance with operational or functional requirements is tested by placing the hardware item in an initial condition and causing it to function in a way that exposes the performance of the item, which is then measured and compared with the required value. Sequences of these actions are organized to evaluate item performance defined in all of the performance requirements in the item specification.

Design constraints are similarly inspected. Interface conditions may be evaluated by testing the interfacing items separately at these interfaces based on knowledge of the mating item static and dynamic characteristics. We can physically combine the interfacing items, as in

a. Installing one onto the other for a physical interface verification or by checking one interface against a match tool created to represent the other side of the interface;
b. Electrically or pneumatically connecting and operating the items in accordance with their item or subsystem requirements.

Environmental constraints will require a means to create the anticipated environmental stimuli whether it be a natural phenomenon (wind, rain, atmospheric pressure, gravity level, or temperature), a hostile action (explosive force and accuracy of placement of opposing weapons, possible nature of the hacker's attack on a new information system, or the potential physical attack that could be staged on a facility to gain access), or the uncoordinated actions of noncooperative systems that operate within the same mission space as our system (electromagnetic interference frequency spectrum, timing, and intensity). This is very similar to the problems faced by those who make movies. If they are very lucky, the area they have chosen in which to make the movie will be characterized by all of the weather phenomena they need within the time frame they need them. Baring that unusual possibility, they will have to manufacture the effects to create local spaces large enough to capture the planned action. So it is with hardware testing for environmental phenomena. You cannot depend on the natural environment to generate the conditions needed, especially at the extreme values needed, within an economically acceptable time framework. Thus we resort to salt-spray test stands, high-intensity lighting (to simulate the effects of the sun) banks, thermal ovens and low-temperature chambers, and rate tables to create the environmental illusions we need when we need them.

In testing hardware environmentally, we should also not omit the subtlety of self-induced environmental impacts. The way this omission most often happens is that these requirements are not identified in the specification so they are never properly tested for. An induced impact occurs when the product includes a source of disturbance that interacts with the natural environment to produce an effect that would not be impressed on the product if the product were not present. This may seem like some kind of false dilemma on the surface, but induced environmental impacts are real. An example of one is found in a space transport rocket firing its powerful engines on launch causing tremendous acoustic noise communicated through the atmosphere to all of the elements of the rocket rising off the pad. If those darn engines were not firing, there would be no acoustic noise to stress the onboard computer or guidance set. At the same time, we would not be rising off the pad if they were not firing. The sources of these induced impacts are commonly energy sources within the product. If they are accounted for in the requirements, they are not hard to test for in the verification process since they simply are natural environmental stresses as they are applied back to the product.

The third kind of constraint, specialty engineering requirements, offer many challenges in hardware testing. Many of these requirements are probabilistically rather than deterministically phrased making for problems in determining the confidence of our testing results. This is overcome by testing multiple items, randomly varying the conditions to reflect reality, and extending the test time span. Many of these requirements are best verified through analysis rather than test or through examination or demonstration. In the case of a remove-and-replace time requirement, for example, we could analytically evaluate the removal-and-replacement of each item with knowledge of the design, accept these times as means, and combine these numbers deterministically in a maintainability model to calculate the system figure of merit. We could also wait until we have hardware and actually perform the work and time each action combining the results using the maintainability model. We might choose to do both for all items or select a few items to be accomplished in a demonstration only to determine the credibility of the analytical data.

Some of the specialty engineering requirements are much more straightforward. Weight can easily be verified through examination by placing the item on a scale. Dimensional requirements can be measured with a rule.

6.6.2 Application-specific integrated circuit (ASIC)

An ASIC is a specially designed concentration of integrated hardware circuits intended to provide a predetermined relationship between a set of product inputs and outputs that may entail thousands of possible combinations. It is designed for a specific product application. The difference between a conventional hardware electronic circuit board and an ASIC is in the ASIC density of circuitry and complexity of input–output relationships. The circuitry may be designed to handle both discrete and analog signals as well as analogs acted upon, selected, or controlled by discrete signals.

ASICs are too complex to test in a reasonable amount of time using a human-oriented procedure. They can be tested most effectively by using a computer programmed to stimulate input combinations and check for appropriate output conditions. Their design can be tested even before fabrication because the design represents logical constructs that can be tested in software. The same design description thus tested is used as the driver for forming the gate structures of the ASIC so there can be good continuity from design to manufacture and the staged testing leads to good confidence in product performance.

The most efficient process for creating ASICs applied in industry applies VHSIC Hardware Description Language (VHDL), a special hardware design language derived from a Department of Defense program to develop very high speed integrated circuits (VHSIC) and it evolved into an IEEE standard for the description of complex electronic circuitry during the design process. This process entails the following sequence:

a. A set of requirements are defined for the circuitry — the inputs, outputs, and relationships between them defined.
b. The designer, using a computer tool such as Signal Processing Workbench (SPW) on a Sun workstation, assembles the functional elements needed to satisfy the predefined requirements on the screen by calling up circuit elements from a library and placing them on the screen. These elements are connected by drawing lines with the mouse between particular standard terminals of the elements. The library contains standard circuit elements such as flip-flops, filters, and so forth whose properties are well characterized in the tool in the context of specific external terminal relationships and their effects on input–output conditions.

   Once the circuit is "designed" on the screen as some combination of circuit entities interconnected by lines, it can be tested with particular input signal characteristics and output behavior and timing studied and compared with expectations and requirements. Adjustments may be necessary to the design to correct observed performance problems. This is a validation process establishing that it is feasible to satisfy the requirements as well as a preliminary design activity.

c. The circuitry is used as the basis for the generation of VHDL code that characterizes the required circuitry. This code can be written directly by a skilled programmer based on the circuit elements shown on the screen or scratched out on paper and the ways they are connected. But it is much more efficient to use the code generator built into tools like SPW to generate the code automatically. Tests have shown that code created manually by a skilled engineer requires on the order of 1/20th the number of lines of code that are commonly generated by automatic code generators but both will result in the same circuit as an end result. Further, the automatic generator creates the code many times faster than the human can do it. So, there appears to be no adverse consequences in the added code complexity from the automatic generation process. The automatic code generators create very generic code including features not essential to every application resulting in added lines of code that do no harm. Should we desire to do so, we may test the results of the coding process at this level before proceeding with the next step.
d. The code is then passed through a synthesizer that converts the code into instructions needed to drive a particular foundry equipment that actually manufactures the ASIC in silicon and other materials. These instructions include a placement file that determines how the active circuit elements shall be laid out on the chip and a gate-level netlist that provides interconnect instructions.

e. The manufactured ASIC must be tested to verify that it is properly configured. The process of creating the ASIC also generates test vectors required to drive the final testing process in terms of input stimulus combinations and expected output responses.

Each stage of this process permits testing of the circuit described in a different and more-advanced form with good traceability throughout the process to the original requirements upon which the design was based. The foundry step resulting in the manufactured chip is sufficiently costly that it is seldom exercised as part of an iterative improvement spiral development process. The tools for developing and testing a new ASIC design are sufficiently advanced and proven that the design can be perfected prior to the production run.

This is a far cry from the days (only 25 years ago) of relay-diode logic and vacuum tube circuits connected by bulky wiring in black boxes strung together with other black boxes via external wire bundles. These designs evolved over extended periods involving the man-hour intensive creation and maintenance of engineering drawings in pencil. Manufacturing involved human wiring of the circuits with the potential for human error. These systems had to be tested *after* construction and circuit problems had to be corrected in engineering drawings, test plans and procedures, manufacturing planning, and on the equipment itself followed by a return to testing.

Circuit miniaturization and computer-aided ASIC design, testing, and manufacture has dramatically reduced the cost of electronic circuitry development and permitted design of very complex circuit elements of low weight and tremendous capability in a brief time span. Chip Express advertised in the February 1996 *IEEE Spectrum* that it will deliver low-volume ASICs in 1 week. Compare this with the time it required to design the circuit, breadboard it, test the breadboard and modify it for desired results, complete the final design on a drafting board including the packaging design, and manufacture the item on a production line following previously developed planning and instructions. This evolution of the electronics circuitry design process offers a shining path for efficiency improvements in other engineering domains by expanding the scope and extending the period during which the design is under the control of a traceable stream of related computer software.

6.6.3 *Firmware*

Firmware is a special variation on hardware called programmable read only memory (PROM). These devices are programmed in a fixed way to accomplish actions predetermined by a stored program commonly burned (or "blown") into the hardware device in a nonreversible or nonchangeable fashion. The software element is permanently introduced into the circuitry. The software can be changed only by removing and replacing the firmware circuit element that has been programmed differently. Electrically erasable programmable read only memory devices (E^2PROM) are available that do permit the program content to be changed.

After manufacture and prior to being blown, the PROM can be tested to verify that it does not have defects in the form of open circuits that will give the effect of blown junctions in operation. Subsequent to programming the PROM, it can be tested in accordance with the instructions used to blow it to ensure that it reflects the desired configuration. The effects of the firmware on the processor it becomes a part of can also be tested to verify system operation in concert with hardware and software elements with which it is combined.

6.6.4 Computer software

Computer software is a series of instructions that control the operation of a hardware device called a computer. The hardware computer is designed to manipulate binary numbers formed by only the numerals "0" and "1" strung together into "words" commonly of 8, 16, or 32 characters in length. These words form both instructions that tell the computer hardware how to perform and data upon which the machine operates to satisfy needed functionality.

In the beginning of the digital computer story, engineers had to write programs in what is called machine language composed of words formed from strings of ones and zeros recognizing the way the machine interpreted these words and the substrings in them. Part of a word might be a memory address from which to obtain data, while another part defined an operation to perform on that data resulting in an output placed in a predetermined location in the machine. The analyst had to understand how the machine functioned in some detail in order to write a program for it, for parts of these instructions were the 1s and 0s that were applied directly to gates and flip-flops to control machine operation.

Today, most computer programs are written in high-order languages (HOL) that are more like English but not quite. The analyst need not understand the detailed functioning of the machine upon which the software will run. The computer programs thus written must either be interpreted as they run or be compiled before they run to create the machine code that will run in the planned machine. Commonly, one line of code in the HOL results in many lines of code in machine code. Obviously, it would be possible to write the HOL statements incorrectly such that the compiler output would not accomplish what was intended. To the extent that the analyst fails to follow the required syntax of the HOL, the program would fail to compile and an error message would be displayed on the development workstation telling how to repair the code. So, just as in the case of ASIC development, software development includes automated error detection features that prevent many errors from ever being introduced into the completed product. It is entirely possible, however, even in the most effective software development environment to create flawed software. These more subtle errors in content must be detected through testing and weeded out to the extent that is possible before the software is delivered to the customer or otherwise placed in operation.

Software development over the period between the 1950s, when mainframe computers began to become commonly employed for nonmilitary applications, and the 1980s focused separately on two aspects of software: processing and data. Attempts to merge the development process using input–process–output (IPO) charts and entity relationship diagrams (ERD) in combination with data flow diagrams (DFD) were seldom effective, especially when two teams, the data team and the processing team, were formed but not required to work together effectively. The object-oriented (the earlier approaches being data oriented and process oriented) approach has attempted to bridge that gap by associating both data and processing aspects with objects that are linked together to accomplish the desired functionality.

The result of applying any of these organized approaches to the development of computer software is a clear view of the requirements for the software as a prerequisite to designing the software in terms of particular computer programs and modules written in a specific language and running on a specific machine. The next step is coding the software, the equivalent of hardware manufacturing. Final acceptance testing verifies that the software satisfies predefined requirements.

It is true that computer program code can be written without having previously done the analytical work discussed above, but the result will most often be disappointing in terms of cost and schedule overruns and a shortfall in performance of the delivered software. By capturing the basis for the software design in terms of models, we also provide those responsible for testing with the information they need to define testing requirements and effective test methods.

The greatest problem in software testing is realized when the software is developed in an undocumented and *ad hoc* environment. This often happens when the development and testing tasks are accomplished by two separate teams of people, the development team and the testing team. When the software is presented by the development team for testing, there is nothing but lines of code with no information on the structure of the code, the requirements it is intended to satisfy, nor the logic used in creating the code. In this worst of all worlds, the test engineers must seek to understand from scratch what the analysts and programmers have created. When the software has been created in accordance with clear requirements and follows a sound written structured development process, the testing is much easier to plan and implement because the logic programmed into the software is available for review and use in test planning.

The most effective and least-cost method for developing software is to merge the development and development-testing functions such that the evolving software is tested at each stage of its development, including incrementally by the developers, with close-coupled feedback of the results. This is possible today because the computer software industry has developed effective environments within which to create and test software. The tools one uses to analyze and understand the problem are often capable of generating code that can be tested to verify the performance defined in the analytical model. Formal acceptance testing is best done by a specialized test function to ensure that testing to verify that the software satisfies the requirements for which it was created is not compromised. Note the word verification was used here in keeping with the definition used in this book. A software engineer may very well refer to this work as validation.

6.6.4.1 *Software test sequences*

Software can be created using top-down or bottom-up methods and the development testing approach should follow that same pattern. While it would be possible to develop the software in a top-down direction and test it in a bottom-up pattern, we would not have the benefit of proving out the created software coordinated in time with the development of it. There would be a similar problem with the other extreme.

When software units are written as separate entities from the bottom up and then combined into higher-level units or components, they can be tested by creating test drivers that present the module the environment it would see as a member of a larger software entity as noted in Figure 6-12a. As higher-tier software components become available, they replace the test drivers and testing is accomplished to verify performance in accordance with requirements. This process can flow right up to the complete software system replacing drivers with higher-level elements as they are completed.

This process can work downward as well, as shown in Figure 6-12b. The system-level software is written, external interfaces verified, and internally tested, playing into software stubs that simulate the lower-tier software not yet created. As the lower-tier elements are completed, they replace the stubs. Obviously, these processes can be combined for middle-out or outside-in development sequences. In any case, these stubs and drivers are software as well as the product being developed and must be thoughtfully developed, configuration managed, and tested every bit as energetically as the product software itself. Figure 6-12 illustrates a simple software architecture suggesting a hierarchical input–output relationship between the modules. Other interface relationships are possible, of course, and common.

6.6.4.2 *Software test methods*

All software testing can be broken down into black box or white box testing. The difference depends on the degree of visibility the tester has into the inner working of the program under test. In the case of white box testing, you can see into the logic of the program, whereas in black box testing the inner workings cannot be or are not taken into account. Another high-level way to partition all testing is based on the purpose of the test. Testing designed to prove that the software works in accordance with the requirements is often called clean testing, whereas tests designed to break the software or expose bugs is called dirty testing. A third testing breakdown is

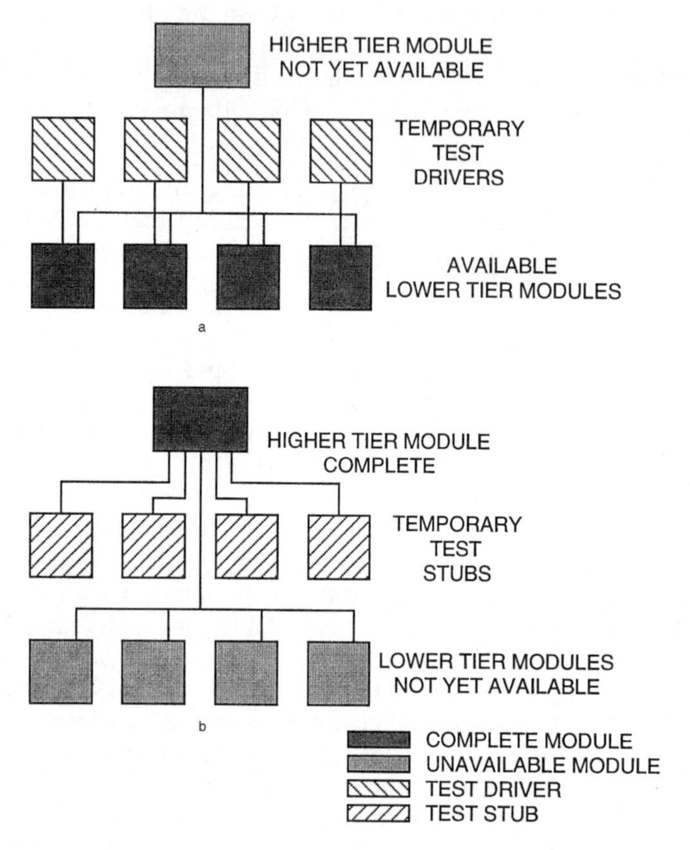

Figure 6-12 Use of test drivers and stubs. (a) Bottom-up development; (b) top-down development.

focused on the level of the software. The lowest testing level is commonly called unit testing. When two or more units are joined together and tested, it is referred to as component testing which may occur through several layers of aggregation. System testing is conducted at the top level. Integration testing is performed at all levels to test two or more elements that have previously passed tests at their own level prior to grouping them into a combined entity. Regression testing is accomplished at any level to verify changes in design accomplished subsequent to prior testing.

Black box, functional, behavioral, or dynamic testing is constrained by a lack of knowledge about the internal workings of the article under test. Such testing is accomplished based on the specification for the article and tends to be input–output data dominated. If the software is developed through *ad hoc* methods with little or no surviving documentation, those responsible for test may be forced to apply one of these methods for lack of knowledge about the internal workings of the program. In addition, if the specification was poorly prepared or nonexistent, the testers may be denied the information they need for white box testing and be forced to either reconstruct the program logic based on the code or focus on black box testing without the requirements. Examples of black box methods include logic based testing, behavioral transaction flow testing, syntax testing, domain testing, data flow testing, and loop testing.

White box, glass box, structural, or static testing is accomplished with full knowledge of the internal workings of the article and tends to be oriented toward testing the logic of the program. Commonly, both kinds of testing are useful because each has its own strengths and weaknesses. When the software is developed in response to a well-written specification and using effective modeling tools with surviving documentation, the testers have the perfect condition for effective testing because they are free to select the types of testing most effective for the situation and best supported in the documentation. Examples of white box testing methods include flow graphs and path testing and structural transaction flow testing.

Different testing methods offer different approaches to exposing bugs, but, in general, testing tries to evaluate the system under test by exercising every possible pathway through the software and every input and output at least once. But, software faces more challenging situations in real use than we can possibly dream up valid test cases for. For complex systems, the numbers of different situations may or may not be infinite but they are very large and difficult to foresee. Testing for software units (at the bottom of the software hierarchy) tends to be static in nature using deterministic values. As we aggregate to higher levels in a bottom-up development and test approach, we have to account for an increasingly complex and dynamic situation. In reality, inputs may be random and require probabilistic definition. Multiprocessing, networking, and a need to respond in real time complicate the software development and testing work.

We tend to develop software and hardware within a frame of reference constrained by our linear outlook on the world around us, but when software is combined in a complex system operating in its real environment, the whole often becomes a nonlinear system where small effects can produce large influences not foreseen during the development of the system. It is probably impossible to fully test a large and complex system to the extent that there is a probability of 1.0 that there are no remaining bugs in the software. We are left with a plateau defined by the amount of risk we can stand. Given that we are confident that the system will not kill or seriously injure anyone nor result in great damage to the system or other things, we will generally be content to locate bugs of less seriousness through normal use of the product. When the user finds these bugs, resources are assigned to correct the problem and develop a new version or patch that does not exhibit the offending behavior.

Some people believe that software development will eventually become fully automated with no need for human-directed testing of the software product. But, as Boris Beizer points out in his *Software Testing Techniques*,[1] as the complexity of software increases and the ingenuity of testing professionals attempts to keep up, the bugs that infest this software will become increasingly insidious and more difficult to discover. As a result, testing will continue to be important in developing quality software and it is hard to imagine that humans will not be involved in the thinking part of the development and testing of it.

6.6.5 *Personnel procedures*

Most systems include a human element in the controller path that observes system performance and makes decisions based on that performance to alter future performance toward achieving predetermined goals. The human normally follows the guidance offered in operating procedures. The design of the procedure is verified by demonstrating that when it is followed in operation of the item or system, required results are achieved. Part of these procedures should be based on a set of standard conditions experienced under what are defined as normal conditions. The challenge in devising these demonstrations is to determine an adequate range of nonstandard conditions that stress personnel performance to determine if there are appropriate steps to cover all reasonable possibilities. Test, analysis, and examination are not normally applied in verifying personnel procedures.

Where these procedures appear in formal technical data on DoD contracts a two-step process is commonly applied. First, the technical data are demonstrated by contractor personnel, generally at the contractor's site. This is referred to as validation. Second, the technical data is demonstrated at the customer/user site by customer/user personnel, and this is referred to as verification.

6.7 *Commercial subset*

The foregoing may seem awfully complex to those not familiar with DoD programs, and it is true that it may be in excess of anything needed on all but the most complex ventures. There are ways to simplify the process for commercial purposes, thereby decreasing cost and schedule impacts while preserving the central notion that it is good business to prove to ourselves as well as our customers that the results of our design labor has actually satisfied the requirements defined to constrain the design solution so that the product will satisfy a customer need. We have the option, of course, of not following the systems approach but, as much fun as that may be for a while, it is not a viable option when we are dealing with other people's time and money.

A sound tailored commercial approach to test planning can be assembled from the previous as follows:

a. Define the requirements for the product as a prerequisite to detailed design.
b. For each requirement, define how it shall be verified that the design does satisfy the requirement in terms of a method and a process.

c. Make the transform between the item verification requirements and verification tasks focusing on a list of items which will have to be subjected to qualification testing for the application.
d. Build an integrated verification plan within which to capture the test planning and procedures.
e. Publish the results of the verification events and review them to ensure that the actions do prove that the product satisfies the pre-defined requirements. To the extent that this is not proved, change something (requirements, design, or verification procedures) within a context of professional and technical integrity and seek once again to establish that the product satisfies its requirements.

References

1. Beizer, B., *Software Testing Techniques,* Thompson Computer Press, 1990.

chapter seven

Item qualification test implementation

7.1 Implementation responsibility and organizational structure

A project should seek to organize all verification implementation work under the management responsibility of a single person just as suggested for the planning work. The selection of this person is a function of how the enterprise is organized to some extent. Where the enterprise is organized as a matrix with knowledge-based functional departments and multiple programs organized about the product architecture into cross-functional teams, the program system or integration team should be made responsible for the overall verification process, its planning, documentation, and coordination. The person chosen to do this work should be from the test and evaluation functional department or skills base because that method of verification is the more complex and difficult to master from a management perspective of the four methods discussed in this book (test, analysis, demonstration, and examination). This person will be referred to as the system agent or coordinating agent for purposes of identification though we may choose to refer to him or her as a system technical agent or a system management agent, both of whom could be the same person, depending on the nature of the topic and assumed organizational structure.

This system agent must manage the overall test program through a series of test task team leaders and principal engineers responsible for each of the test tasks identified in the plan in accordance with the planned budget, schedule, and plan content. Some of these test tasks may be under the general responsibility of a product development team (where the enterprise uses teams) and others under the responsibility of the system team. In this arrangement, the test responsibilities may be effectively matrixed where the test principal reports to the product development team manager for performance of team work related to testing and to the system agent for coordination of team test work with overall program testing activity.

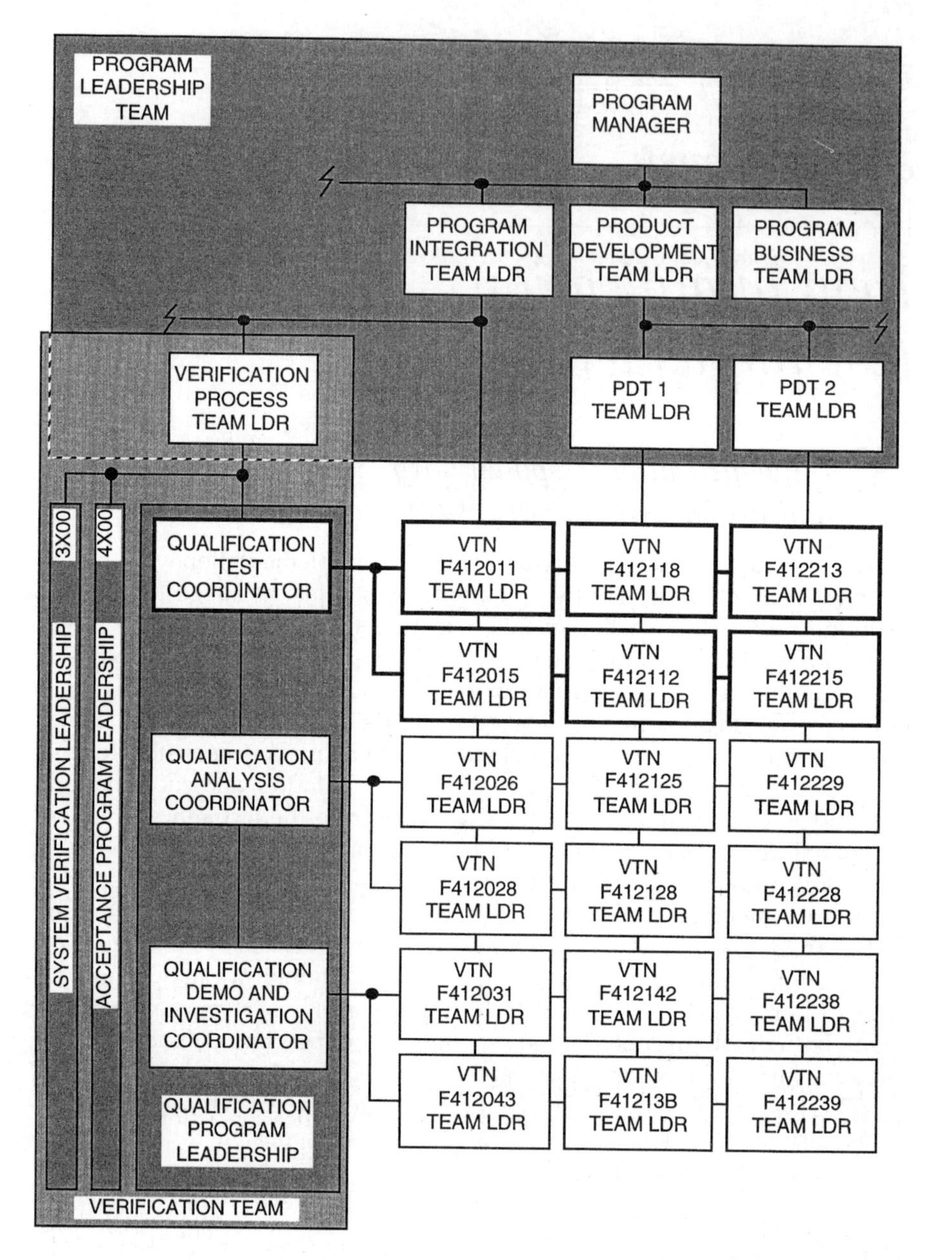

Figure 7-1 Organizational structure for qualification.

Figure 7-1 offers one suggestion for organizing the qualification force assuming the enterprise employs the matrix structure. There are many other ways to structure this force. Indeed, some critics of the matrix structure might truthfully express concern that Figure 7-1 actually shows a third axis of the

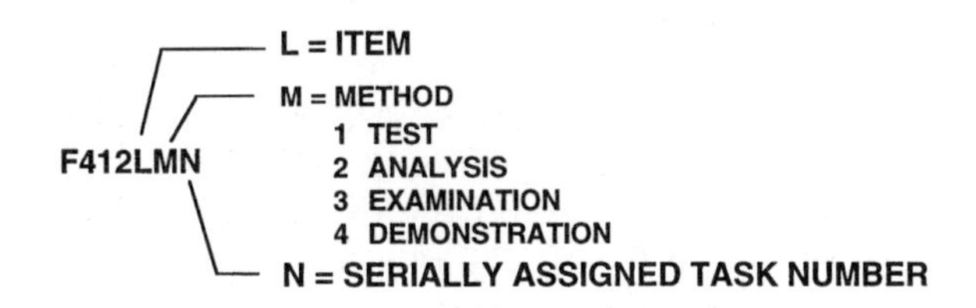

Figure 7-2 Universal task identification numbering system.

matrix in that the functional department structure has been suppressed in order to focus on the qualification process leadership roles and that workers are often confused by reporting paths even in a two-axis matrix. That is a valid objection to the matrix arrangement where the people staffing the program have not made the leap to organizational flexibility comfort and management people have not made the leap to management for the greater good of the enterprise rather than their individual career opportunities. The reality is that enterprises need a much more agile workforce than was common in industry in the period 1950 through 1980 and programs need a lot of flexibility to restructure as a function of the program phase needs.

In the structure shown in Figure 7-1, all of the qualification work has been collected under a verification process leader responsible for all qualification, acceptance, and system verification work. Earlier in the program this same role could have been made responsible for the validation work that would have been accomplished within all of the teams. This diagram arranges leadership under the verification leader by verification method and this is only one possibility. On a small program, the verification leader may be able to do all of the coordination work with no staff at all. You will note that the verification leader has been hooked into the organizational structure reporting to the program integration team along with possibly other process-oriented teams.

The test verification work, as noted above, would be accomplished by members of the product and system teams and the work of those people in their test work would be managed on a day-to-day basis within the context of their team leadership. The test qualification coordinator should work with the principal engineers and team leaders (larger verification actions) responsible for specific verification tasks (identified on Figure 7-1 by their verification task numbers or VTNs).

In Figure 7-1, the author appears to have violated his own suggestion against smart numbering systems for the VTNs in the interest of standardization within an enterprise. The reflective reader will recall from Chapter 5 that F412 is the generic task identifier for "Accomplish Qualification Process" which includes all of the qualification tasks on a program. The remaining characters have the meaning indicated in Figure 7-2. The F412 component will be the same for every program for qualification tasks. The method identification is obvious. As we identify the many items that will be subjected to qualification, we assign them an identifier N using a base 60 system. If there are more than 60 items, we might have to use a two-digit N where base 10 would work for up to 99 items. If more than that are expected base 60 would provide for up to 3600, many more than any program would ever require. In the bottom row of Figure 7-1 the task F41213B is an example of the use of the base 60 system for N. The numbers 1 through 9 and A have apparently already been exhausted making B the next available character. The N characters would simply be assigned in serial order within each F412LM category. The example in Figure 7-1 assigned the L characters as: 0 = System, 1 = Team 1 Item, and 2 = Team 2 Item.

The use of a partial smart numbering system was used to emphasize the subset relationships among all of the VTNs, teams, and methods. If, of course, smart numbering systems are useful in this case, it could be argued that they may be more generally useful than the author claimed in Chapter 6. The VTNs related to test are characterized by a heavy border in Figure 7-1 to focus on the purpose of this chapter. On a fairly complex program, there may be hundreds of VTN for qualification testing spread across many teams organized into as many as three tiers of teams. The more complex the program, the more important the need for a verification program integration and management focus coordinated with the management granularity applied to the program. The reader should note how easy it is to identify the tasks in different categories in Figure 7-1 once you break the code defined in Figure 7-2.

Some readers may wish to add a leadership string for development evaluation testing (DET) but the author has refrained from doing so considering that part of the validation process rather than verification. A really adventurous engineer might take the integration process one step farther than the author and work toward an integrated validation and verification program, however.

On a program with many items and many verification tasks per item, it may be necessary to expand the numbering system to recognize up to three, or even four, unique tasks within any one verification category (such as qualification) and method (such as analysis). For example, task F412L2005 might be the fifth analysis task identified in all of the qualification work for item L. Some people might prefer to delimit the 005 in this case as F412L2-005 or F412L2.005. As noted above, we can also apply the base 60 numbering system to avoid multiple place value systems and delimiting characters.

7.2 General coordination activities

Verification implementation coordination work begins with assurance that all needed verification tasks are identified in approved planning for all product items or for those that lead the schedule, at least. All tasks should be listed in the verification task matrix in terms of the name of the task, the product item to which it applies, the method to be applied, the responsibility for planning and implementing the work, and the date that the task must begin and end. These entries act as the direction for work that must be accomplished, and it should be recognized that the work may be accomplished or not. So, the system-coordinating agent must ensure that the person or team identified as responsible accepts it and agrees to work toward the indicated schedule and goals.

Each party with a test verification responsibility must produce the planning associated with the task and cause it to be included in program verification planning data, and this planning data should be reviewed to ensure that it defines the necessary resources, schedules their availability, and provides procedures that will be effective in producing valid evidence of requirements compliance. The coordinating agent must arrange for peer review of a tremendous flow of planning documentation or ensure that it has happened. Your enterprise or your customer may also insist that this planning data be formally reviewed and approved. You may even have to deliver test planning data for formal customer review and approval.

Given that each individual task is planned at the appropriate time and that the individual task planning is consistent with proving whether or not the design satisfies the requirements, there is a grander level of coordination also required. Two or more verification tasks may require the same resources at the same time. Conflicts must be identified in a timely way and resolved. Two or more test tasks may also be interrelated or hooked into the results or inputs needed for an analysis task. The planning data should identify any of these verification strings and the coordinator must ensure that these connections are made in the review of the test results leading up to authorizing the beginning of testing.

Throughout the run of the qualification test program, the coordinator must maintain, and communicate to the program in general, the ongoing program status with respect to schedule, cost, and major issues unresolved. Where issues surface in the performance of test tasks that cannot be easily resolved by those responsible for the test or the product team, the coordinator must assist in acquiring the resources to resolve the issue in order to maintain forward progress of the testing while not compromising on test effectiveness.

Verification task results must be documented, reviewed, and approved as the tasks are completed. There is a natural tendency on the part of the engineers who must do this to put it off as long as possible as more-pressing matters come to the fore. The coordinator must insist on the timely publication of the test results for they must be reviewed in the near term to determine if test goals were satisfied. We must be able to reach a conclusion about whether or not a test has proved that the design satisfies the requirements and is therefore qualified for the application. As all of these tasks come to pass, we should annotate our Verification Task and Compliance Matrices to reflect current verification status and to signal the availability of test results for review by others.

7.3 Individual test coordination

Every test involves preparation actions followed by test actions and, finally, posttest actions. These actions must be undertaken in accordance with the plan by the persons assigned responsibility. The test conduct entails accomplishing planned test work and, when necessary, resolving problems that inhibit continued testing. Preparatory steps lead to a state of readiness to execute the test. Post test actions entail completing reporting requirements and disposing of test residuals. Figure 7-3 offers a generic test process diagram. All of the work within this diagram is the responsibility of the individual test lead engineer (principal or team leader). To the extent that this person has any difficulty making ready for the test he or she should appeal to the test coordinator for assistance.

All of the qualification test work external to all of the individual test tasks is the responsibility of the test coordinator. Table 7-1 lists the steps related to an individual test task and notes the responsibilities and exit criteria for each step.

7.3.1 Authorization to proceed

At least 30 days prior to the scheduled beginning of a test, longer if it is a very complex test, the coordinator should inquire of the designated test leader if there are any reasons why the test should not proceed as planned. Given a negative answer, the coordinator should enable the budget associated with the test and make sure the test leader has any other support needed to proceed as planned. If the answer is that there are problems, then the coordinator must assist the test leader to resolve those problems with a schedule change, acquiring the resources needed to proceed on schedule, or other action that will minimize the negative impact of the problem on the program. With the budget authorized, the test leader can now charge to the test and begin bringing onboard the people and resources needed to prepare for the test.

Please recognize that commonly there are long lead problems in acquiring test resources. The comments about beginning test readiness ramp up 30 days prior to when the test is scheduled does not suggest that there is nothing happening between approval of the planning data and 30 days prior to the test. The person responsible for the test may require some sustaining budget throughout this period to keep things on track.

7.3.2 Test preparation

The test leader must bring together all of the necessary test personnel, resources, and facilities as planned and make ready to start testing. When a state of readiness is achieved, the test engineer should notify the coordinator of a preferred date and time for a test readiness review (if these are required on the program).

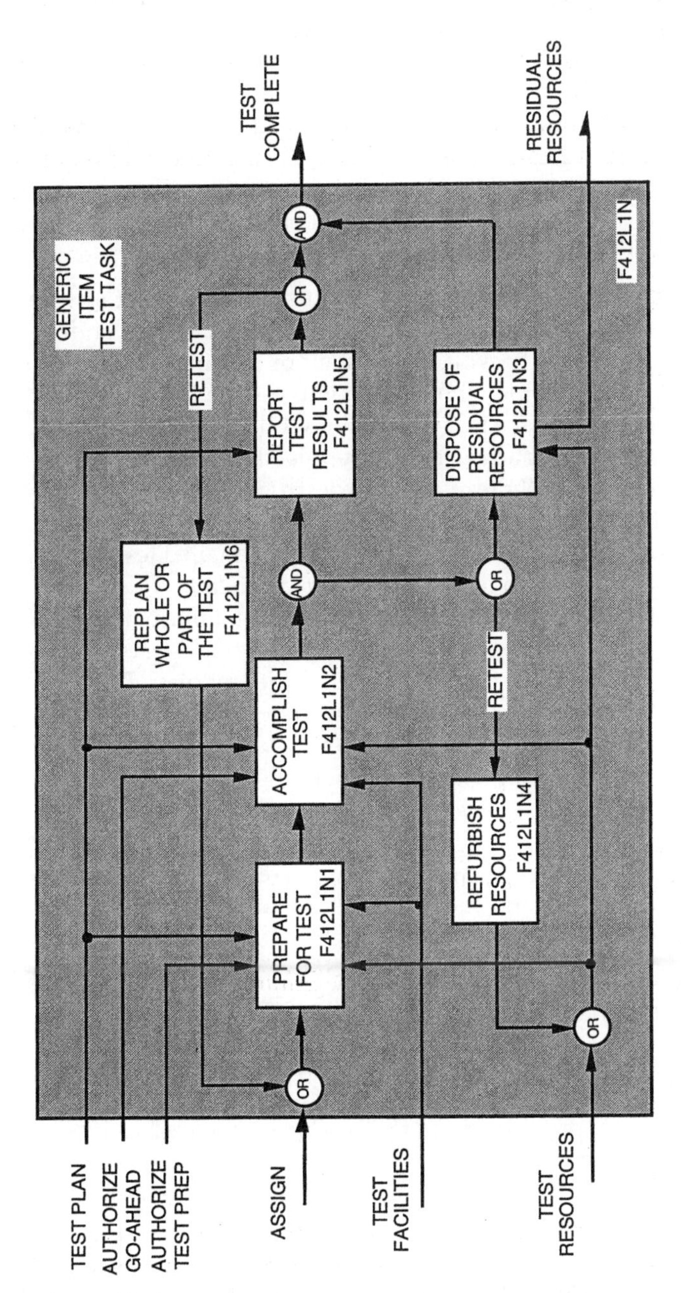

Figure 7-3 Individual test implementation model.

Table 7-1 Test Task Responsibilities and Exit Criteria

Test task	Responsibility	Terminal event
Authorize Test Preparation	Test Coordinator	Assign Responsibility
Prepare to Accomplish Test	Test Leader	Receive Test Go-Ahead
Run Test in Accordance With Plan	Test Leader	Test Complete
Report Test Results	Test Leader	Report Approved
Declare Test Complete	Test Coordinator	Test Complete

7.3.3 *Test readiness review*

Overall test planning should identify a need to reach a state of test readiness and gain some form of go-ahead approval before actually conduct of the test. This acts as a final checkoff of the state of readiness and provides an opportunity to notify people who have to witness the test. If this step is not respected, it is possible that a test will be run while not in a state of readiness resulting in a waste of budget and schedule that may be impossible to retrieve. If the test normally results in the destruction of the item under test, the result will be more serious, possibly requiring the development of another item for the test rerun. So, the more costly a failed test run, the more care that should be taken in approving a test go-ahead.

To satisfy this sensible requirement, formal or informal test readiness review should be staged by the test engineer to demonstrate that the team is fully prepared to begin testing as covered in the test plan. The coordinator should be the decision maker reviewing the state of readiness unless this is a major test involving considerable program risk, cost, and customer sensitivity, in which case, the program manager should probably be the reviewer.

The term *test readiness review* is often applied on programs only for software testing, but it is a valid review for all testing, however informal the review may be. For a simple test it may only require a brief conversation between the test engineer and test coordinator to reach a mutual understanding of readiness. Here follows a test readiness review checklist motivated by the items in Appendix F of canceled MIL-STD-1521B extended to the more general product situation than originally intended:

a. Have there been any changes to the requirements for the product item since the most recent major review? If so, how have the product, test planning, test procedures, or test resources changed as a result?
b. Have there been any design changes aside from requirements changes since the most recent review of the product and its test planning? If so, how have these changes influenced or driven test planning, if at all?
c. Have there been any changes in the test plan or procedure since it was last reviewed and approved? If so, what was the rationale for those changes?

d. Are all of the planned test resources available in a state of readiness to begin testing at the end of this review? If not, will those that are not yet available be available when their need is first planned if a go-ahead is given at the end of this review?
e. What were the results of any validation testing that has previously been completed on this item?
f. What is the configuration of the item to be tested with respect to the current approved configuration? If there is any instrumentation included, how can we be sure that it will not compromise test results? If the configuration is not precisely the same as the current approved configuration, why should testing be approved at this time?
g. What is the condition of traceability between requirements that will be verified in this test and the test planning and procedures data?
h. Is any customer furnished property, equipment, software, or information to be used in any way in this test? If so, how do we know that it is in the correct configuration?
i. Are there any valid limitations that we must respect in testing for safety reasons or other reasons?
j. Are there any current design problems with the item under test that have not been resolved? What is the solution schedule and how will those solutions impact the results of tests run in the current configuration? Does it make better schedule and cost sense to run the planned test now and subsequently run any needed regression testing or to hold up this test until changes are made and run the test only once?

As noted earlier, this review should be a formal review for a large and costly test with a lot of leverage on future program activities. In this case, the customer may insist on participating and it may even be in your contract that they will cochair the review. In this case, you will also likely be required to deliver to the customer after the review is complete the minutes of the review and any action items assigned during the review. If there are action items assigned by the customer, the contractor may be inhibited from starting testing until they have been answered to the satisfaction of the customer. The test coordinator should maintain track of these situations and work with the persons responsible for responding to close all test-related action items as scheduled.

7.3.4 *Test task go-ahead*

Based on the results of the test readiness review, the test coordinator should authorize the test leader to proceed with the planned test. The status of the task should be changed in the verification task matrix as well. If there are specific persons who are required to witness the test from the customer's ranks or from an associate contractor, independent validation and verification contractor, regulatory agency, or supplier company, it should be made clear in the go-ahead instructions that it is contingent upon those parties being in attendance. To the extent that those parties cannot or will not attend in a timely way, the test coordinator must assist the test leader to gain authorization to proceed without their attendance, with attendance by an alternate, or with their expedited attendance under duress.

7.3.5 Test conduct

Given that the test has been authorized to proceed in accordance with an approved plan with everything in a state of readiness, there is very little for the test coordinator to do with respect to that particular test. Throughout the test activity, the coordinator should maintain track of how each test is proceeding and give support when asked for or when the coordinator sees or senses that there are problems not reported. The organization that has perfected the use of its networked computers to communicate test status will offer the coordinator an efficient way of tracking performance.

7.3.6 Resolving test problems

Since all planning, including test planning, must be accomplished before the fact and often a long time before the fact, test plans are prepared based on some beliefs about the future. It may or may not evolve that the conditions predicted in the plan actually come to pass. An accident could cause damage to the plant, an unseasonably bad stretch of weather could frustrate the outdoor testing schedule, or a strike could deprive the test organization of the personnel it needs to implement the planned tests. For this reason, all test planning should be subjected to a risk analysis with efforts made to mitigate the effects of disadvantageous events. But despite our best efforts, these kinds of problems may still come to pass.

In addition to what we might term planning failures, the item under test could fail during testing leading to identification of a significant design problem that must be quickly resolved and testing restored. The failure in question could, of course, be a simple random failure that happens to occur during qualification testing or it could be a design problem that has to be corrected followed by a return to testing. The failure could also be a matter of failure to satisfy the required performance rather than a part failure. In the latter case our design is deficient in that it fails to satisfy required performance. The possibility remains, of course, that the item failed because it was being stressed beyond its design limits.

On the C-17 transport program, McDonnell Douglas agreed to test the wing for two lifetimes and they came within a very small fraction of passing this very demanding test. There was some negative press at the time that clouded over the real achievement but some modifications were made and wing testing moved through three lifetimes while this was being written. When the wing structure failed before two lifetimes, one of the first suggestions was to fix the problem in software. This is, of course, a common hardware engineer offering when big problems are exposed late in a program, but — fix the wing with software? Well, it was proposed that ride control be employed as on the B-1 bomber such that the wing would not be exposed to the stresses then being experienced, yielding a longer life. A structural fix was selected, but this situation suggests that there may be a very wide range of options available to us and we should search for them before leaping to the first apparently viable solution that comes to mind.

When these kinds of problems occur, the contractor and customer must consider the most appropriate action. It may be that the requirement is unrealistically demanding, beyond anything that the product will experience in actual use in its life. If this can be proved, the requirement should be changed and the product tested for the reduced value unless testing already accomplished can be shown to prove that the product satisfies the reduced requirement. Alternatively, it may be that the requirement is perfectly valid and the design simply cannot satisfy it. This is, after all, why we run tests, to find conditions like this. In this case, the design should be changed and tested to the requirement.

There is a third possibility that might be considered where the item fails a test and it is concluded that the design is not adequate for the requirement. The customer and contractor may reach an agreement that it will be very difficult to satisfy the requirement fully, meaning that it will add considerably to program cost and delay availability of the product and that the current performance proven in the test is adequate for initial operations planned by the customer. In this case, the customer may be willing to grant the contractor what DoD calls a deviation that will allow the delivery of the product that meets a reduced requirement for a limited period of time or number of product articles. Between the time the immediate problem is resolved and the end of the deviation time or article period, the understanding is that the contractor will find a solution to the problem, change the product design, and prove it through testing so that the product does satisfy the requirement.

The decision on what to do about these situations is often conditioned by the precise contractual terms, the availability of new funds, and the relationship between the contractor and customer up to that point. If the contract up through development has a cost-plus basis and verification work is accepted as part of the development process, then the contractor will be well disposed to make any changes requested by the customer so long as it is covered through an engineering change proposal (ECP) that carries with it new funds to fix the problem. If the contractor foolishly agreed to a fixed-price basis for the development period of the contract, they will be very opposed to making any design changes especially if the program is already in trouble financially.

There is a lot of gray space between these extremes. The author discussed one such problem, while this chapter was being written, with a program manager and the company's lawyer prior to them going to court with the government. The central issue, in a very messy situation involving failure to satisfy requirements that were poorly defined by the customer very late, was when the development period ended and production started because the development period was cost-plus and the production period was fixed-price. The customer position was that the development ended at CDR and ushered the contractor into CDR before the design was ready. As a result, many failures were observed in qualification testing that required fixes prior to production. The contractor claimed they should be funded under the cost-plus contract period and the customer that they were to be corrected under the fixed-price contract period. With millions of dollars involved and bad feelings on both sides, the issue was on the way to court.

This is a worst-of-all-worlds situation for both the contractor and customer and it highlights the need for crystal-clear contract language (a possible oxymoron) augmented where necessary by contract letters between the parties to clarify uncertainty as early in the program as possible. It also emphasizes the need for adequate time up-front to define the requirements clearly and do the verification planning work. It is very difficult to make up for all of the past sins on a program during the verification work. At this point, you are on the tail end of everything — schedule time, money, and possibly civility. Lost time cannot be restored, money spent badly cannot be retrieved, and some strings once tightened cannot be relaxed.

We have considered the problems of reality unfolding in ways not planned and failures to meet test requirements. A third source of discontinuity in the test process is the unavailability of test apparatus or consumables due to damage during testing, unexpected delay in acquisition of the resources needed, or expenditure of all of the available budget or schedule time prior to test completion. The time and budget problems are of a programmatic nature that can be solved through program reserve, if that risk abatement practice had been applied on the program and there was any reserve left at the time (late in the program it may all be gone). If there was no program reserve or it is all gone, the funds must be located through replanning, acceptance of the cost increase by the customer, or an acceptance of reduced profit on the part of the contractor. If the test is on the critical path, the planning network must be searched for ways to get the test off the critical path.

The author recalls working on several unmanned reconnaissance aircraft programs at Teledyne Ryan Aeronautical in the 1960s and 1970s as a field engineer where there was no available funding for overtime by the time field engineering was called upon (the end of the line) but there were many things that had to be done to make up for schedule problems. Luckily for the company and those programs, field engineering, flight test, and other tail-end groups were staffed by mission-oriented people who got the job done regardless. You can't always bet on having these kinds of people on the tail end who will be understanding about the results of earlier mismanagement.

Test resource problems may require expediting the needed resources, reaching agreements with competing users of those resources, development of alternative test planning or apparatus, or replacement or augmentation of the test in part or in total by another method of proof.

So, there are many things that can result in stoppage of planned test work. There should be a process in place as a fundamental part of the test program to review the cause of any real or potential test stoppage and to decide upon a corrective course of action. In these reviews, the test people, engineers responsible for item design, and management must understand the real problem and evaluate alternative solutions to those problems as a prerequisite to selecting a course of action with the best chance of satisfying program needs. Some examples of this process at work have been included in the foregoing.

7.3.7 Test results documentation and review

As a test comes to a close, those responsible should produce a draft summary of the results to give management an immediate indication and understanding of actions that may have to be taken in the near term prior to completing formal documentation, that could take some time for a complex test. This summary may indicate a need to rerun some parts of the test or change the design followed by more testing. Normally, a good design and sound test procedure will encourage a favorable result that still needs to be communicated to management.

The overall verification plan should require a maximum time between the end of a test and the publication of the formal results. The coordinating agent should track performance here and note for management scrutiny and action any overdue reporting cases. Overdue reports should be treated very seriously and their number managed to zero.

When the raw information from the test is available, it must be reviewed by the people responsible for the test and the people on the design team for the item. If the test produced a great deal of raw instrumentation data, that data will have to be reviewed, analyzed, and digested to reach a conclusion about the item design adequacy. A conclusion is needed as quickly as possible on whether or not the item design satisfies its requirements. If the test was well planned, the results should provide credible evidence of the item satisfying or not satisfying the requirements. The criteria should be that the item must satisfy all of its requirements in all reasonable combinations. Failure to satisfy only three requirements out of 123 should not be looked upon as some kind of probabilistic victory.

7.3.8 Residual disposition

Often a test will be completed with considerable residual material and it is difficult to decide what to do with it. The test apparatus normally can be returned to its source whether the property of a company test and evaluation laboratory, provided by the customer, or rented for the duration of the test. The item under test is more problematical. The kinds of tests run during verification most often render the item under test unsatisfactory for normal use as part of the actual delivered product and these items should not be installed for delivery. At the same time, these items may be very costly and there will be a lot of reluctance to simply put them in the trash.

Test residuals can sometimes be used in test or integration laboratories or simulators where they will not have to endure the full stresses of normal operation, but one must be careful that posttest changes in design are introduced into these items to avoid compromised results in activities involving their use. They can be used for preliminary training purposes where the differences between test and final items are not great. Units of this kind should be clearly identified for these purposes beyond the normal controls imposed through effective configuration management techniques. Some organizations paint these units a special color or paint a brightly colored orange stripe on them.

Other items may not be useful even in a reduced operational application on the program but may have value for historical or marketing purposes. A program manager may want the guidance set test residual to mount in the program lobby or his or her office as an example of the quality products produced by the program.

The test item may not be usable in any practical way because of its test history but may contain valuable material such as gold and silver that should be extracted before disposal of the item. Other parts may be recycled by salvage agents for aluminum, steel, and other materials where they involve a substantial amount of the material or can be included in a stream of material from other sources. The worst problem arises when the item under test includes hazardous materials.

The author recalls, while he was a field engineer supporting SAC on unmanned aircraft at Utapao Air Base, Thailand in 1972, when the troops had to empty a vehicle fuel tank in preparation for repairing damage or otherwise working within the fuel tank the troops would sometimes open the fuel drain cock and tow the vehicle around the base taxiways until empty. They found this was faster than calling for a defuel truck. No doubt equivalent actions have been taken by industry in the past as a way of disposing of posttest residuals but the rules are very clear today (they were at Utapao in 1972 as well) that hazardous materials must be disposed of properly. There are national, state, and local environmental protection regulations that must be considered in planning, implementation, and especially during post test actions.

The disposition problem is more pronounced when the item is classified or proprietary in nature. It may have to be altered prior to disposition even to the extreme of smashing it into pieces from which the original could not possibly be re-created or any of the classified features discerned. In the case of classified material, the customer may insist on being the disposition agent.

The test plan should include instructions for disposition of test residual. In so doing it may refer to company or customer procedures or call for the application of commercial or governmental standards for guidance. This may be done within the context of different responsibility levels, of course. The test plan may only require that the item be routed to the company material disposition organization and the final disposition determined based on their rules after an evaluation of the item.

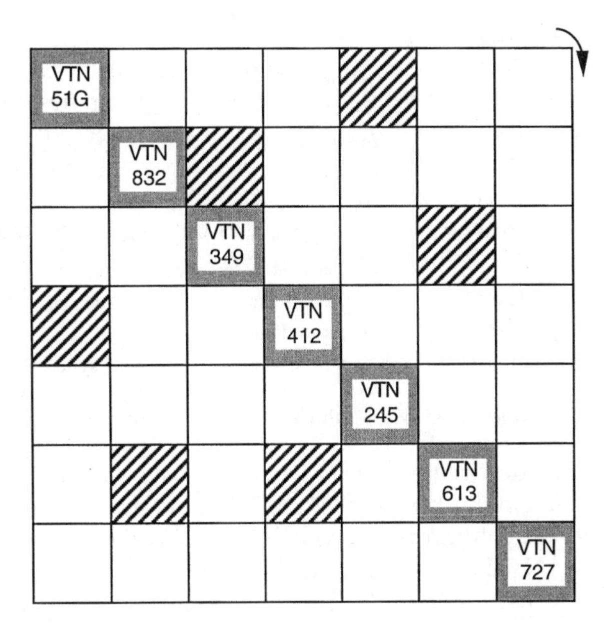

Figure 7-4 VTN N-square diagram.

7.4 Intertest and intermethod coordination

Each of the test leaders will properly focus on optimizing their work about their assigned tasks. There may be a larger story that requires coordination that they cannot be expected to remain focused on that will require the system level attention of the test coordinator and other parties or the qualification test coordinator working with the other qualification coordinators (in the context of the responsibility patterns exposed in Figure 7-1). To the extent that each verification test task is independent from all other work, this coordinating conversation will not be very voluminous. Commonly, the more complex the program, the more interactions that must occur between test tasks and between test and other method tasks. These could be identified in a VTN N-square diagram. Alternatively, it could be shown in process or IDEF0 block diagram format where the arrow-headed lines connecting task blocks reflect a need for flow of results. In a simple program either of these diagrams will show very little intertask interaction but on a more complex program it may be worth the time to construct one. Figure 7-4 illustrates such a diagram in N-square format.

This diagram uses the same numbering conventions applied in Figures 7-1 and 7-2. The prefix F412 has been omitted in the interest of space. The marked blocks are telling us that test VTN F41251G must supply something to demonstration VTN F412245 and test VTN F412412 must supply something to test VTN F41251G. The arrow at the top right corner defines the directionality intended on the two sides of the diagonal that identifies

SOURCE VTN	DESTINATION VTN	INTERFACE
F41251G	F412245	Test results that show it safe to demonstrate max velocity
F412832	F412349	Measurement of thrust link length
F412349	F412613	Mechanical demonstration of latch manipulation in preparation for test of explosive operation
F412613	F412412	Proof that pressure can be sustained in maneuver
F412613	F412832	Proof that adjustment has full range
F412412	F41251G	Computer qual in prep for avionics system tests

Figure 7-5 VTN interface tabular listing.

all of the VTNs included in the diagram. An alternative way to do this would be to craft a simple tabular list as in Figure 7-5 using the full task identifiers in this case. The left column is the source VTN and the second column is the destination VTN. The third column simply tells what it is that must be supplied from one VTN to another. This tabular listing, which contains the same information shown in Figure 7-4, can be implemented with a word processor, spreadsheet, or database software application. This format does make it easier to capture the interfaces.

7.5 Qualification test sequencing

We have discussed the qualification test in a very detached way up to this point only requiring that the test prove that the requirements have been satisfied. It should be noted that there are two principal elements of these tests. We must prove that the item functions in accordance with the performance, interface, and specialty engineering requirements, and we must prove that the item can withstand the environmental extremes defined in the specification. It is generally the environmental requirements that stress the item such that we would not be comfortable subsequently using the item in an operational way.

Obviously, it will not be possible to test everything simultaneously so our planning data must identify priorities and sequences for testing various combinations of requirements. A common approach is to perform functional tests interspersed between each of several environmental tests as suggested in Figure 7-6. Note that we are not requiring the functional test to be successful during the environmental testing in this case. Rather, we are running the environmental tests interspersed with functional tests to see if the unit survived the previous environmental test. If it did not, then this would be cause for concern. We would want to know why the unit failed and consider how we can change the design such that it will pass the test on the next run. One could require that the functional test run concurrently with the environmental tests and that would, of course, be a more severe case. At the same

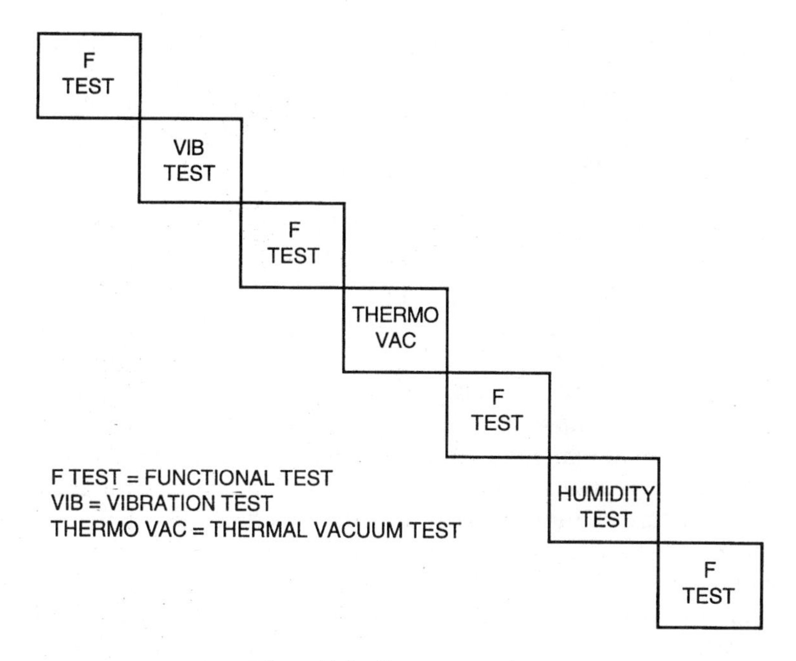

Figure 7-6 Test sequencing.

time, this kind of testing can be set up to give clear information about the cause of the failure whereas in the interspersed functional testing we have to try to recreate the situation intellectually based on the failure evidence.

Given a failure during any of these tests, it implies that the unit cannot satisfy the item requirements. This need not necessarily be the case. It could be that the item simply failed due to a random part failure. So, our first effort must be to find out exactly what the failure was and try to reconstruct the cause. An effective failure review and corrective action system (FRACAS) is a necessary prerequisite to qualification testing. Qualification testing also offers a development program an opportunity to wring out the problems in such a system before entering production.

If it is necessary to test for functionality and environments concurrently, it may be necessary to develop special test equipment and interfaces such that they are compatible with environmental test chambers and stands. The environmental test equipment may have to be modified as well. For example, if you must operate the equipment while it is undergoing thermal vacuum test, you will need a chamber which permits external interfaces with power, signals, or plumbing sources as appropriate to the item. A third alternative would be to run environmental test under power but without an organized stimulation of inputs or monitoring of outputs. Functional test between tests would still be required to ensure that the item had survived the test.

Figure 7-7 offers one view of a FRACAS paralleling the qualification test process. We enter the functional test initially and, given a pass, we move on to the first environmental test. We will then repeat the functional test to verify that the item still functions after the environmental exposure. At this point there are several possibilities. A pass will enable an entry into the next environmental test whereas we must find the cause if the functional test fails. The cause is going to be a random failure or an inability to survive the previous environmental test without failure.

Given a failure to satisfy the environmental requirements, we must evaluate the best solution to the problem. One alternative is that the test stimulated the item in excess of the requirements leading to a need to repair the item and retest it within the requirements values. Even though the item failed to pass the prescribed requirements, we might still conclude that the requirements should be changed to more realistic values. A repair and retest would also be in order at the new values.

A more common situation would be that the item failed because it could not satisfy valid requirements leading to development of a fix in the design. All of the documentation must be completed for the design drawings, test planning, logistics, manufacturing, and quality data. The unit must be repaired and subjected to retest followed by completion of the complete item qualification process. It is possible that it will be less costly and a faster route to build another article rather than repair and refurbish the item that failed in the prior test.

7.6 First production article integration

We have considered so far in this chapter that the qualification test activity is isolated from all other work on the product. In reality, as the first article moves through the manufacturing process in the form of all of its parts, assemblies, subsystems, and end items, the production process may be mixed with qualification test tasks building qualified subsystems from qualified components and qualified end items from qualified subsystems. The exact arrangement between these two activities is a function of several factors. A program with very few high-value production articles may use the first article as the qualification article, followed by refurbishment and shipment later in the production run, or the refurbished components may be included in spares with the hope that they never have to be used operationally. On a program with a long production run and an extensive qualification program, the first set of components may go into the component qualification stream while the second set goes into the subsystem qualification process, possibly involving some computer simulation and software integration steps, and the third set may then go into a system test article tested to acceptance test levels rather than qualification levels.

Other steps that will tie up product that may also have to be integrated into the early product stream include technical data validation, flight test or customer personnel training, mass properties measurements for the individual items and the end item in different configurations, special compatibility tests, and special maintenance, safety, and logistics demonstrations.

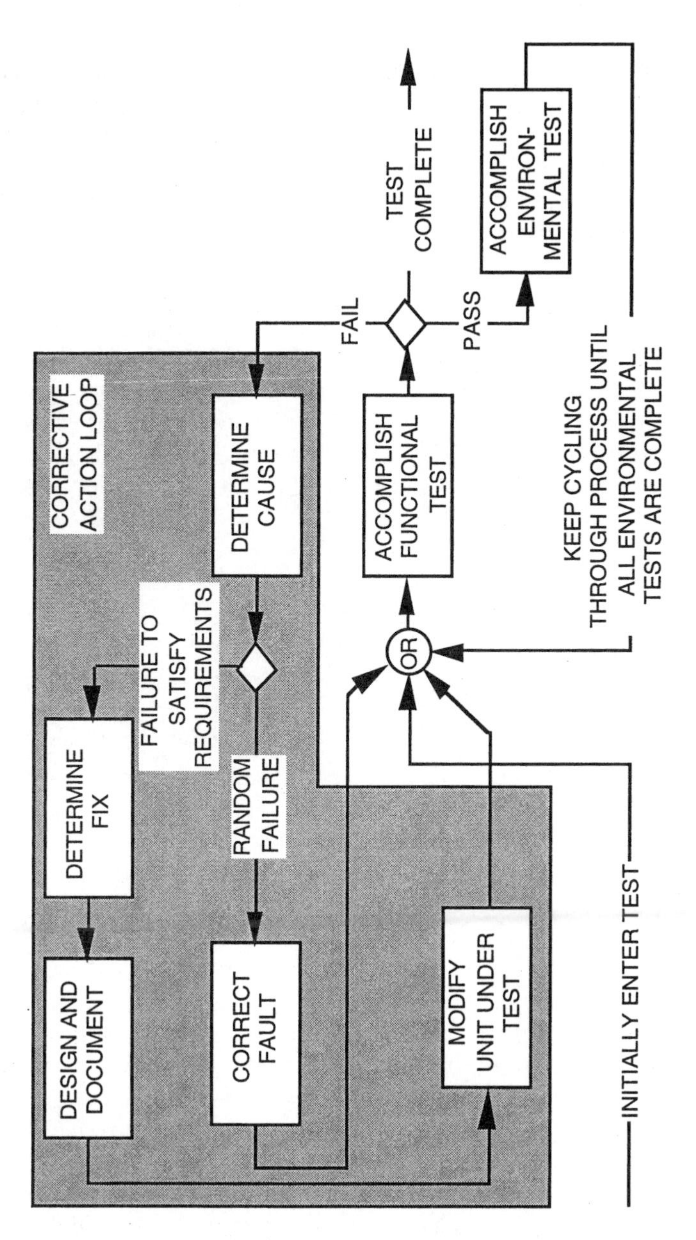

Figure 7-7 Corrective action loop.

chapter eight

Nontest qualification methods coordination

8.1 Other methods and meanings

Some engineers believe that the term *test* includes analysis, demonstration, and examination as well as test. MIL-STD-961D, on the other hand, suggests that all four methods are examples of inspection. This is another semantic problem along with the V words that can cause confusion among people and organizations during the verification process. In this chapter we continue the differentiation of all verification work into the four separate methods as previously defined. Some engineers prefer also to identify a fifth verification method termed *based on similarity* (BOS) where it is determined that the item is so similar to another item previously qualified that it is unnecessary to endure the cost and time needed for another verification cycle. Successful application in a prior use proves that the item is qualified for the new one. The author would prefer to simply include this as part of the analysis method because, in order to properly reach that conclusion, you must do some analysis comparing the characteristics and applications of the two items and evaluate the earlier test data relative to the new situation. This analysis may take all of 5 minutes, but it is, nonetheless, an analysis.

This chapter deals with the three nontest methods and the work necessary to coordinate their implementation. This work is generally very different from the test method from the organizational responsibility perspective because in most enterprises test and evaluation is a single organization that accomplishes most or all of the testing. All analysis, demonstration, and examination work commonly does not fall under the responsibility of the people from any single functional organization. Therefore, the coordination of these activities is more difficult than testing since there is commonly no central organizational agent.

8.2 Nontest verification responsibility

In the case of the test method, it is difficult to distribute the responsibility because of the common need for specialized facilities best provided through the critical mass of a single organizational entity. On a program organized in accordance with the cross-functional integrated team concept encouraged in this book, the test responsibility can best be taken on by what has been called the program integration team (PIT). It may be that the actual work is accomplished within test laboratories run by a functional organization serving the whole company and the work accomplished by people remaining as members of the functional department working in accordance with program-prepared test plans and procedures, but the control and direction must come from the PIT under whatever name. There may be tests with very limited resources and extent that can be assigned to specific product teams (especially if a functional department is actually doing the work under program direction), but, generally, the testing is going to best fit under the PIT umbrella, at least as far as program coordination of the work is concerned.

Other methods of verification may be farmed out to the teams and members of those teams and should be. We should be careful, in the process, to recognize valid system-level concerns, however. For example, if we must assign reliability requirements and subsequently prove that the product satisfies those requirements, this work can be done in piecemeal by the reliability persons on each team (if the project is so large as to warrant this arrangement) accomplishing the analysis corresponding to their own system element. The results of these analyses should be collected and combined into end item and system-level analyses by the reliability function of the PIT. In so doing, we may find that the team numbers do not combine to satisfy the higher-level goals and that we need to reallocate the reliability figures, consume margin, or redesign the product to conform to the requirements.

In all verification work, we must clearly identify all elements of the verification work and assign responsibility for performance of that work. In this chapter we will continue the rationale of prior chapters referring to the atomic verification work structures as tasks. Each task is created through fusion of a finite number of verification strings (identified by what we called verification string numbers or VSN) into a coherent activity. A person, and possibly a team, is then assigned responsibility for planning, developing, and accomplishing that task at the appropriate time in accordance with a written procedure.

8.3 Nontest documentation

In a few companies you might find nontest verification planning in an integrated test plan but this is not thought to be very common. More often you will not find this planning and procedural information documented anywhere. The reason for this is that the nontest verification work is commonly not as well organized as the test verification work because it does not have a central organizational structure responsible for all of it.

Normal practice aside, the analysis, demonstration, and examination tasks should be planned and have procedures associated with them. Some of these plans and procedures sections for analysis, demonstration, and examination may be satisfied by reference to specialty engineering plans. Examples include reference in the integrated verification plan (IVP) to the system safety program plan, reliability program plan, and logistics program plan. These documents should provide general and specific plans for performance of verification work in those fields and can be generally referenced rather than including all of the content in the IVP. In other cases, the IVP should include the plans and procedures. This need not be extensive or costly. It may only require a few paragraphs to define nontest verification task controls clearly.

8.4 Analysis

8.4.1 Specialty engineering requirements

The requirements for an item will flow into the specification from many sources during the development process. Just in the specialty engineering area you may find any or all of the kinds of requirements listed, in no particular order, below. Many of these requirements can only, or most economically, be verified through analysis due to the probabilistic nature of the value statement. Others can be verified most effectively through demonstration or examination while others may be most responsive to the test method.

Reliability	Availability	Guidance Analysis
Maintainability	Survivability and Vulnerability	Nuclear Analysis
Stress Analysis	Aerodynamics	Thermodynamics
Thermal Analysis	Mass Properties	Structural Dynamics
Human Engineering	Logistics Engineering	Corrosion Control
Materials and Processes	Operability	Producibility
Quality Engineering	Electromagnetic Interference/ Electromagnetic Compatibility	

All of these disciplines, when performed well, follow a similar pattern during the development of the product and the verification work. First, they should participate in the definition of requirements for the product from their own specialized perspective. They should assist the designers, and other team members, in understanding those requirements and what it means to be compliant. It may even be necessary to hold a brief school for the team members on the fine points of their specialty as they relate to the product line involved in the program. At some point the designer must synthesize his or her understanding of these requirements into features of the design that are believed to be compliant across all of the specialty requirements as well as all of the other kinds of requirements (performance or functional, interface, and environmental). Subordinate paragraphs explore how the analytical process works for several specialty engineering disciplines. A similar pattern is applied in others.

It is very difficult with the engineering tools available prior to and at the time this book was being written to establish the relationship between the specialty requirements and the design. It is, therefore, often difficult for the specialty engineers to establish analytically that the design complies with the requirements. Perhaps at some time it will be possible for the designer to embed in the CAD model some form of traceability of the features to specific requirements that drove their inclusion. This would result in a much easier job for specialty engineering verification if these features satisfy the requirements. We would simply have to check the traceability condition between their requirements and the design features, review those features, and reach a conclusion about the nature of the compliance condition achieved based on the principles of that specialty discipline. We might call this a microapproach to analytical verification. It might work with safety requirements, but requirements like reliability are distributed throughout the design in a way that is hard to isolate. Therefore, the most often applied technique is a macroapproach. The specialty engineer builds an intellectual model of some kind and assesses the design within the context of that model.

8.4.1.1 System safety

A safety engineer uses a process model of system operation to look for situations where, during normal operation of the system, things can happen that injure personnel or damage equipment that is in the system or its environment. In addition, the analyst considers all of the ways the system can fail and determines the effects of those failures. This is often called a failure modes effects and criticality analysis (FMECA). This work may be done by reliability specialists and subsequently reviewed by safety engineers to check for safety concerns. In both of these cases, the safety engineer may have to devise ways the adverse action can be avoided through design changes or procedural steps or find some way of attenuating the negative aspects of the response.

The safety engineer maintains a list of safety hazards uncovered during the analysis and maps these hazards to actions taken to mitigate those hazards. To the extent that the safety engineer can remove all of those identified hazards from the active list through design changes, procedural changes, use of protective or safety equipment, or other means, one can say that the system is safe; that is, it satisfies the safety requirements. If the analysis has comprehensively examined all of the possibilities, a difficult thing to prove, then we may have confidence in the analysis. The analytical evidence, composed of tabular data identifying all of the hazards and documentation evaluating each one, can be presented to the customer as proof of compliance with safety requirements.

The strength of the analytical results can be reinforced by following a very orderly method of exposing potential safety hazards. This will encourage confidence in the results, reducing the chances that anything of consequence has been missed. The safety engineer needs two models to satisfy this goal. First, a model of the product is needed and can be provided by the architecture and schematic block diagrams. These two diagrams tell what is in the system and how they are related interface-wise during operation of the system. The other model defines how the system shall be used and can be provided by a system process diagram consisting of blocks representing the processes connected by directed line segments showing sequence, a process flow diagram.

The things in the architecture should then be mapped to the blocks in the process diagram in a process-architecture map or matrix (PAM). This tells the safety engineer what things must be considered during the analysis of each process. The safety engineer then studies each step in the process flow diagram using his or her imagination, and any procedures that have been created to date, to create a picture of what will happen in that process. Throughout this thought process, the safety engineer must ask what would happen if this or that transpired or how might it be possible for things to go wrong. The assumptions upon which the analysis are based must be questioned as well.

This analysis should first be done assuming that the product is in perfect working order. The safety engineer remains alert for ways human operators and maintenance personnel may be protected as the steps in the process are sequenced in his or her mind. Also of interest are ways that the equipment may be damaged simply from normal operation, ways that it may be damaged by things in the system environment, or ways the system environment may be negatively influenced by the system. Here, the work of the safety engineer and the environmental specialist overlap to some extent. It may be useful to make clear on programs and in our functional organization where the boundary conditions lie to ensure that the same job is not being done twice.

After the analysis has been completed based on proper operation of the system, it should be repeated in the context of failures defined in the FMECA that relate to the things mapped to this process block. We should not omit the human failure modes in this analysis.

This study may uncover possibilities that were not considered in the FMECA and these should be added as they are uncovered. An alternative would be to first do the failure modes analysis based on the product architecture and design information available at the time. This would identify all of the ways the system can fail. Then, we could complete the effects and criticality analysis in the context just described. This might save some budget while not taking advantage of the cross-check that the two separate activities would provide.

In some government domains a safety engineer must be certified through some formal process. Alternatively, the customer may require some particular criteria for a person doing this work such as a degree in an engineering discipline and/or 5 years of experience in safety work. Companies may have similar requirements as a defense against the potential for product liability litigation.

8.4.1.2 *Reliability*

Reliability engineers begin their process by first recognizing a system mission reliability figure, which may be specified by the customer. The mission model must recognize the amount of time required and the stresses applied to the product. We must determine how each end item is affected by these factors leading to an end-item reliability, mean time between failures (MTBF), or failure rate figure. The resultant end-item reliability figures are further allocated to lower-order items in the architecture in the context of the reliability mathematical model. The predicted reliability figures are based on an assessment of the evolving design and are obtained from prior experience, substantiated supplier figures, standard lists, or estimates based on design complexity and maturity. The predicted reliability figures can be assembled from piece part analysis or a parts count, in both cases an analytical method.

These predicted figures are compared with the previously allocated figures yielding a conclusion that the requirement was satisfied or it was not depending on which figure is the greater. Where the analysis reveals a failure to meet the allocation, we should study alternative solutions such as reallocation of the figures, consumption of margin, exchange of margins or reallocation of excess capability in other aspects of the product design, or redesign the product to meet the previous allocation.

The reliability engineer should be using precisely the same model of the system used by the safety engineer, which, of course, is the same one used by all of the other specialists. The reliability engineer may express it differently in a reliability model illustrating the serial and parallel functions, but the same things should make up the system. The system should not be cut up differently in the different specialty engineering models.

As the product items become available, they may be subjected to special reliability testing to substantiate failure rates. These tests tend to be very costly because, if the equipment is reliable, the equipment must be operated a considerable period of time before failure occurs. It is possible to reduce this time by testing at elevated stress levels, but it is often difficult to establish the relationship between the amount of excess stress and an acceptable MTBF or failure rate. The results of these tests can be used to verify compliance with the reliability figures, but more often the predicted figures are accepted in the interest of cost.

Once product becomes available, the FRACAS should be used as a source of failure data for reliability verification. Within the contractors sphere of influence, this can cover all qualification and acceptance testing plus similar supplier experience prior to shipment of product to the customer. Some customers, such as the military, maintain maintenance records that may be acquired under the contract to calculate a demonstrated reliability figure. Alternatively, where the contractor maintains field engineering services at user sites, failure data can be acquired through this route.

8.4.1.3 Maintainability

Maintainability engineers establish end-item corrective and preventive maintainability figures consistent with system maintainability requirements defined by the customer. These corrective maintenance figures may be as simple as remove-and-replace times for selected items in the system or extended to model a more complex and realistic corrective maintenance situation entailing supply time, administrative time, and other time factors. These numbers are first allocated to items in the system and as designs become clear, the maintainability engineer must analyze those designs in the form of a step-by-step process which the engineer steps through mentally making estimates of times required for each step. As in the predicted reliability figures case, if the allocations are not satisfied in these figures, the maintainability engineer should first try to find ways of accomplishing the work more rapidly or in a simpler fashion. If that does not close the gap, then reallocations, margin consumption, or design changes should be considered. These analytical figures may be corroborated through timed demonstrations of the removal and replacement of these items.

Scheduled or preventive maintenance actions can be evaluated analytically by establishing a process flow diagram (the same one used by safety and everyone else) recognizing all required activities. Step-by-step procedures can then be defined and each step associated with an estimated time figure. By accumulating these figures in accordance with the process model, it is possible to determine analytically how long it will take to accomplish this work.

8.4.1.4 Mass properties

Mass properties engineers begin their work like other specialists by allocating weight figures to the items in the system architecture. As the designs begin to mature, they analytically determine weight estimates (predictions). As qualification test items become available, they actually weigh them, a form of examination, and compare the figures with required values. If all items satisfy their weight requirements, then the whole system should. The mass properties engineer may also have to plug the actual weight figures into a center-of-gravity (CG) calculation to verify CG for some items analytically. The mass properties person may also be responsible for space utilization and have to verify by inspection the volume and form factor of items and that they actually fit into the assigned spaces with adequate clearance.

On a prime item that must function within several mission scenarios, the mass properties problem will be much more complex than in a case where there is a single operational possibility. For example, a multirole aircraft like an F-16, AV-8B, or F-15 will have to maintain a safe CG throughout many kinds of missions as fuel is burned and weapons are released or fired or not under a wide range of possibilities. Each of these mission scenarios must be run on the computer and the results subjected to analysis.

Adjustments may have to be made in the design of the fuel tankage configuration or the fuel transfer sequence to move the CG trajectories into compatibility with safe controllable flight conditions.

8.4.1.5 Specialty engineering verification integration

There are many opportunities for the specialty engineering disciplines to cooperate, which also leads to many potential development failure modes when they fail to do so. For example, the maintainability math model requires reliability figures as well as maintainability numbers from analysis and/or demonstration work. If the engineers from these two disciplines do not coordinate their work, the numbers will not be available in pairs for the items in the system leading to some inefficiency. Also, they can assign values in conflict. If the reliability engineer allocates a high reliability figure to an item and the maintainability engineer allocates a low remove-and-replace time to the same item, they will have wasted availability. In that the item will seldom fail, a rapid remove and replace time will not provide much benefit from a maintainability perspective. It would result in a better system solution to reduce the reliability allocation or increase the maintainability figure and take advantage of the unused allocations where they could realize a more harmonious combination and better availability effect.

8.4.2 Interface requirements verification analysis

Interface requirements are most often going to have to be verified by test, demonstration, or examination, but there are some that will surrender to analysis at least in part. It may not be possible to arrange for some interfaces to exist in reality prior to their first mating, thus making it impossible to verify them through methods other than analysis. Any time there is a probabilistic element to an interface, it will probably be necessary to complete some of the verification work via analysis. It may be possible to apply one or more of the other three methods to good effect, but, as a minimum, some of those results will have to be subjected to analysis.

8.4.3 Environmental requirements verification analysis

There are several kinds of environmental requirements as noted in Chapter 1. The five different elements include: (1) natural, (2) cooperative, (3) uncooperative, (4) induced, and (5) hostile. Most natural environmental requirements are verified through test but these actions sometimes must be supported by analysis. Suppose, for example, that we are going to take subsurface samples on Mars. We have some knowledge of martian surface characteristics and atmosphere and can test the penetrating capabilities of our sampler on material similar to that anticipated. But, we cannot be certain that our lander will place us in a place that precisely matches the requirements driving the design of the sampler. Perhaps this represents a flawed set of requirements, but we are convinced the probabilities are in our favor and we cannot afford to provide a sampler that will work in 100% of all possible situations. In this case, our verification of the sampling capability may have to include an analysis of the accuracy of placement of the lander relative to plan and the credibility of our surface estimates.

Cooperative and induced environments will generally yield to test or demonstration in that we can converse with those responsible for the other terminal and gain access to the design details of those mating interfaces. Uncooperative interfaces, like electromagnetic interference, can be subjected to test but they will often have to be preceded by some form of analysis accomplished to determine the range of possibilities for those characteristics. It is uncommon where we will have the mating element of a hostile interface. Our antagonist would do well to do everything possible to deny us access to the details of their capabilities in terms of the magnitude of the burst pressure and its rate of increase and decrease as well as their accuracy of placement of the weapon relative to our location. So, we may have to resort to analytical methods in verifying some aspects of this environment.

We will also often have to apply analysis to the results of environmental testing. The raw data may not be immediately intelligible by people and have to be reduced or annotated so that other persons can make some sense out of it.

8.4.4 Analysis strings

Each of these analyses should be identified in program planning and assigned a VTN as discussed under the test method driven by specific requirements. In the test chapter, however, the implication was made that test VTN would often be in a one-to-one correspondence with the items in the system. In the case of analysis, this often will not be the case, as illustrated in Figure 8-1. It may be more sensible, for example, to map the analysis VSNs related to reliability requirements for all of the items in the system to a common VTN for reliability analysis of all of the items in a centrally maintained model. Some customers prefer a combined reliability and maintainability report that could be the output from a systemwide analysis of both of these factors. We may also choose to assign parts of these analyses to product teams such that in a system with five product teams there may be six reliability analysis tasks, one for each team and a system-level analysis to integrate the results of the product team analyses.

8.5 Demonstration

8.5.1 Specialty engineering requirements

Many specialty engineering requirements will yield to verification by demonstration. One can demonstrate the remove-and-replace time for an end-item maintainability repair time requirement; the safe operation of refueling controls on a flight line refueling truck, as well as how the safety features

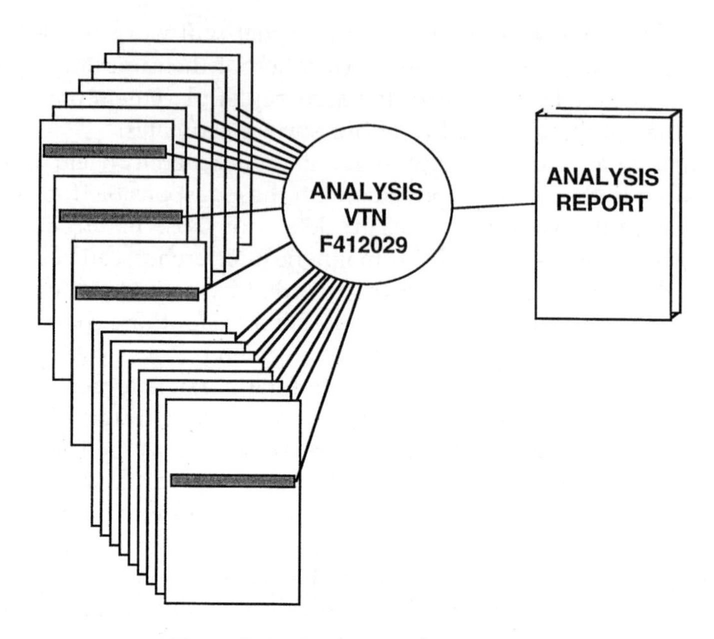

Figure 8-1 Analysis task grouping.

encourage safe operation; the fault isolation time for a new radio transmitter; and the ease of human operation of a new telephone design.

8.5.2 Interface requirements

It is very commonplace to verify interface requirements by demonstration. The two interfacing elements are brought together and fit is checked, cable assemblies mated, plumbing lines joined, and bolt patterns matched possibly as a prelude to running active functional tests on the interface. In some cases where it is very difficult to bring the two units together because of schedules, distances, and transportation difficulties, the physical fit of one item to another can be verified by specially created matched tooling jigs. One of the interface partners creates a jig that accurately reflects product configuration and ships it to the other partner where it is used to check the interface. This process could be cross connected by both contractors shipping fixtures to the other, but it is generally adequate to consider one side of the interface as the master where the interface is rigorously controlled by one contractor who supplies the jig to the other.

8.5.3 Software verification by demonstration

Computer software does not possess a physical existence. It has an intellectual character. Granted, it appears in paper documents as lines of code, is recorded on disks and tapes in combinations of magnetic states, and appears in computer memory as combinations of 1s and 0s, but it has no real physical existence as does the hardware computer within which it runs, a piston in an automobile engine, or a whole Boeing 747 jet transport. Therefore, since it has no physical existence, one might conclude that it is impossible to test software. This is, of course, not true. Test is the principal method for verifying software. Testing is implemented by running the software in the intended environment under controlled conditions and observing the results. There are, however, applications for the other methods, including demonstration at particular places in the development cycle for software.

Software that is very interactive with a human operator's observations on the screen and actions on the keyboard can be effectively verified by the human demonstrating it. The operator runs through a planned series of actions and results are monitored with flaws noted for analysis. One could argue that the alpha and beta testing approaches involving hundreds of potential users of a new Microsoft product fall into this category.

8.6 Examination

This book uses the word examination rather than inspection primarily because it is used in MIL-STD-961D, Appendix A, which reserves the word inspection for a more-general meaning encompassing all methods. That is, a test, demonstration, examination, or analysis would all be termed examples of inspection in MIL-STD-961D. For consistency, the author has used the word *examination* for software as well even though the word *inspection* is more commonly applied within that field. The difference in the meaning for a few common words between hardware and software development continues to cause needless confusion as in the case of the words *validation* and *verification*.

8.6.1 Software application

Examination is a popular method for verification of software, but is not adequate by itself. As a prerequisite to testing, the code for a software module can be subjected to a peer review, walk-through, or examination by people with knowledge of the application, who possess the right language skills, and who understand the company's software development process. The principal software engineer explains the code followed by an opportunity on the part of the reviewers to study the code and offer critical comment focused on improving the code. The reader is encouraged to study the book *Walkthroughs, Inspections, and Technical Reviews*[1] by Daniel Freedman and Gerald Weinberg for details on this activity.

8.6.2 Hardware too

Examination is commonly applied in hardware acceptance verification as in quality assurance inspection on the production line. The quality inspector studies product features against a standard for imperfections using the unaided senses or simple tools (such as a ruler, caliper, gauge block, or magnifying glass) and comes to a conclusion based on a pass–fail criterion. During hardware development this method can also be applied, however. We must produce a part using a new fiberglass resin because of changes in the state environmental protection regulations, so we make samples and subject them to a visual and tactile surface smoothness examination. The fiberglass payload fairings for the General Dynamics (now Lockheed Martin) Atlas Centaur space launch vehicle at one time were examined by a person thunking the surface with a quarter listening for delaminations indicated by the resultant sound.

References

1. Freedman, D. and Weinberg, G., *Walkthroughs, Inspections, and Technical Reviews*, 3rd ed., Dorset House, New York, 1990.

chapter nine

Item qualification verification management and audit

9.1 Setting the management stage

In previous chapters we have identified ways to collect and organize information pertaining to the work that must be done to verify that the product does, in fact, satisfy its requirements. This information has been organized into a series of documents, ideally, within the context of a relational database.

This chapter covers the overall management of the item qualification verification process using the integrated verification plan (IVP), integrated verification report (IVR), and verification management report (VMR) introduced in earlier chapters. All of the work grouped together under qualification is focused on a major review at the downstream end of the qualification process called by DoD the functional configuration audit (FCA). At this review, the customer is presented with the evidence accumulated through the qualification program to prove that the design satisfies the requirements in what would now be called in DoD the performance specifications. In past years these specifications would have been called the development specifications or the Part I specifications, but MIL-STD-961D changed the terms for DoD in 1995.

Given that we have captured the correct data and organized well for implementation, the remaining problem is one of doing the planned work and management of that work. Management involves bringing the right resources to bear on the problems we have defined in accordance with the schedule, also predefined, within the context of limited economic resources consistent with the planned work. We desire to accomplish the planned work that will result in evidence of a condition of compliance. Where that evidence does not support the conclusion that the product satisfies its requirements, we must take management action to bring us to a condition of compliance. The possible actions we can take entail changing the requirements, changing the verification planning data and reverifying, changing the product design and reverifying it to possibly new planning data, or obtaining a temporary waiver for the noncompliant design features.

Management skill was required throughout the preparation for verification. This included insisting on solid verification requirements as well as sound planning and procedural data, following the development of needed verification resources and providing guidance and direction where needed, and arranging for availability of product to test in accordance with planning data. As the product, in its several representations (engineering drawings, parts lists, physical product, and lines of code), becomes available for test, analysis, demonstration, and examination, the verification management focus must shift to taking action to cause planned work to occur as scheduled, maintaining status of planned work, overcoming conditions that cause work stoppages, and taking action to prevent or correct for budget and cost overrun conditions.

As the item qualification verification work begins to come to a close, verification management must focus on communicating the results to the customer through a formal review of verification data and achieving customer acceptance of a successful item qualification verification effort. The prize at the end of this run is authorization to engage in system-level verification activities potentially involving actions that are hazardous to life and property such as occurs in the flight or ground testing of weapons systems. Progress and incentive payments may also be tagged to completion of item qualification.

9.2 *The management agent*

Within a program that applies the cross-functional team organizational approach with staffing from lean functional organizations, the overall verification management responsibility should be assigned to the program integration team (PIT), by whatever name, as described in Chapter 1. All item verification tasks should be assigned to one of the product teams with a clear definition of resources and timing in published planning data as discussed in prior chapters. As the time planned for these tasks comes into view, the PIT verification leader should ensure that the responsible team is moving toward readiness and that the necessary resources are being provided by supporting personnel. An effective means should be in place for all verification actions in progress to encourage persons responsible for those actions to make available to the PIT accurate status information. The PIT must review this information periodically and take action where appropriate to maintain the verification process on track.

The system verification management agent (the PIT in this book) must take direct responsibility for all qualification verification actions accomplished at the system level including any tasks focused at a higher level than the product teams, which may include tasks below the system level depending on how the product teams are assigned. The responsibility for a particular task could be delegated to a team, of course.

In prior chapters we identified all verification work in the context of VTNs that were assembled from VSNs and oriented about items in the system architecture. Each VTN must be assigned to a particular engineer or team in our verification task management matrix. Each task should have associated with it a task plan and task procedure captured within the integrated verification plan and should result, when implemented, in a component of the IVR. The tasks provide a sound basis for verification management but there are groupings of these tasks that also prove useful.

The sum total of all VTNs associated with a particular system item is referred to as an item verification task group and the sum total of all of the item verification task groups assigned to a single person, group, or team is a responsibility verification task group. There will be one or more item verification task groups subordinate to every responsibility verification task group. If a team is responsible for verification work for two items, its responsibility task group will consist of two item task groups. These verification task groups consist of all of the verification tasks, of potentially all four methods (test, analysis, demonstration, and examination), associated with the items involved.

The person from the PIT leading the item verification process should form a virtual organization of all of the persons responsible for these verification task groups which may involve tiers of groups and individuals rather than a simple single-layer fan structure. This responsibility is that of a project engineer working across organizational structures to achieve overall program goals in this specific area. This is an integration responsibility not strictly a management responsibility. But, this is not an unusual condition as it is the role of most people in a system-level team. The system reliability engineer is looking for the optimum reliability condition across the system. The system-level mass properties engineer is similarly interested in prime item mass properties rather than that of the items upon which the teams are focused.

Figure 9-1 offers a way of illustrating this grouping of individual verification tasks into groups oriented toward architecture items and teams. Given that product development team 1 is responsible for architecture items A11 and A12, the responsibility verification task group for this team includes all of the verification tasks in all of the item verification task groups mapped to that team. In the verification task management matrix, all of the tasks indicated under team 1 would show a responsibility assignment to that team or someone reporting to that team.

We now have the makings of a manageable situation. We have identified a comprehensive series of tasks clearly defined in plans and procedures, driven by a clear set of requirements, hooked into budget and schedule figures, and assigned to specific persons and/or organizations. The remaining problem is to track the progress of the responsible persons in satisfying

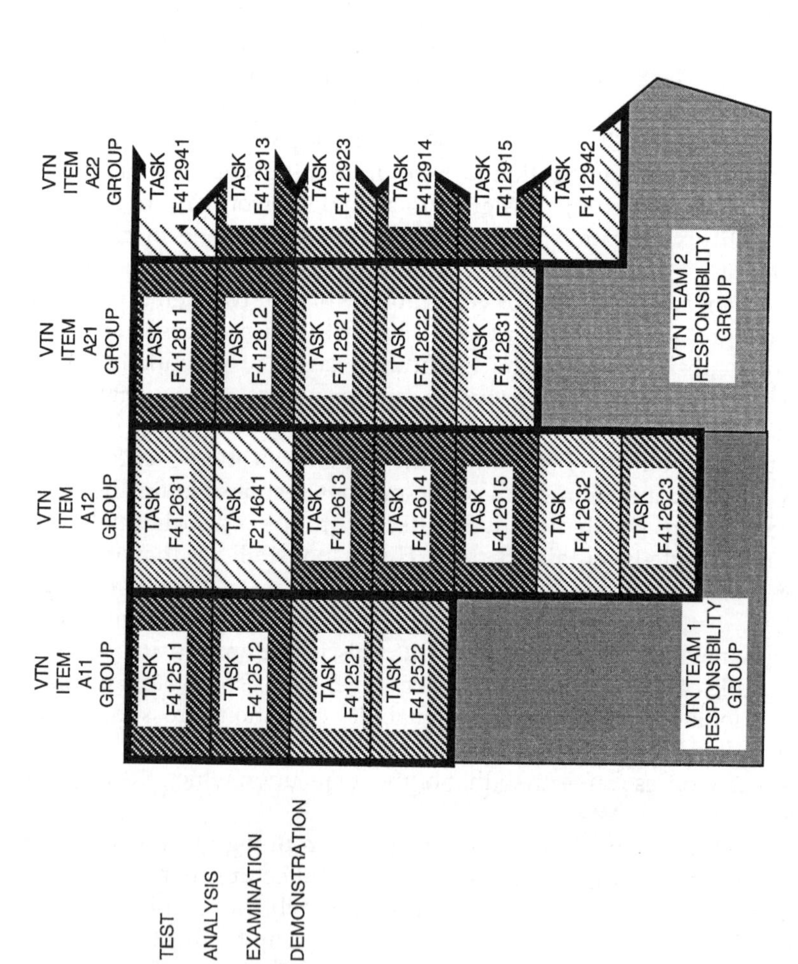

Figure 9-1 VTN groups.

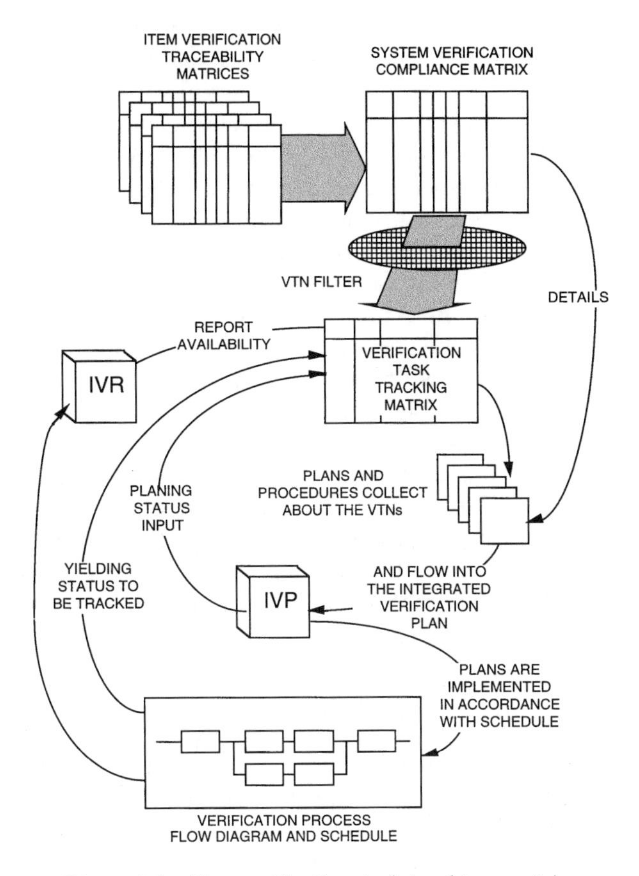

Figure 9-2 The verification task tracking matrix.

task goals within assigned cost and schedule figures and take such actions as required to encourage the planned outcome.

9.3 Documentation overview

Figure 9-2 illustrates the general management pattern during the item qualification process. Those responsible for verification tasks are stimulated to implement planned work identified in the verification task tracking matrix. Corresponding planning data is implemented by assigned personnel in accordance with the schedule and budget, yielding results documented in the corresponding integrated verification report component. The tracking matrix is updated and the whole program continues to follow this pattern toward completion of all planned verification work.

The line items in the verification task tracking matrix are VTN derived from the detailed listing of the verification compliance matrix which is the union of all of the verification traceability matrix in all of the program specifications. Table 4-1 shows a fragment of a verification traceability matrix that appears in each specification Section 4. It parses the requirements if necessary and maps item requirements to verification requirements and methods to be applied in verification. These entities form the atomic structure of the verification process called VSNs. Table 4-2 illustrates a fragment from a verification compliance matrix composed of the union of all verification traceability matrices and it maps every VSN to a VTN that relates to the specific tasks in the qualification program. As shown in Figure 9-2, the content of the compliance matrix is filtered into the verification task tracking matrix which has one line item for each VTN.

This is the matrix that the verification coordinator should use to manage the program and track status. If any program had to maintain all of these matrices as independent entities, it would tax available resources. The only way to maintain this data is through the use of relational tables in a database.

The verification plans and procedures are focused on the VTNs and collect in the sections of the integrated verification plan as outlined earlier. The test coordinator uses the test plan, test schedule, and verification task tracking matrix to determine when to assign and authorize tasks and tracks those tasks in work as they mature and complete their cycle in accordance with the IVP content.

9.4 *Task integration and scheduling*

As the atomic components of the verification process become available from teams or individuals in the form of specific verification tasks of the several kinds (test, analysis, examination, and demonstration) they must be coordinated with respect to place of implementation, the time frame within which they will be accomplished, and other tasks upon which they depend and to which they must supply results. Ideally, this will be possible within the requirements tool applied on the program, but if a truly comprehensive system is not available, a good alternative for this part of the process is offered by the many good network tools capable of program evaluation and review technique (PERT) or critical path method (CPM). Microsoft Project, Mac Project, and Primavera are examples. The tasks can be entered and linked to time and resources in these tools and schedules generated for use by those who must implement the qualification tasks. The critical path is also clearly identified in these tools.

The tasks (VTN) do not stand alone generally. They are interconnected by information dependencies. These dependencies should be noted in the planning data and used to help determine the impact of problems in completing particular tasks on schedule. Where a task has no downstream dependencies, it will be less of a problem to allow it to overrun schedule-wise. Where it has many downstream dependencies, it will commonly appear on the critical path and signal a need to pour in sufficient resources to expedite completion.

9.5 Task readiness review and task authorization

In the case of tests, there is a generally accepted notion of reviewing readiness to execute the test as a prerequisite to beginning the test. This reflects an appreciation of the cost and schedule consequences of implementing a test prior to a condition of readiness. In any action, if we begin work before a condition of readiness is achieved, we may have to repeat the work. This could be called a first-degree risk where the spent money and time characterizes the extent of the risk. In some qualification tests, the item under test becomes damaged as a result of the stresses applied and this unit thereafter is not acceptable without refurbishment for subsequent retest when it is discovered that a condition of readiness was not in effect and retest is required. This results in a second-degree risk involving funds and time in addition to those required to simply repeat the work.

As a minimum, the verification task readiness review concept should be applied where there is a second-degree cost and schedule risk. Alternatively, we could require a review prior to every verification task recognizing that most of these would require no more than 10 or 15 minutes by the principal engineer and the team leader. Each VTN listed in the verification task tracking matrix should be annotated to indicate the kind of readiness review required from a set of possibilities such as the following: (a) none, (b), team leader, or (c) program verification coordinator.

In the case of verification tests on high-dollar items involving large cost and schedule increments, the customer may wish to be present and voice an opinion on readiness. The matrix should also note these cases so that there will be no mishaps where the customer is not notified of such meetings.

9.6 Evaluation and refinement of evidence

The final review of the qualification verification evidence will take place at the FCA. But this evidence should have received a thorough going-over long before then by several parties. It does happen in some organizations that considerable money is spent planning for and accomplishing qualification tests only to have the results spark no interest. This is especially true for supplier test data but also occurs for one's own test reports. Verification actions should be undertaken for good reasons and the information produced as a result is important deserving of careful scrutiny. If the work isn't important enough to read the report, it probably wasn't important enough to have done the work in the first place.

9.6.1 In-house review of in-house verification reports

The verification task tracking matrix should identify who in the plant must review and approve the results of each verification task. This may be the principal engineer, the product team leader, or the program chief engineer and leader of the PIT. In some cases, we may wish to have several people collectively review a report and meet together to reach a consensus on the results. Where one concludes that the report does not provide clear evidence of compliance, the problem should be elevated to a management level where a decision can be made about what to do.

Each verification task listed in the verification task tracking matrix should have an assigned review responsibility and an indication of the status of that review. The verification management agent must ensure that these reviews take place and that the results of these reviews are acted upon.

9.6.2 In-house review of supplier verification reports

A large company working on a complex product will procure many parts of the system from other companies. Where the product is purchased off the shelf, the buyer must simply accept it or appeal under a warranty or guarantee where it is defective. Where you thereafter alter the item, you have to accept the responsibility to verify that the resultant product satisfies your requirements. Where you procure an item built to your blueprints you should do the qualification testing yourself since they are your requirements, in which case the verification work simply falls under your qualification program. Alternatively, you could require the supplier to test it to your requirements, possibly with you witnessing the tests. In any case you should expect some kind of written report of findings, and it should be reviewed for findings, and the veracity of those findings.

Where you procure an item developed in accordance with your procurement specification and statement of work, the verification work is best done by the supplier. Your subcontract should call for review and approval of the qualification plans and reports. On a large program with many suppliers, the amount of evidence to be reviewed and retained for future reference can be almost staggering. General Dynamics Space Systems (GDSS) Division (prior to sale to Lockheed Martin) accumulated so much supplier data by 1993 for the Titan Centaur Program that the file cabinets within which this data was stored became so numerous that data management had to move people to other office space to make room for them. Later in the program, data management brought in a system provided by Hershey Technologies that largely solved the storage capacity problem and went a long way toward solving some other problems in routing the data for review and approval.

This system included a scanner to convert paper documents received from suppliers into bit-mapped images that were temporarily stored on a server where they were available for review by persons contacted by data management. As a result, it was no longer necessary to run copies of these documents and physically distribute them to a list of persons. The people notified by E-mail that they had to review these documents could open them up during a scheduled review window and respond with comments. When a sufficient body of data was available on the server, data management loaded it onto a CDROM using a CDROM maker built into the workstation.

This CDROM was then stored in a file cabinet. Subsequently, an engineer could call data management and request that a particular document be uploaded to the server from a CDROM and the engineer could make a paper copy of the parts needed or view it on the screen of his or her workstation. In a company so equipped, this document could be called up across the network to a computer projection unit in a meeting where the content of the document could be discussed. This is a tremendously powerful capability that can repay the small cost of implementation many times over in efficiency or productivity.

Other companies have come up with alternative ways of making supplier data available for review and approval as well as use. One could put the CDROMs in a carousel or disk storage bank to avoid the manual upload problem, but there is a cost-effectiveness trade that should be run here. If the data will seldom be requested after initial review and approval activity, it can be more cost-effectively provided using the verbal request method discussed above. The problem with scanned data is that it is very consuming of computer memory even where special compression algorithms are used. You may conclude that it would be better to scan the incoming data as text and graphics to reduce the memory space and also to provide the opportunity to edit it subsequent to receipt. You should be aware that even with the very best scanner and optical character-reader software there will be errors in the resultant data. There will be no errors in a bit-mapped image. Further, we should not modify data received from our suppliers even if we could. This is formally delivered data across the contractual boundary and changing it internally after receipt could lead to many ethical and legal problems.

The workstation for supplier data receipt should also include a modem for receiving data submitted by electronic data interchange (EDI). Over time, this pathway will increase as more and more companies become EDI capable. The paper submission method will become less popular but will likely never become a complete void. One step beyond the paper and E-mail or fax delivery of data we find database delivery. Every company today creates its documentation on a computer and eventually more and more companies will create their data in relational databases. This data could be delivered to customers by refreshing the content of a customer version of the database in accordance with a schedule or at specific milestones. Internet, ever evolving, will also offer useful data delivery options in the future.

9.6.3 Associate interaction and interface verification resolution

Interfaces pose one of the most serious development challenges on a program. Where the two sides of the interface are represented by different companies with no contractual relationship, the contractors are referred to as associate contractors. This happens where a common customer separately contracts for several segments of a system. Each of the contractors has a contract with the common customer but no contract with the other associates. The customer solves this problem by requiring each contractor to submit a memorandum of understanding citing that they will cooperate with other associates in accordance with the associate contractor clause in their contract and jointly agree to work in accordance with an interface development plan and interface control document that one of them will be called upon to create and maintain as the integrating contractor.

A team is formed called an interface control working group (ICWG) cochaired by the common customer and the integrating contractor and including representatives from all of the associate contractors as members. This group is responsible for developing an interface development plan and developing compatible interfaces across all of the associate interfaces in accordance with this plan. The first step in this process is to identify each necessary interface and define interface requirements for each of them and capture them in an interface control document (ICD).

An ICD can be used in one of two ways. First, it may be a living ICD which remains active throughout the life of the program. Alternatively, it can be used only as a means to define the requirements which are moved to the two specifications for the two interfacing elements. In the latter case, the ICD passes from the scene when the interface is mature and defined compatibly in the two terminal specifications.

In the case of the living ICD, the verification requirements should be captured in the ICD Section 4. The verification traceability matrix in Section 4 should list all of the requirements and identify the methods to be applied in verification, the level at which they are to be verified, and the responsibility for the verification work. An interface plan and procedure prepared by the ICWG should provide all of the necessary data in support of verification work to be accomplished. When this work has been completed, the responsible party should be required to prepare a report of results referenced in the interface compliance matrix prepared and maintained by the ICWG. An interface FCA could be held or the results could be audited at the system FCA.

If the passing ICD is applied on the program, it is consistent to require the associate contractors to prepare interface verification planning and procedures in their integrated verification plan with traceability to the ICD. If this work is done well, the ICD verification traceability matrix tells responsibility and points to the responsible contractor for each requirement. The interface verification data developed on the two sides of the interface should be reviewed in-process by the ICWG. A joint FCA could be called or the two sides of the interface audited separately by the same group of people close together in time.

9.6.4 IV&V witnessing and review

The customer may not have assigned to its program office the personnel required to interact technically with the contractor(s). To satisfy this need, they may contract with another company to provide technical, independent validation and verification (IV&V), or system engineering and integration

(SE&I) services. One of several companies that have done this kind of work in the past is Teledyne Brown Engineering in Huntsville, Alabama. The customer may require the contractor supplying product to invite the technical services contractor to in-plant meetings for qualification test planning and test readiness purposes. In these meetings the technical services company represents the customer technically but cannot direct that changes be made that change the contractual scope or alter previously agreed-upon formal baselines.

The agent may have to be invited to witness all or specific qualification tests or be provided with the results of qualification testing for review. The agent may interact with your company over technical details of the test and results and with the customer about their conclusions on product suitability for the application as exposed in the qualification test.

Some development organizations feel insulted when their customer calls in an IV&V contractor because they feel that the customer does not trust them. This is a very immature attitude. The facts are that very complex systems are not easy to create and an independent view is a healthy thing. The responsible contractor cannot be totally removed from the thoughts and emotions involved in the development of such a system. GDSS Division decided to produce space launch vehicles for the commercial market as the NASA and military launch market weakened in the late 1980s.

Some people in GDSS were relieved when it was no longer necessary to deal with the IV&V contractors that NASA and the U.S. Air Force brought into their programs. It was not long, however, before GDSS realized it was a healthy thing to have an IV&V contractor reviewing its development work and hired a firm to provide that service for its commercial programs.

9.6.5 Customer coordination

As verification reports become available, they should be made available to the customer incrementally for review. Ideally, the customer should have access to updates of the tracking matrix, which shows the building status of this evidence, and should be able to obtain computer access to the evidence directly. In the best of all worlds the appropriate members of the customer organization will remain tuned in to the availability of this evidence, review it as it becomes available, and notify the appropriate person in the contractor's ranks about their conclusions. To ensure that this does happen, the principal engineers identified for each verification task or the team leaders responsible for the accumulating evidence should talk to their counterparts in the customer's organization and encourage them to review newly available reports and discuss their findings with them. There should be some way for the contractor to capture in a formal way the results of a customer review of the reports but at least a verbal comment should be requested.

One of the possibilities is that the customer will be convinced that the requirements were satisfied based on reading the one or more verification reports related to the item. In this case, the contractor should be able to move on to other things of importance and consider that the item in question is ready for FCA. It is also possible that the customer will not be convinced by the evidence. In this case, a mutual understanding must be reached between customer and contractor to take one of several actions:

a. Agree to disagree until the FCA where management action will be taken to resolve the dispute. This course of action may have schedule risk problems in that if it is finally concluded that the design is deficient, it will require some time to correct the problem and verify the fix, possibly adversely affecting other schedules such as the beginning of flight test, for example.
b. Agree that the evidence does not support the conclusion that the item satisfies its requirements and take one of the following actions:
   1. Change the design and reverify.
   2. Request a waiver which would authorize use of the current deficient design for a limited period of time or limited number of articles until such time that the problem can be corrected. This will require formal work through the contract to achieve and should be complemented with the necessary work to understand how the waiver condition can be lifted and when this can be done.
   3. Change the requirements to reflect the level that was satisfied in the verification work. This course of action may entail the contractor giving the customer consideration for permitting a relaxation of the requirements. This consideration may be in the form of accepting a lower price for the product, agreeing to fix some other problem that is out of scope but desired by the customer, or paying back an amount of money to the customer.
c. Reach an agreement that the evidence really does support the conclusion that the product design satisfies the requirements.

If this kind of work is not done during the qualification process it will all pile up at the FCA leading to a lengthy meeting and numerous action items dragging out the final qualification approval. The FCA should be a relatively short meeting where agreements are quickly arrived at between company and customer regarding the evidence of compliance because both parties are familiar with the item requirements and verification evidence. It should be possible for the cochairs from customer and contractor to move the agenda quickly through the items being reviewed. There may be a few items where real disagreement remains and those few actions can be the subject of splinter sessions yielding outbriefs and direction from the co-chairs on a course of action.

The tracking matrix should include information on who the customer contact is and the status of their review of the report. The corresponding contractor representative, the VTN leader the item principal engineer, or corresponding team leader should recognize it as one of his or her tasks to

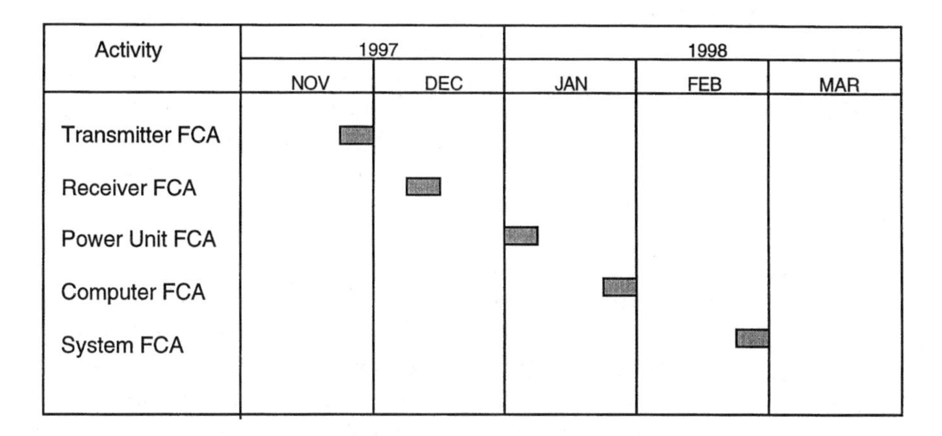

Figure 9-3 Multiple FCA schedule.

coordinate with this customer person to encourage the person to obtain and read the report, send it to them if necessary, and discuss his or her conclusions after having read the report. The results of this conversation should then find their way into the tracking report. If the customer reviewer finds cause for concern about the report, those concerns should be considered by the contractor and a mutual agreement derived for action to remove the concern. In the worst case, this could mean rerunning some parts of the qualification work to improve the evidence of compliance. By the time the FCA takes place, this matrix should be full of okays for customer review and acceptance.

9.7 Staging the FCA

9.7.1 Single or multiple audits

All major reviews, including FCA, are intended to be held at the configuration or end-item level. Where a contract only involves a single item, a single FCA is clearly called for. Where the system entails several end items, a separate FCA may have to be held for each item. Alternatively, all of these audits could be combined into a single FCA. Where two or more separate end-item FCAs are held they should be followed by a system FCA as indicated in Figure 9-3.

9.7.2 Audit preparation

9.7.2.1 Completion criteria

Generally, the FCA will be scheduled for a specific date but the customer and contract may respect an event-driven schedule where the audit is not held until the contractor is completely ready to stage it. In the latter case, all of the tasks that have to be complete in order to support the FCA will have clearly defined completion criteria. This may extend to all of the qualification test, analysis, examination, and demonstration tasks including publication of the results in each case. As we begin to prepare for the FCA, we should also clearly understand the completion criteria for the FCA. Customers should have clearly defined those conditions but if they have not done so, the contractor should ask for them or offer a closure criteria for consideration. This criteria could include the following conditions:

a. All verification tasks have reports available describing the results. This could be confirmed by the customer reviewing the task tracking matrix and noting the references to reports and their status and further confirmed by spot-checking the availability of several.
b. Customer review of the task reports for some subset (possibly all) of the tasks with a conclusion that the evidence supports the conclusion that the item satisfies its requirements in each case.
c. All action items stemming from the FCA closed. This may entail rerunning some tests and reporting the results or simply responding to specific questions.

MIL-STD-973, Configuration Management, includes instructions for FCA for DoD programs and offers a set of forms for formally closing out the FCA. The reason this is so important is that successful closure will enable important program activities, such as system testing, and it may very well enable one or more progress payments to the contractor. If the closure status of the FCA simply drifts along without decisive action, it will also result in inefficient use of personnel and possible program loss of key personnel to other programs within the contractor's organization. If failure to close out the audit precludes forward motion on the program, then the program must reduce personnel to prevent a cost overrun since many of these people cannot possibly be efficiently employed. If the program drops some critical persons for lack of budget support, the program may never get them back when sufficient funding is made available again.

9.7.2.2 Audit agenda

The contractor should prepare an initial agenda for the audit and submit it to the customer for review and comment. The contract may require that this agenda be submitted as a contract data requirements list (CDRL) item. Figure 9-4 offers such an agenda based on the assumption that we will only be auditing one item.

The task audits could be reviewed in sequence at a single meeting (as suggested in Figure 9-4) or broken up into subsets for auditing by teams composed of the appropriate disciplines from company and customer staffs. The latter will permit more work per unit of time but require a final integrating session (not included in Figure 9-4) where each audit team reports back to the committee of the whole.

1	AUDIT OPENING	20 MIN
1.1	Host Co-Chair Welcome	
1.2	Customer Co-Chair Comments	
1.3	Host Chief Engineer Discussion of the Agenda and Audit Plan	
2	ITEM OVERVIEW	40 MIN
2.1	Item Requirements Discussion	
2.2	Item Design Features Summary	
2.3	Item Risk Coverage	
3	VERIFICATION OVERVIEW	60 MIN
3.1	Verification Traceability Matrix	
3.2	Verification Requirements Coverage	
3.3	Verification Planning and Procedures	
3.4	Summary of Verification Tasks	
3.5	Item Verification Schedule and Process Flow	
3.6	Verification Documentation Cross Reference	
4	VERIFICATION TASK AUDITS	60xN MIN
4.1	Verification Task 1	
4.1.1	Task Objectives	
4.1.2	Review of Task Resources	
4.1.3	Task Personnel and Qualifications	
4.1.4	Requirements To Be Audited	
4.1.5	Exposition of the Evidence of Compliance	
4.1.6	Recommendation	
4.1.7	Discussion	
4.1.8	Conclusion	
4.N	Verification Task N	
5	EXECUTIVE SESSION	120 MIN
5.1	Disposition of Action Items	
5.2	Closure Criteria Review	
5.3	Closure Plan	
6	AUDIT FORMAL CLOSING	60 MIN
5.1	Summary of Action Items	
5.2	Review of FCA Closure rules	
5.3	Planned Closure Schedule	
5.4	Customer Co-Chair Remarks	
5.5	Host Co-Chair Remarks	

Figure 9-4 FCA agenda.

If the audit will cover multiple items then the agenda should reflect this by including the structure within Section 4 shown below. We have simply added another indenture for the item.

9.7.2.3 *Agenda responsibilities assignment*

The customer may respond with some recommendations for agenda revision and those changes must be worked out with company program management resulting in a final agenda. Each of the agenda items must be assigned to a presenter and these people asked to prepare their presentation. The responsibility for Sections 1, 5, and 6 of the outline is obvious. Section 2 should be presented by the principal engineer or team leader for the item. Section 3 should be presented by the program verification manager or one of his or her people. Each task in Section 4 should be audited under the company responsibility of the engineer named in the task tracking matrix as the person responsible for that task. This person should make the necessary documentation available for the task audit, guide the discussion during the task audit, and report out the results to the whole if the parallel auditing approach is used. Each task audit may also have assigned a customer cochair where parallel meetings are employed.

This principal presenter need not necessarily be from the contractor. Where the customer and major suppliers have been active team members throughout the verification process, these other parties may be the best qualified to make the presentation in some areas and should not be denied that opportunity based on contractual boundaries. This depends on the customers, of course, and their tolerance for these ideas. Coordinate with each responsible party and let them know what is expected and negotiate the period of time they have to present their topic. Let them know what presentation resources are available (overhead or slide projection, computer projection, video, film, audio, etc.) and inquire if that will be adequate for their needs.

9.7.2.4 *Audit venues and resources*

You should then plan for the necessary resources including:

a. Main meeting room and smaller caucus rooms;
b. Presentation equipment;
c. Presentation materials generation, coordination, printing, and distribution;
d. Meeting administrative services for attendee check-in, attendee telephone and computer use, meals (if provided) and refreshments, and incoming telephone calls for attendees.

It may be possible to decrease travel costs by holding the conference with videoconferencing equipment, but you will have to provide for tighter discipline in the meeting rooms and brief local chairmen that they must brief their participants on the need for more formal identification of speakers and questioners than is necessary when all are physically meeting in one place. A simple meeting etiquette sheet would be useful to this end. You should determine whether or not there is any benefit is recording the meeting and, if so, notify the participants of the decision to tape. You should recognize that taping will probably change the character of the meeting. Generally, it will amplify the normal differences between introvert and extrovert attendees. Introverts will tend to be more quiet and extroverts will tend to be more active. As a result, the chairman will have to make an extra effort to bring out the opinions of the demure and control the energy of others.

Presenters may also find it useful to bring the reviewers to a particular test cell or location in close proximity to the review site in order to convey in some detail the significance of ongoing system verification work or to emphasize the validity of past testing. You can show slide after slide on equations and environmental parameters at the review to little effect on the attitudes of some doubters, but they may be convinced completely by a brief trip to the test cell where the item is put through a representative round of tests involving thermal, vibration, shock, rain, and salt-spray cycling. Thereafter, the briefing materials will likely be more believable to everyone. Still photos and video can also be very effectively used for the same purpose where the test site is geographically removed from the review site.

9.7.2.5 The presentation materials

The biggest job in preparing for the audit will be the acquisition of the presentation materials from the presenters, the review and approval of these materials, and the integration of them into a final presentation handout for attendees. Presenters should be given a style guide in electronic media, a page count maximum, a clear goal for their segment, and a date when their contribution must be turned in for internal dry run and approval. If this material is all created in electronic media you should give some thought to a computer projection presentation. Otherwise, the materials will have to be converted into overhead transparencies or slides only to be thrown away after changes are made in these materials in preparation for the final review materials.

9.7.2.6 Audit dry run

You may choose to dry run the audit to make certain that everything is in order for the real thing. You need not take as much time as the planned audit. Rather, the first three sections can be discussed to the extent that the program manager has confidence in the readiness of presenters. Then each task leader can briefly preview his or her plan and recommendations. In some cases, the program manager may want a full dry run of the evidence.

9.7.2.7 Audit data reproduction

Paper copies of Sections 1 through 3 should be provided to customer personnel in the form of a paper handout. You will need to make at least one paper copy of the Section 4 data that will be reviewed. It is unreasonable, however, to make a copy of every document for every member of the auditing team. The amount of documentation can be very voluminous and therefore very costly to reproduce. The ideal approach, where all of this data is on line on a server, is to provide a workstation and local printer in the area of the audit such that if anyone wants a copy of a particular document, it can be called up and printed. Each room used for the audit should be equipped with computer projection capability such that any of the data may be projected in the room for all to see. This combination will dramatically reduce the amount of data reproduction required in support of the meeting. A few paper copies of the report data should be made available in the meeting room for use by anyone not computer literate or to use in combination with data viewed on the screen for comparison purposes.

9.8 Presentation of the qualification results, the audit

Refer to Figure 9-4 for a list of audit topics for each task to be reviewed. As noted above, all of the data should be available for computer projection from a central verification library consisting of many text and graphics documents or the content of a relational database covering the whole process from requirements in specifications through verification reports. In some cases, the customer representatives will have read the related task reports and accepted the conclusion that the item design satisfies its requirements. If the company principal engineer has maintained contact with a customer counterpart and assertively but politely encouraged customer readership as the reports became available, this is a more likely result than if the FCA is the first time the customer has seen the documentation.

The contractor and customer representatives may be able to agree at the outset of a specific task audit that everything is acceptable except one or two particular areas where the conversation can focus, avoiding wasting time discussing things that both parties agree on completely.

Where the customer representatives are reluctant to accept that the contractor proved that the item satisfied its requirements, the contractor task lead may find it useful to take the customer representatives to a nearby test laboratory to see the test apparatus or go over the simulation or modeling software and ways they have validated them. The discussion about the physical resources used to verify the requirements may be more convincing than the data produced as a result.

From the contractor's perspective, the audit should provide convincing evidence of the following:

a. Test procedures comply with specification content.
b. Procedures were run as documented.
c. Results reflect design complies with the requirements.
d. All requirements were verified (completeness).

The evidence may not be convincing to this end. If there were areas where verification action was not taken, could not be taken, or was taken and the results were either inconclusive or proved that requirements had not been satisfied, these facts need to be presented along with a recommended action. This action may entail retesting, suggested changes to the requirements, design changes followed by retesting, or a deviation approved by the customer permitting some number of product deliveries that are less capable than required while the problem is being corrected.

Unless clear evidence has been uncovered during the audit that the item failed to satisfy its requirements, the company representative should make a recommendation that the customer accept that the requirements have been verified for qualification purposes. Before leaving the audit session, a clear statement of the customer's conclusions should be solicited if it is not forthcoming. If the two parties cannot come to a common conclusion, they may have to jointly brief the complete audit team of their different opinions. The cochairs will have to agree upon a way to finally resolve the dispute.

9.9 Postreview action

Subsequent to closure of the review the contractor must work off all of the action items and track the status of this work as it is driven to null. Some of these action items may entail additional qualification work to close out customer concerns expressed during the review. During the review, the customer could finally grasp that the product does not satisfy the current wants. The product may perfectly satisfy the current requirements as expressed in the customer-approved baseline but the customer has discovered that this item could, but does not currently, satisfy other requirements not identified earlier in any formal way. This may trigger engineering change proposal (ECP) action that will define new requirements and design changes leading to fabrication of the changes and qualification of those changes. Any ECP action should not be allowed to interfere with closure of the current FCA. The changes can be the subject of a delta FCA or regression-testing FCA at a later date.

The contractor should keep track of each specific task that must be completed to gain FCA closure, assign someone responsible for closure of each item, and manage the closure to completion. This is especially true of all of the action items assigned at the audit.

9.10 Information resources

As the reader has observed, this process is very documentation intensive, for it seeks to capture formal evidence of compliance in the form of test reports and analysis reports. Therefore, the management process must rely very heavily on sound information management techniques and systems. Not only must we ensure that the reports are created and reviewed, but we must be able to access any of these reports on short notice whenever we or the customer wants them.

Engineers on programs that were not well run from a verification aspect have experienced the stomach-churning realization that the documentation for particular verification tasks was not available during or shortly before the FCA. Some programs permit the principal engineer for a verification task to retain the corresponding documentation providing copies as required on the program. It has happened on such programs that one or more of these engineers has retired, died, or left the company for other reasons and the contents of his or her desk and files were disposed of in the trash or to the retiree's garage in a pile of boxes never to be opened again. Sometimes you might be lucky enough to locate the box in the engineer's garage in one of these cases and be able to retrieve the only copy of the report in time for the FCA, but don't bet on it.

These goals can be achieved through the use of verification documentation captured with word processors with good discipline, but the optimum way to do this is to use a relational database. Some commercially available computer tools (RDD-100, Doors, RTM, and Slate, for example) include many good features supporting the verification process and the schema of many of these tools can be adjusted by the buyer to improve and tailor the capability for specific enterprise needs. The ideal tool you select or build for your own purposes should capture the requirements completely in a relational database from which a specification can be printed and within which traceability can be established within and among the requirements and between the requirements and the verification process information.

The verification requirements should be captured in the same relational database keyed to the same specification they relate to with traceability between the requirements section (commonly 3) and the verification requirements section (commonly 4). Further, it should be possible in this database system also to assign these verification requirements to a method and collect them into predetermined verification tasks. The integrated verification plan content should be created in this same database system with its paragraph numbers, titles, and text also included in records just like the requirements. The planning content should be traceable to the verification requirements from which they were derived.

Finally, the verification reports should also be database captured with traceability back to the planning and procedures. With this kind of structure, all of the planning and reporting data is fully traceable and specific documents can be called up or printed in whole or in part upon demand. It should also be possible to status the whole process in some automated fashion based on the content of the database. Figure 9-5 illustrates this structure in IDEF1X format explained in Department of Commerce *FIPS Publication* 184.

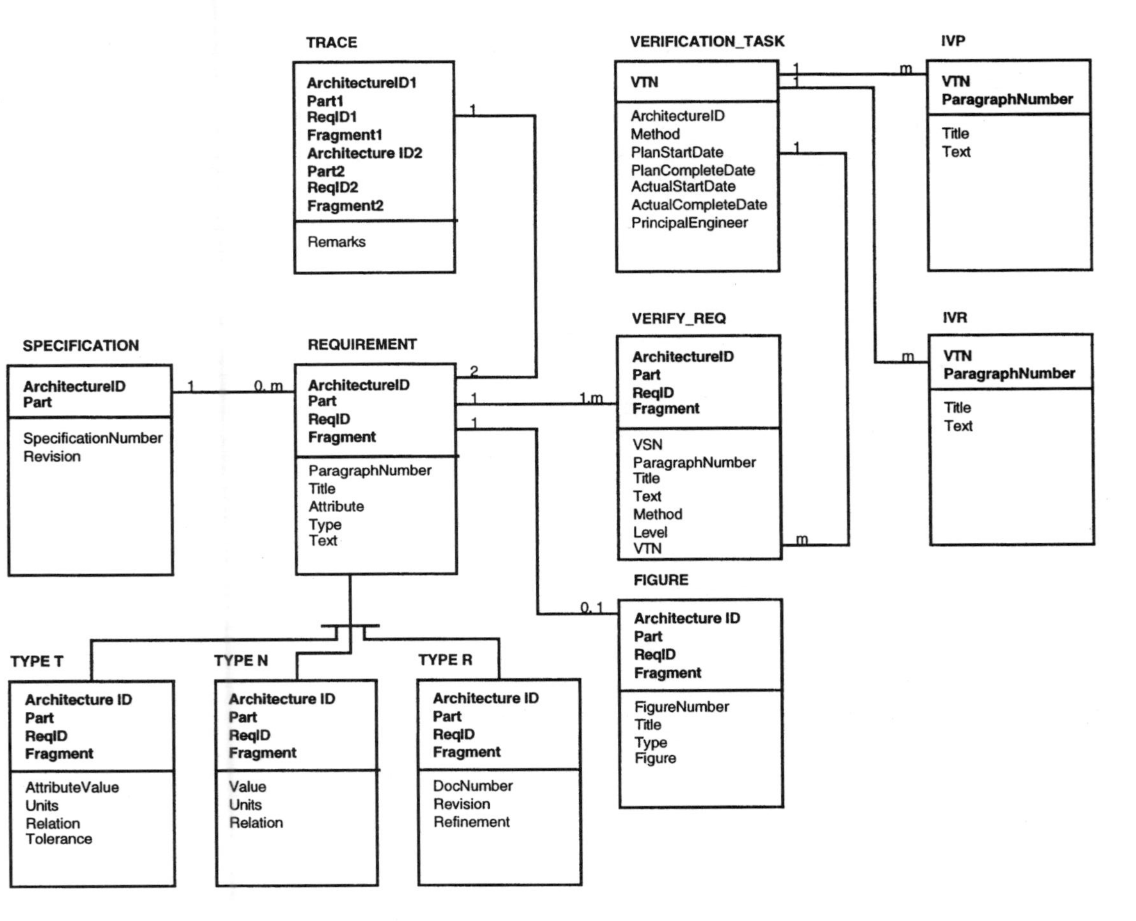

Figure 9-5 Specifications and verification support database structure.

chapter ten

System verification

10.1 System verification overview

The qualification process properly occurs at the end-item level, possibly supported by lower-tier component tests and analyses. These activities are accomplished within the context of the item verification process as discussed in prior chapters.

Most customers of complex systems will also require system-level tests and analyses to show that the end items will interact as planned to achieve system requirements. This is primarily accomplished through the use of the system in accordance with its planned use profile. If the prime item is an aircraft, we call it flight test. Land vehicles may be tested through on- or off-road ground tests depending on the kind of vehicle involved. A missile system will undergo test firings under a range of conditions anticipated in actual use against targets that perform like the planned targets. This system testing is not thought of as part of the qualification process that culminates at FCA because qualification is a component activity that is a prerequisite to operation of complete end items in which they are installed.

If we have had to conduct a segmented FCA process because of the complexity of the system and availability of end items distributed in time, the customer may also require a system-level audit to go over the results of all of the item audits and ensure that all prior action items stemming from item FCAs have been closed. This system audit acts to link together all of the individual FCAs into an integrated activity. It is uncommon to coordinate the system FCA with the end of any system testing that is required since the system testing may require a very long time, years on a new fighter aircraft, for example, and both the customer and the contractor have used the FCA process only to ensure that the end or configuration items are qualified for the application.

The system FCA, like the sensible use of multilevel hierarchical testing, illustrates a point about which system engineers must remain consciously aware because of their perfectly normal and useful mania for decomposition of complex problems into series of related smaller problems. In the opening chapter we noted that because we decompose large problems into small ones in the interest of focusing our energy onto problems consistent with our technical and managerial capacities, we must also integrate and optimize at the higher levels. It is necessary but not sufficient, for example to verify the operation of five items independently that will form a subsystem in the end item. There are very likely higher-level subtleties that cannot be exposed by lower-level work. In space programs it is normal to test at every level of indenture in the system architecture. The reader may recall the unfortunate case of the Hubble Space Telescope program where an integrated test of the optics was not done because of facility scheduling and cost difficulties and the belief that the item testing had been adequate. With systems that will be more easily accessible after deployment, we may properly be more tolerant of holes in the testing string logic but each gap in the string is an invitation for risk to enter the program. As in all cases, the technical risk of not doing work must be balanced against cost and schedule risks of doing it possibly unnecessarily.

We have several goals in testing the complete system and combining the results of the end-item verification results. First, we wish to establish the readiness of the complete system for deployment or delivery and operational employment. We also hope to validate the models and simulations used in development so that they may be subsequently used to predict performance in later studies and change analyses. In a commercial situation, we may also be interested in proving that the product is safe and reasonably free of flaws that might stimulate legal action.

There are many product lines where government regulations require new products to pass the scrutiny of special testing laboratories or criteria. The FAA requires new passenger aircraft to be certified for revenue-producing flight. Road vehicles must pass or be prepared to pass tests focused on safety regulations. New drugs must wind their way through lengthy evaluation before being allowed on the shelves of pharmacies. Many other products must pass similar hurdles en route to public sale or distribution. In some cases, the responsible agency will accept your test results for review. In other cases, the agency insists on their own tests at their own facility or conducted by a contracted licensed test laboratory.

The system verification process depends on both the results of the item verifications and on unique system-level activities that cover requirements that could not be verified at the end-item level. The latter consists of all of the verification tasks that were identified in the compliance matrix for verification at the system level. These tests and analyses must be characterized progressively in order to approach incrementally conditions where the maximum risk to life and property are realized. We also need to focus on comprehensiveness or completeness over the full range of system operational and logistics actions. Actual completeness may entail an infinite space and thus be practically impossible to achieve in a sufficiently complex system, so system testing is also a great compromise requiring careful consideration by experienced engineers to determine what subset of completeness will be sufficient.

10.2 System verification audit

Where item verification audits are held, it is useful to terminate the verification audit process with a system review even if there are no system-level verification activities. The system audit in this case acts to tie together any loose strings that were identified at the end-item audits and ensure that all action items identified at those audits have been properly addressed and been closed out. Where the system includes both hardware and software elements, as is so often the case, the system verification audit also permits an integrated consideration of the isolated results in these two areas.

If only a system audit is held, where the system includes two or more complex items and that audit must accomplish the intent of the item and system audits, you would need to apply the content of Chapter 9 as well as the content of this chapter. The system audit should be broken down first into a series of items audits (following the pattern discussed in Chapter 9) followed by an audit of system verification in accordance with guidance offered in this chapter.

10.2.1 System audit planning

Prepare an agenda in cooperation with the customer and other interested parties. Determine who is chairing the audit. Is it the customer, an IV&V agent of the customer, or some third party? Decide on some goals for the audit which might include the following elements:

a. Review and approve the actions taken at all previous item audits.
b. Ensure that all past action items have been addressed and closed.
c. Review the results of any system-level verification actions that have occurred and place earlier item audit results in context with them. The system test work will commonly not be complete when the system functional audit takes place.
d. Review the effects of any engineering or manufacturing process changes that have occurred since the completion of item qualification work and ensure that any and all effects on previous qualification conclusions have been considered and, where appropriate, have resulted in additional verification work which has also been the subject of FCA action either independently or within the context of the system FCA.
e. Formally conclude the qualification process with a formal notice of completion.

The latter may be a conditional conclusion subject to the completion of a small number of remaining actions, but there should be closure on this issue rather than allowing it to remain a lingering sore for an indefinite duration.

10.2.2 The audit

If there has been no significant changes in the product since item FCAs were completed, the system FCA should only focus on the results of those FCAs and action items that lingered after meeting close. So, there should be an agenda section for each item FCA. If there have been significant changes since item FCA closures, the customer could have elected to stage delta FCAs for all or some of the engineering changes where the differences were subject to review. These delta FCAs simply become add-ons to the corresponding item FCA discussion.

There may be changes where delta FCAs were not held with the understanding that the matter would be taken up at the system FCA. In this case, the agenda should include a full-blown audit of the changes rather than a review of the prior audit results.

10.2.2.1 Item verification summaries and integration

The current status of all item audits should be reviewed with special focus on any retesting that had to be accomplished subsequent to the item audit. Engineering change proposals that have been identified and processed subsequent to the item audit should be discussed in terms of their effects on the design and results of the qualification process for the item. It should be agreed that subsequent to closure of the system verification audit, any change proposals should carry with them the responsibility and costs for any requalification action that may be necessary.

There may be some item requirements that were not verified wholly at the item level, but rather at a lower or higher level. In all of these cases, we should connect up the verification actions focused on items and review the adequacy of the collected work for each item. Ideally, this should have been done for each item audit where promotion or demotion occurred, but some of these actions may not have been complete at the time of the item audit in question. For example, if we completed the verification work for item A123 and held an item audit on February 21 and some of the requirements were going to be verified at the parent level (item A12) in testing that was not scheduled to take place until June 13, our A123 FCA would be incomplete. In this case, there should have been an action item assigned to report the results of A12 testing related to item A123 and offer a recommendation regarding closure of the A123 audit. In cases where there are many promotions and demotions, it is possible that not all of the intended integration actions occurred. At the system audit all of these strings should be traced and proper closure assured.

10.2.2.2 Item audit action item closeout status

It is likely that some action items were assigned at each of the item verification audits. It is hard to imagine that all item audits were held without any action items having been assigned. So, any open action items should be reviewed for status and ways found to close them out. Customer and contractor should jointly review those closed out previously and concur that they were completely addressed in their closing action. MIL-STD-973 provides forms for action items and verification certification that can be used as a guide even with a nonmilitary product.

10.2.2.3 Interface verification audit

When you arrive at the system audit, item interfaces have probably previously been verified at the item specification pair level. In this case, we should compare the verification results for both ends of each interface and ensure that the results were satisfactory and complementary. At higher levels in the system, some interface requirements may be documented in interface control documents and have to be verified with respect to that documentation. The only difference in interface verification results is that they relate to an interface rather than an item in the system. The evidence is developed in the same way in test, analysis, demonstration, and examination reports. The review of system-level interfaces could be viewed as part of the system-level work but is singled out to preclude it from being overlooked.

The interfaces may have to be physically verified quite late in a program. Space systems often first come together completely only at the launch site. So, many interfaces have to be verified as the first vehicle is built up in preparation for launch. As a result, adequate time has to be built into the schedule for this work that is in addition to normal launch processing. Many of the problems that can occur in this work can be eliminated long before this time through the use of matched tooling for mechanical fit checks at the factory of one or both mating items and special test equipment reflecting the design of the mating item.

10.2.2.4 System testing results

If the system FCA covers system testing as well as the integration of the item FCAs, we should check to ensure that the system-level test and analysis work has produced a comprehensive body of knowledge covering the whole system. This work will have consisted of a finite number of tasks accomplished at system level leading up to the system review. Each of these verification tasks should have had associated with it one or more goals or purposes and specific actions defined in a verification plan and procedure driven by requirements in a specification. Most of them will be based on the system specification, of course, but lower-tier requirements may be included that were to be verified at a higher level. The review must determine if the results of the testing and analysis has proved that the requirements have been satisfied. Each system verification task should be reviewed to determine if the results cover the planned verification space and that the whole leaves no planned verification work undone. Refer to Sections 10.3 and 10.4, below, for details.

10.2.2.5 Audit follow-up and closeout

The contractor should create a final report of the system-level audit noting any decisions arrived at, action items assigned, and any remaining steps necessary to complete the whole qualification verification process. Any remaining action items should be assertively worked to completion and submitted to the person or organization requesting them. If actions were assigned to the customer, associate contractors, a prime contractor (if you are a major subcontractor), or team members, a lot of telephone and E-mail contact will be required to encourage closure on time. As soon as all remaining actions have been completed, that fact should be communicated to the customer, sponsor, or IV&V contractor identified by the customer and formal closeout action requested.

The final action in this long and difficult process is that the contractor should encourage the customer to send them some kind of letter or form formally indicating successful completion of the qualification verification requirement. On many programs completion of the verification process is required to begin any system-level testing that may entail risks to damage or loss of scarce development resources (like flight test aircraft) and death or serious injury of personnel involved in the testing. It is a good idea for contractual and legal purposes to acquire some kind of formal agreement that the program was ready to move to this stage before doing so.

10.3 System test planning

System test planning should start way back with the development of the system specification. The V diagram illustrates this very well. As noted in Figure 10-1, the content of the system specification should drive system test requirements crafted at the same time as the specification is written. Ideally, these requirements will be included in the verification section (4 in a DoD specification) of the specification. Some organizations place these requirements elsewhere or fail to go through this step, and the result is a poorer specification. If you have to write the verification requirement as the product requirements are being written, you will write better requirements. There will not be any requirements statements like "... must work well and last a long time ..." in the specification because one cannot craft a way to test or analyze whether or not the final product satisfies this kind of requirement.

All of those system requirements mapped to test in the system specification traceability matrix should come to rest in the hands of the person responsible for creating the system test plan. These may be augmented by verification requirements, or verification string numbers in the context of

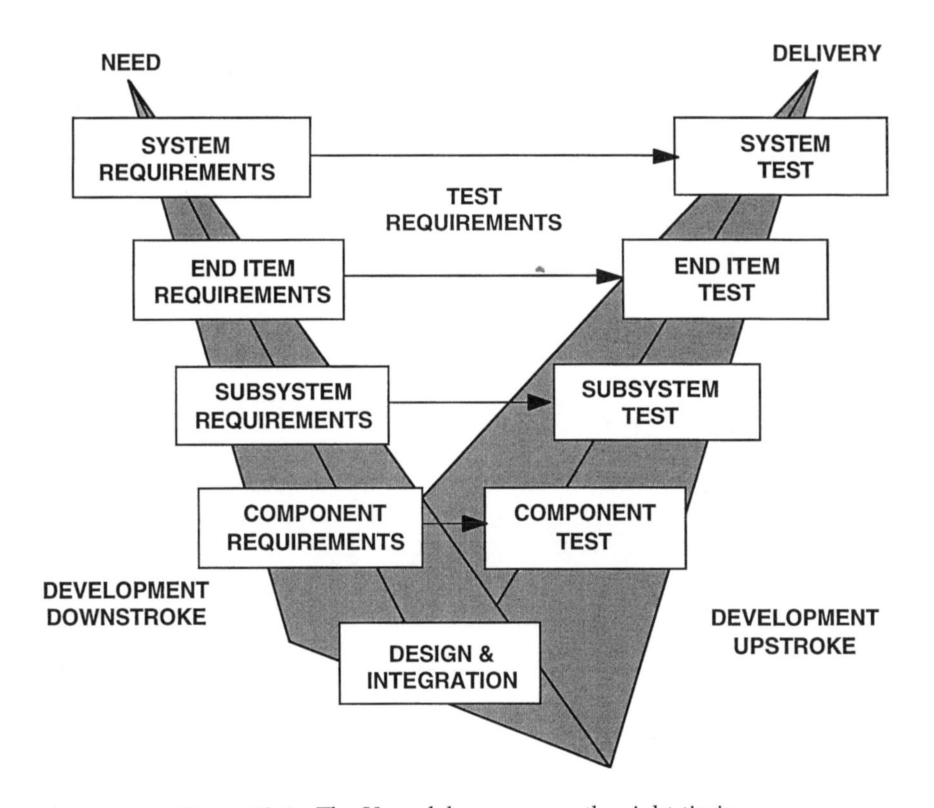

Figure 10-1 The V model encourages the right timing.

item verification work, promoted from subordinate items to parent-level verification. As described for items, the system test-planning process must determine how to verify that the system satisfies these requirements least expensively and most effectively.

The first step is to synthesize the test requirements into a set of test tasks. On a simple product, this may only require a single test task. On a new aircraft, this can entail many flights over a long period of time preceded by many ground tests and taxi tests. The first flight might be a takeoff followed by an immediate landing followed by a takeoff and several touch-and-go landings with all flight within visible distance from the flight test base or under surveillance from chase planes. Based on the results of these tests and analysis of any instrumentation data recorded during them, engineering should provide guidance on subsequently planned flights. It may be necessary to alter initial flight plans restricting certain flight regimens or loading conditions or moving them into later flights until data can be further analyzed.

Generally, it is sound practice to test from the simple toward the complex, from the small toward the large, from the specific toward the general, from the safe toward the more hazardous, and from the easy toward the more difficult. Our system testing should progressively build confidence in future testing by proving the soundness of increasingly larger portions of the design.

10.4 System-level testing and analysis implementation

10.4.1 Product cases

The nature of the system testing required during development is a function of the product. An aircraft, helicopter, missile, or unmanned aircraft will likely have to undergo some form of flight test to prove that it can actually fly under all required conditions and achieve mission success. A ground vehicle will have to undergo some form of ground testing. A tank, truck, or other kind of vehicle should be required to perform over real or specially prepared terrain of the kind covered in the system specification while verifying, in a military product, offensive and defensive capabilities. In all cases, the end items subjected to system test should be composed of components, subsystems, and systems which have previously been qualified for the application as discussed in earlier chapters.

Because we have tested all of the parts, there may be a temptation to conclude that it is not necessary to test the whole. This conclusion is sometimes correct but more often not. The problem with complex systems is that it is very hard to intellectualize everything. This is true especially in space systems. Many readers will recall the problem with the Hubble Space Telescope that could not focus properly when deployed because the optics were not correctly configured. Cost, schedule, and security problems during development with respect to the use of facilities capable of adequately testing the telescope resulted in the rationalization that major elements tested separately would work properly together. Since the telescope was placed in a low Earth orbit it was possible to modify it while in orbit using the space shuttle but only at tremendous cost.

Space systems are commonly tested throughout the product hierarchy starting at the bottom and working toward the whole up through each family structure. This is true during development to qualify the system as well as when assembling a particular deliverable end item for acceptance. The logic for this is pretty simple. Shuttle repair of the Hubble notwithstanding, it is not possible to routinely solve design problems that were not observed in testing during development after launch. We simply do not have the same flexibility in space systems that we have in ground, aeronautical, or ocean-going systems, systems with which we can interact with relative ease after they have departed their manufacturing site.

The intensity with which we will choose to pursue system testing is a function of the planned use of the product, its complexity, and available resources. This should not be a matter of chance or personal preference. One of the common failures in system test planning is to select those things that can be tested based on the current design. System testing should be based on system requirements. We should be trying to prove that the system performs as required in its system specification within the context of prior successful component and end-item test results. During the design development, based on specification content, we should be concurrently designing the system test. Where it is very difficult to accomplish a particular verification action at the system level, and it cannot be accomplished at a lower level, we should consider a design change to make it possible. This is, of course, the value of concurrent development. If we can catch these problems early enough it is economically feasible to develop the best possible system that is also consistent with the right test, manufacturing, logistics, and use processes.

There are relatively few good written sources for the testing of purely hardware systems or the hardware elements of grand systems. Those that the author has found useful are military standards. One of these offers a good superset for testing guidance within which most test situations will be a subset. This is MIL-STD-1540C, Test Requirements for Space Vehicles. MIL-STD-810D, Environmental Test Methods and Engineering Guidelines, and MIL-STD-1441, Environmental Compatibility Requirements for Space Systems, also provide useful content relative to environmental testing, a significant portion of the test problem for systems and lower-tier elements.

There is a good deal of good literature on the testing of computer software in the form of books written by Boris Beizer[1] and William Perry,[2] to name two respected authors in this field. Software cannot, of course, form a complete system in that it requires a medium within which to run, a computer of some kind. Given that the machine, its compiler, and the selected language are fully compatible, the software can be treated as an isolated system, however. The software must be thoroughly tested internally and integrated with the hardware system within which the computer operates.

Grand systems, composed of complex arrangements of hardware, software, and people performing procedures based on observations and circumstances (sometimes referred to by the troubling term *peopleware*), involve very complex system testing demands because it is so difficult to intellectualize all of the possibilities. We are commonly left with few alternatives but to select some small finite number of alternatives from the infinite possibilities and extrapolate across these few test cases. Given that the hardware, software, and procedural elements are each examined carefully throughout their range of variation, there is a reasonable expectation that system testing, so constituted, will be effective in proving compliance with system requirements. Whether or not we will succeed in uncovering all of the possible fault modes undiscovered analytically through failure modes and effects criticality analysis is another matter. In a complex system, we will be hard pressed to imagine all of the strings of activity that can occur in actual operation especially where the user is intent on getting every possible scintilla of capability out of the system and is not averse to risk taking to satisfy some borderline operational opportunity.

10.4.2 System test categories

DoD recognizes several categories of system test that can also be applied to a broader range of customers. The category boundaries are based on who does the work and the conditions in effect rather than on military parameters. In all of these cases, the understanding is that the components and end items that make up the system have been qualified as a prerequisite to system test.

10.4.2.1 Development test and evaluation (DT&E)

DT&E is accomplished by the contractor in accordance with a test plan that focuses on engineering considerations. It should be driven by the content of the system specification, and the testing accomplished should produce convincing evidence that the system to be delivered does satisfy the requirements in that specification.

10.4.2.2 Operational test and evaluation (OT&E)

OT&E is accomplished by the customer in accordance with mission needs that dictated the procurement of the system adjusted over the development life of the system for its evolving characteristics.

10.4.2.3 Interim operational test and evaluation (IOT&E)

If the DT&E reveals a need to make significant changes to the system and these changes will take some time to develop and implement, the residual assets from DT&E and any other end items produced to satisfy the customer's initial operating capability (IOC) requirement will not be in the correct configuration to use as the basis for the OT&E. This may lead the user to alter the original plan for OT&E to focus on those parts that are not affected by the changes early on and postpone test elements that are affected until the changes can be implemented. The resultant testing may be called IOT&E.

10.4.2.4 Follow-on operational test and evaluation (FOT&E)

As the final resources begin to become available to the user subsequent to the DT&E changes discussed above, the user may choose to transition to a final set of test-planning data that corresponds to availability of fully operational assets leading to identification of the test activity in the context of a follow-on activity. Another way to get into this kind of T&E program is that during OT&E user testing uncovers problems that did not surface during DT&E because of the different perspectives of the people and organizations responsible for DT&E and OT&E. DT&E plans are structured based on an engineering perspective, whereas the OT&E is structured based on a user's perspective. The user will set up test sequences focused on the missions that are known to be needed, and these differences may expose the system to modes not evaluated during DT&E. These tests may uncover system problems that have to be fixed and follow-on testing accomplished after the fixes are installed.

The cost of making the changes driven by OT&E testing may or may not be properly assigned to the contractor. In order to allocate the cost fairly, it must be determined whether or not the testing attempted is appropriate for a system as described in the system specification. It is possible that the user has extended the mission desires beyond the mission capabilities they procured. This problem can easily materialize where a procurement office acquires the system and an operational unit operates it. The people in the latter commonly have very little interest in the contractual realities within which the procurement agent must function. They are strictly focused on their mission and, if the system will not satisfy some portion of it, the system should be changed. It may even be the case that the system in question provides better capabilities than required but it still falls short of the capabilities that the user has extrapolated from the capabilities observed during DT&E that suggest other opportunities not yet satisfied nor covered in the contract. Users march to a different drummer than procurement people, and contractors must remain alert to ensure that they are not steamrollered into making changes that are actually out of scope for the current contract.

10.4.3 Test results applications

The results of system testing are first useful in verifying that the system complies with system requirements. In addition, the results of system testing are useful in confirming the validity of models used to create the system. Throughout the development process, while the system was an intellectual entity defined in a growing list of engineering drawings, models and simulations were crafted as part of the intellectual system description. These models and simulations were built based on some assumptions as well as rational assessments based on science and engineering. We should have done our best to validate these models and simulations as early in their life as possible, but the availability of the complete system is the first time that some of the features of these objects could be fully validated.

If system testing confirms that the models and simulations predict performance that is actually observed in operation of the system, then we have a very valuable resource on our hands. The reader may conclude that since the system is already built it is of little consequence whether or not the models agree with its operation so long as we are satisfied with system operation. Why don't we simply get rid of all of the other representations of the system? We are finished.

The reality is that this system will have a long life and in that long life it will experience phenomena that we may not understand immediately. If we have available to us a simulation that has been validated through system testing, then we can manipulate the simulation to help us understand the cause for the problem observed in a very controllable situation. Without these resources, we may be forced to conduct dangerous flight tests to observe performance aberrations. Also, the simulation will permit us to examine a wide range of situations in a much shorter period of time than is possible through actual operation of the system.

So, we should make a conscious effort to correlate the results of system testing with any models and simulations we may have future use for. In some cases they will not agree, and the cause could be either the simulation is incorrect or we could have interpreted the results of system testing incorrectly. In any cases where there is not good agreement, we should look for a cause on both sides of the divide.

10.5 Other forms of system testing

10.5.1 Quality and reliability monitoring

On a program involving the production of many articles over a long period of time, the customer may require that articles be selected from time to time to verify continuing production quality and reliability. This is especially valuable where the product cannot easily be operated to develop a history. A good example of this is a cruise missile. It would be possible to configure selected vehicles off the line for flight test purposes involving adding command control and parachute recovery systems, but, while these changes would permit the very costly vehicle to survive the test, they would invalidate the utility of the test. In order to fully test the missile we would have to cause it to impact a representative target and explode (conventional warhead assumed here, of course).

Military procurement of missiles for air-to-air, air-to-ground, ground-to-ground, and shipboard variations commonly involves purchase of blocks of missiles. Since the military cannot determine how many wars of what duration they will be required to fight, these missiles end up in storage bunkers and warehouses. The service may require that missiles be pulled from stock at random and actually flown to destruction from time to time. These actions, where the missiles work, encourage confidence in the store of hardware. These events also offer user units a great opportunity to conduct live fire training so these two goals can be connected to great benefit.

10.5.2 System compatibility test

Teledyne Ryan Aeronautical developed a complete unmanned reconnaissance system for the U.S. Navy in the late 1960s. This was a shipboard rocket-assist launched aircraft using an adapted ASROC booster installed on the vehicle located in a near-zero length launcher angled at 15 degrees to horizontal. Ryan had been supplying the Strategic Air Command with this equipment for air launch from DC-130 aircraft for several years at that point but had never completely outfitted a new user. Past operations grew into their then current capability in the presence of a complete Ryan field team with adjustments over time lost in the corporate knowledge. The program manager correctly sensed that there was a risk that the development team might not be capable of detecting problems from a purely intellectual perspective so he required the performance of a system compatibility test. The author was asked to design and manage this test because he had experience in the field with the system and all of its support equipment.

The series of tests uncovered many small problems with vehicle and support equipment design but did not disclose several very serious ones that were only exposed in the subsequent flight test. The principal value of the test was to confirm the soundness of the maintenance process as a prerequisite to entering flight test. Most of the problems that survived until flight test entailed operational aspects that could not be examined during the compatibility test.

Some things that were discovered:

a. The system was to be operated on an aircraft carrier at sea continuously for weeks at a time. When the vehicle returned to the ship after its mission it was recovered through parachute descent to water impact where it was picked up by the helicopters otherwise used for air–sea rescue on the carrier and placed on deck for postflight work that included freshwater decontamination of the aircraft and engine. An engine decontamination tank was provided to flush the external and internal parts of the engine with freshwater to remove the corrosive saltwater. When an attempt was made to place the engine in the tank, it was found that it would not fit. The problem was traced back to a conversation between the lead designer and the new engineer given the responsibility of adapting a previous tank design to this engine. The original tank design had been made for a J69T29 engine common to the Ryan target drone. The lead designer told the new engineer to go down to the factory floor where there were a number of the correct engines awaiting installation in vehicles. The new engineer, not knowing that there were several models in production with four different engines happened upon a group of larger Continental engines used in a different model that did not have the same external interface and dimensions and used this engine as the model for his design. Therefore, the adapter the new engineer created did not fit the J69T41A used in this model. It had to be modified.
b. A number of other incompatibilities were uncovered but the author has forgotten the details.
c. The test also acted as a technical data validation and numerous problems were discovered in technical data and corrected.

Some things that were not discovered:

a. The Continental engine representative said that the J69T41A engine would not accelerate to launch thrust after start without an engine inlet scoop. The launcher design team located a retractable scoop on the front of the ground launcher such that it could be retracted out of the way subsequent to engine start and not be struck by the rocket motor moving past during launch. During the compatibility test (which included a launch simulation but not an actual launch), the scoop worked flawlessly while the vehicle sat in its launcher with the engine running at launch rpm. Several weeks later, the system went to flight test operated out of Point Mugu, California. On the first launch, the rocket motor blast tore the engine inlet scoop completely off the launcher and threw it some distance. It was quickly concluded that the inlet scoop was not required and the system worked flawlessly for many operational launches from the USS *Ranger* in that configuration.

b. A special circuit was designed for this vehicle such that if the vehicle lost control carrier for a certain period of time, the circuit would restore to a go-home heading so as to regain a solid control carrier at shorter range. The only time this circuit was ever effective during operational use it caused the loss of the vehicle rather than the salvage of it. The remote pilot during that mission was experiencing very poor carrier conditions and kept trying to issue commands to turn the vehicle back toward its control aircraft (an E2 modified with a drone control capability). What was happening was that every time the carrier got into the vehicle momentarily, it reset the lost carrier go-home heading timer so that it never was triggered. Poor control carrier conditions caused the vehicle to go into a turn only during the very brief moments when the controller carrier got through. Since remote turns were not locking, these turns were almost nonexistent and the vehicle continued to fly farther from the controlling aircraft. The controlling aircraft could not follow the vehicle to reduce the distance, because the vehicle was overflying North Vietnam. Finally, the vehicle fueled out over Laos and crashed.

c. In preparation for launch, the jet engine was first started and brought up to launch rpm while other prelaunch checks were conducted. There was only a very brief time before too much fuel would be burned resulting in an unsafe launch so the launch procedure was stressful for the launch operator. Upon launch command with the jet engine running at launch rpm, the ASROC rocket motor was fired, shearing a holdback. The jet engine thrust vector was supposed to be balanced with the rocket engine thrust vector in the pitch axis so that the vehicle departed the launcher at the 15 degree angle at which it rested in the launcher. On the first launch at Point Mugu, California, the vehicle departed the launcher and roared straight up through an overcast and descended shortly thereafter behind the launcher in a flat spin crashing fairly harmlessly between the launch control building and the launcher. System analysis had incorrectly estimated the pitch effect of jet engine impingement on the rocket unit and elevator effectiveness at slow speed so the balance condition was beyond the control authority of the flight control system. It was unfortunate that a few weeks before a Ryan target drone had been ground launched from this facility and the ASROC unit, with an undetected cracked grain, had stutter fired blowing up the drone just off the launchpad on its way to falling through the roof of a hangar on the flight line some distance away. But more problematical was the fact that the base commander was in the control building watching the launch of the new vehicle and his wife and child were parked in the parking lot between the building and where the drone impacted. This probably made it a little more difficult to regain range safety approval for the next launch.

d. The rocket igniter circuit was supposed to have been completely isolated from the rest of the electrical system. It tested okay during the compatibility and EMI tests. On the last launch of the flight test program from the USS *Bennington,* the igniter was set off when the launch control operator switched on the fuel boost pump in preparation for jet engine start. The rocket fired without the jet engine operating and the vector sum drove the vehicle into the ocean just forward of the launcher. Subsequent EMI testing revealed a sneak circuit, which had not showed up in the previous launches and tests, that had to be corrected.

e. In loading a vehicle onto the launcher, one had to hoist the vehicle from a trailer in a horizontal attitude and then rotate it to 15 degrees nose-up to place it on the launcher rails. A special beam was designed with a traveling pickup operated by an air-driven wrench. This worked well during the compatibility test. On one launch on the USS *Ranger,* however, the user experienced a failure of the rocket motor to fire during an operational launch. This is one of the worst things that can happen to you with a solid rocket since you don't know if it may choose to ignite subsequently. The rocket could not be removed while the vehicle was in the launcher so the whole apparatus had to be moved into the hangar deck, where there were many aircraft parked loaded with fuel, and the aircraft hoisted from the launcher with the hoist beam pickup point set for a 15 degree angle. The beam was supposed to permit movement of the pickup point under load to a horizontal position but in this case it did not move. With the pickup point jammed too far forward, the vehicle had to be manhandled into a trailer where the misfired rocket could be removed and pushed overboard.

References

1. Beizer, B., *Software Testing Techniques,* Thomson Computer Press, Boston, MA, 1990.
2. Perry, W., *Effective Methods for Software Testing,* John Wiley & Sons, New York, 1995.

chapter eleven

Acceptance test planning analysis

11.1 The notion of acceptance

The customers of large systems are, or should be, very careful to assure that the products they acquire are precisely what they intended to purchase. Therefore, they should insist on some form of proof that the product does satisfy some criteria of goodness prior to accepting delivery of the product. The requirements that should have driven the design effort are a sound basis for that criteria. We have seen in the prior discussion of qualification how the requirements are the foundation for the whole qualification cycle. Those requirements were development, Part I, or performance requirements, depending on what standard you refer to. The requirements we are interested in for acceptance are variously called product, Part II, detailed, or design requirements. The performance requirements should be solution independent to the maximum extent possible in that they should describe the problem to be solved. The detailed requirements should be solution dependent in that they are describing features of the actual design.

The testing work done in association with development, that is, qualification, is normally accomplished only once unless there are significant changes that invalidate some portion of the results such as a major modification, change in the source of an end item or major component in the system, reopening of a production line with possible subtle differences in the resultant product, or an extension of the way the product is going to be used operationally not covered in the initial development. Acceptance testing, on the contrary, is commonly accomplished on each article, at some level of indenture, that becomes available for delivery, and successful completion of the acceptance test is a prerequisite to delivery to the customer.

11.2 Where are the requirements?

A program may use two-part specifications, one-part specifications, derivative specifications, or some combination of these. In the two-part scenario, the requirements that drive the acceptance test process are in the Part II specifications. In a one-part specification, the document includes both qualification and acceptance requirements where the qualifications requirements are generally more severe. The specification scenario in vogue within DoD at the time this book was being written involved preparation of a performance specification for the development process which drives the qualification testing followed by preparation of a derivative specification called a detail specification by adding content in a paragraph called design and construction and changing the flavor elsewhere. The performance specification content is subject to qualification while the detail content is subject to acceptance. Depending on how the document is prepared, the performance content may have to have associated with it a pair of verification requirements to cover both cases.

11.3 How do detailed requirements differ from development requirements?

Development requirements define the performance requirements and constraints needed to control the design process. They define the problem that must be solved by the design team. They should be solution independent so as not to constrain unnecessarily the solution space or encourage point design solutions. They should define the broadest possible solution space such that the resultant design solution will satisfy its assigned functionality contributing effectively toward satisfying the system need.

Detailed requirements are design dependent written subsequent to conclusion of the preliminary design. They should identify high-level conditions or measurements rather than defining redundant or detailed requirements. If the product satisfies these requirements, it should indicate the whole item is satisfactory. Table 11-1 derived from SD-15, Defense Standardization Program Performance Specification Guide, describes the differences in specification content in several major subject areas.

11.4 Conversion to verification requirements

The content of Section 3 of the detail specification must be converted into verification requirements for Section 4 just as in the case for performance specifications. And, here too, we should write the pairs together to encourage better Section 3 content. It is more difficult to write bad requirements when you must also define how you will prove that the design satisfies them. Also, as in performance specifications, it is possible to include the verification requirements in Section 4 of the detail specification, a special test requirements document, or in the acceptance test documentation. The author encourages including them in Section 4 of the specification, once again, to encourage good Section 3 content.

Table 11-1 Requirements Comparison

Specification requirements	Performance specification	Detail specification
Section 1 — Scope	No difference	No difference
Section 2 — Applicable Documents	As a rule, performance specifications have fewer references: they refer to test method standards; interface drawings, standards, and specifications; and other performance specifications	Design specifications use materials and part component specifications; manufacturing process documents; and other detail specifications as references
Section 3 — Requirements	Biggest differences between performance and design are in Section 3	—
1. General	States what is required, but not how to do it; should not limit a contractor to specific materials processes or parts when government has quality, reliability, or safety concerns	Includes "how to" and specific design requirements; should include as many performance requirements as possible but they should not conflict with detail requirements
2. Performance	States what the item or system shall do in terms of capacity or function of operation; upper and/or lower performance characteristics are stated as requirements, not as goals or best efforts	States how to achieve the performance
3. Design	Does not apply "how-to" or specific design requirements.	Includes "how-to" and specific design requirements; often specifies exact parts and components; routinely states requirements in accordance with specific drawings, showing detail design of a housing, for example
4. Physical Requirements	Gives specifics only to the extent necessary for interface, interoperability, environment in which item must operate, or human factors Includes the following as applicable: overall weight and envelope dimension limits; and physical, federal, or industry design standards that must be applied to the design or production of the item	Details weight, size, dimensions, etc for item and component parts. Design specific detail often exceeds what is needed for interface etc

Table 11-1 (continued) Requirements Comparison

Specification requirements	Performance specification	Detail specification
	Such requirements should be unique, absolutely necessary for the proper manufacture of the item, and used sparingly; an example would be the need to meet FAA design and production requirements for aircraft components	
5. Interface Requirements	Similar for both design and performance specifications; form and fit requirements are accep- table to ensure interoperability and interchangeability	Same
6. Material	Leaves specifics to contractor, but may require some material characteristics; e.g., corrosion resistance Does not state detail requirements except shall specify any item-unique requirements governing the use of material in the design of the item Such requirements should be unique, critical to the successful use of the item, and kept to a minimum; an example would be the mandated use of an existing military inventory item as a component in the new design	May require specific material, usually in accordance with a specification or standard
7. Processes	Few, if any, requirements	Often specifies the exact processes and procedures and procedures to follow — temperature, time, and other conditions — to achieve a result; for example, tempering, annealing, machining and finishing, welding, and soldering procedures
8. Parts	Does not require specific parts	States which fasteners, electronic piece parts, cables, sheet stock, etc. will be used
9. Construction, Fabrication, and Assembly	Very few requirements	Describes the steps involved or references procedures which must be followed; also describes how individual components are assembled

Table 11-1 (continued) Requirements Comparison

Specification requirements	Performance specification	Detail specification
10. Operating Characteristics	Omits, except very general descriptions in some cases	Specifies in detail how the item shall work
11. Workmanship	Very few requirements	Specifies steps or procedures in some cases
12. Reliability	States reliability in quantitative terms; must also define the conditions under which the requirements must be met; minimum values should be stated for each requirement, e.g., mean time between failure, mean time between replacement, etc	Often achieves reliability by requiring a known reliable design
13. Maintainability	Specifies quantitative maintainability requirements such as mean and maximum downtime, mean and maximum repair time, mean time between maintenance actions, the ratio of maintenance hours to hours of operation, limits on the number of people and level of skill required for maintenance actions, or maintenance cost per hour of operation. Additionally, existing government and commercial test equipment used in conjunction with the item must be identified Compatibility between the item and the test equipment must be specified	Specifies how preventive maintainability requirements shall be met; e.g., specific lubrication procedures to follow in addition to those stated under Performance; also, often specifies exact designs to accomplish maintenance efforts
14. Environmental Requirements	Establishes requirements for humidity, temperature, shock, vibration, etc. and requirement to obtain evidence of failure or mechanical damage	Similar to performance requirements
Section 4 — Verification	Must provide both the government and the contractor with a means for assuring compliance with the specification requirements	Same as for performance specifications
1. General	Very similar for both performance and detail; more emphasis on functional; comparatively more testing for performance in some cases	Very similar for both performance and detail; additional emphasis on visual inspection for design in some cases

Table 11-1 (continued) Requirements Comparison

Specification requirements	Performance specification	Detail specification
2. First Article	Very similar for both performance and detail; however, often greater need for first article inspection because of greater likelihood of innovative approaches	Very similar for both performance and detail. Possibly less need for first article inspections
3. Inspection Conditions	Same for both	
4. Qualification	Same for both	
Section 5 — Packaging	All detailed packaging requirements should be eliminated from both performance and detail specifications; packaging information is usually contained in contracts	

Since the Section 3 requirements in the detail specification will be much more physically oriented toward product features than those in the performance specification, it stands to reason that the verification requirements will be more closely related to specific equipment or software code features and measurements. We may call for specific voltage readings on particular connector pins under prescribed conditions, for example.

11.5 Acceptance test planning, procedures, and results data collection

The acceptance test plan and procedure for an item should be an expression of the verification requirements. A test condition will commonly call for a series of setup steps that define the input conditions followed by one or more measurements that should result. Many of these sequences strung together provide the complete test procedure. This process of requirements definition, verification requirements definition, and acceptance test planning forms a sequence of work moving from the general to the specific, from a condition of minimum knowledge toward complete knowledge which conforms to a natural thinking and learning pattern for people.

The verification requirements give us knowledge of what has to be done and with what accuracy. The test plans give us detailed knowledge of test goals and identify required resources (time, money, people and skills, material, supporting equipment, and the unit under test) to achieve those goals. The test procedures provide the persons responsible for doing the test work a detailed process so as to ensure that the results provide best evidence of the compliance condition. The data collected from the test should reveal that best evidence useful in convincing us and our customer that the product satisfies its detail requirements, if in fact it does. This process should not be structured to produce supportive results regardless of the truth, rather produce truthful results. If the results yield a conclusion that the product does not comply, then we have the same problem we discussed under qualification except that we are farther down the road. The possibilities are

a. If we conclude that the requirements are unnecessarily demanding, we could change the requirements and procedures to reflect a more realistic condition. This may require the cooperation of your customer, of course. This is an easy conclusion to reach because it can solve a potentially costly problem late in the development process but there is a very thin line between sound, ethical engineering and unethical expediency in this decision. We may have to retest in this case but if it can be shown analytically that the current design will satisfy the new requirements based on its performance in the failed test it may not be necessary.
b. Change the design to satisfy the requirements and retest.
c. Gain authorization from the customer for a waiver allowing delivery of some number of articles for a specific period of time that fail to satisfy the requirements in one or more specific ways. This waiver condition could exist for the whole production run, but, more commonly, it would extend until a predicted cut-in point where design changes will have been incorporated that cause the product to satisfy the original requirements.

11.6 Associate contractor relationships

The verification work across these interfaces for production purposes may entail special matched tooling jigs. For example, the builder of an upper stage of a space transport rocket that interfaces with the payload manufactured by another contractor may be required to fit a special payload tooling jig, supplied by the payload contractor, to the upper stage to verify that when the two come together at the launch site the mechanical interface will be satisfactory. Other tests will have to be accomplished by each contractor to verify electrical, environmental, and fluid interfaces in accordance with information supplied by the other contractor. Each mission may have some unique aspects so this interface acceptance process may include generic and mission-peculiar components.

The acceptance test requirements for these interfaces should be defined in the interface control document (ICD) and agreed upon by the associates. Each party must then translate those requirements into a valid test plan and procedure appropriate for their side of the interface. Ideally, each contractor should provide their test planning data to the other for review and comment.

11.7 Manufacturing or test-and-evaluation–driven acceptance testing

In organizations with a weak systems community and a strong manufacturing or test and evaluation community, the route to acceptance test planning and detail specifications may be very different from that described up to this point in this chapter. Manufacturing or test and evaluation may take the lead in developing the acceptance test plan prior to the development or in the absence of the development of a detail specification. In this case, the test requirements are defined based on experience with the company's product line rather than written product-specific requirements. Many companies do a credible job in this fashion but the author does not believe it offers the optimum process. In some companies the detail specification would be created based on the content of the acceptance test plan and this is backwards. The value of the work accomplished in creating the detail specification in this case is probably not very much.

11.8 Information management

As in qualification testing, there is a substantial need for data in the acceptance testing process. This stream is similar to that for qualification testing including:

a. A detail (or product) specification containing product requirements, a verification traceability matrix defining methods, level, and reference to the corresponding verification requirements, ideally located in Section 4 of that same document;
b. A test plan and procedure; and
c. A test report that may include nothing but data sheets.

Just as in the qualification testing situation, all of this documentation can be captured in the tables of a relational database. The data collection process could include inputs directly from the factory floor through integration of the verification database with the test support equipment.

11.9 Coordination between acceptance testing and special test equipment (STE)

The design of the STE that will have to be created to support the production line should be based on the acceptance test requirements. As discussed earlier, the acceptance test measurements should be end test points rather than intermediate points. Therefore, the STE may require additional capability where the acceptance testing is only a subset. This additional capability would be introduced to permit troubleshooting and fault isolation given that a particular acceptance reading did not measure up to the required value.

STE may also require the capability to provide the right conditions for factory adjustments or calibrations.

11.10 Relationship between technical data and acceptance

The acceptance testing process may be very close to what is required in the field to verify the product is ready for operational use after a prior application. The test procedure would simply be captured in the technical data with some possible changes reflecting the differences between STE and field support equipment and the hard use the product may receive between test events. The measurement tolerances should reflect a formal tolerance funneling plan including item and system qualification as well as acceptance test and as many as three levels of field test situations. Tolerance funneling calls for a progressively less-demanding tolerance as you move from the factory toward the ultimate user. The purpose is to avoid possible problems with product that fails field testing only to be declared acceptable at depot level (that could be the factory).

The qualification testing should be more severe than acceptance testing and require a tighter tolerance. A unit completing qualification testing generally is not considered acceptable for operational use. That is, it should not normally be delivered to the customer. The acceptance test levels are not intended to damage the item, obviously, in that they are a prerequisite for customer acceptance of the product. All testing subsequent to acceptance should apply the same measurements but accept a widening tolerance as indicated in Figure 11-1. Reading 1 taken at acceptance of the unit, if taken at any level of testing, would be acceptable. Reading 2 would be acceptable at field intermediate level and on the borderline for depot test, but it would not be accepted if the unit were shipped to the factory for repair and testing.

A story is offered to dramatize the problems that can go undiscovered. The author investigated several unmanned aircraft crashes while employed in the 1970s by Teledyne Ryan Aeronautical. On one of these, the AQM-34V electronic warfare unmanned aircraft had been launched from a DC-130 over a desert test and training range in Arizona. Fifteen seconds after launch it had dived directly to the desert floor and buried itself in the desert. When the author, then a field engineer for Ryan, reached the crash site he asked the troops to see if they could find the flight control box in the black hole because this was the unit that was supposed to have placed the vertical control axis on an altitude control mode 15 seconds after launch. Shortly, the box was found in surprisingly good condition. The unit was analyzed at the factory and the altitude transducer opened for inspection. It was found that there was no solder in one of the three electrical pin pots inside the transducer. The history on this serialized unit revealed that it had been sent back to the transducer manufacturer more than once for a problem reported in the field. It could not be duplicated in the factory so it was returned to stock

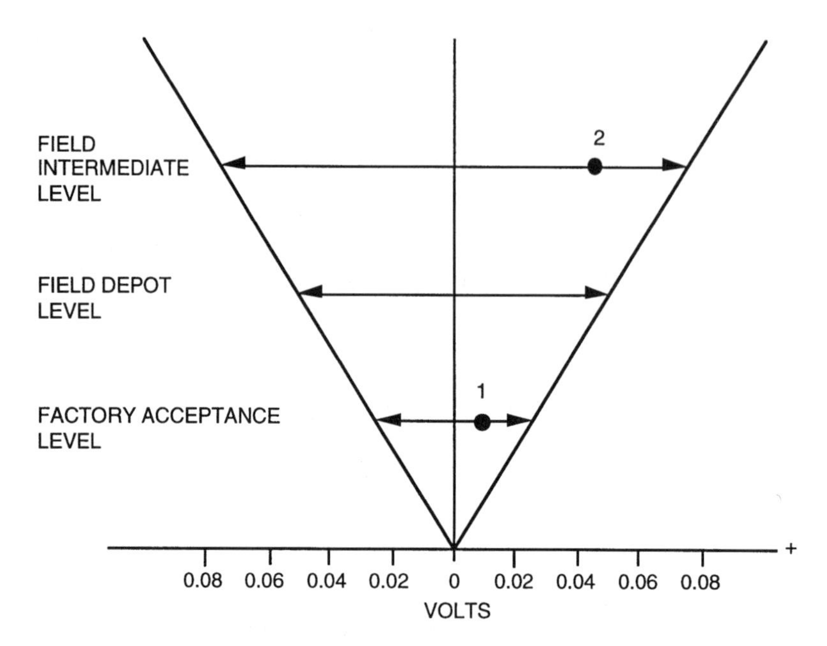

Figure 11-1 Tolerance funneling.

and eventually used in the flight control box in question. Apparently, the wire made sufficient contact with the pin in a static situation to pass all ground tests but when the vehicle hit the dynamic condition of an air launch pullout with its engine running, the contact of the unsoldered pin failed. The logic of the vertical control axis concluded that the altitude command was 0 feet above sea level and the vehicle's race to achieve that level was interrupted by the desert floor considerably above sea level.

The final answer to this specific problem was provided by vibration testing the flight control box and the uninstalled transducers as part of their acceptance test process. Granted, this is more of a binary condition and tolerance funneling is more properly concerned with a graduated response progressively less severe as noted in Figure 11-1. But, the reader can clearly see from Figure 11-1 that if the unit under test in the field fails to satisfy the field test tolerance, then it is unlikely that it will pass the factory or depot test. This preserves the integrity of the aggregate test process.

The example above also highlights another concern in acceptance testing. During qualification testing, it is normal to test items exhaustively for environmental conditions using thermal, vibration, and shock testing apparatus. It is not uncommon to forgo this kind of testing when we move to acceptance testing based on the assumption that the design is sound, having passed qualification. We should eliminate all environmental testing in acceptance only after considering whether or not there are manufacturing or material faults, like the unsoldered pins in the example above, that could go undetected in the absence of all of these environmental tests.

11.11 Postdelivery testing applications

It may prove useful to rerun the acceptance test on some products periodically subsequent to delivery. If the product may be stored for some time after delivery and before operational employment, there is going to be some maximum storage time after which certain components must be replaced, adjusted, or serviced. These are often called time change items. The design should group all of the time change items into one or some small number of cycles rather than require a whole string of time change intervals for individual items to reduce the maintenance and logistic response to a minimum.

When one of these time change limits on a particular article is exceeded, the user may be required to make the time change within a specified period of time subsequent to the expiration or be allowed to make the change whenever the article is pulled from storage in preparation for use. This is primarily a logistics decision unless failure to change out the time change items can result in an unstable or unsafe condition as in the case of some propellants and explosives. Subsequent to replacement actions, the product should be subjected to some form of testing to verify the continuing ready status of the article. This testing could be essentially the same as the acceptance test or some subset thereof.

11.12 Intercontinental acceptance

Some companies are beginning to apply networking and E-mail concepts to acceptance testing. One computer maker develops acceptance test scripts in its facility in Texas and these scripts are sent out over its lines to factories in several foreign countries where they are installed on test hubs and used to test computers coming down the line.

Imagine how these ideas can be applied in the future when Internet II becomes commercially available with its greater speed and versatility. Currently, a customer feels comfortable when all of the elements of its product come together at their prime contractor and everything is tested together in the same physical space. Creative contractors will find ways to apply data communications to link together the elements at different locations around the world in an acceptance test that is virtually the same as if all of the things were physically together on one factory floor. The payoff will be reduced cost of shipment of material and the time it takes to do it.

This same concept can be applied to the qualification process as well where computer simulations may interact with software under development and hardware breadboards distributed at development sites located in Cedar Rapids, Los Angeles, Yokohama, Moscow, and Mexico City. Concurrent development will then leap its current physical bindings. It is hoped, by then, we will have learned how to get the best out of the human component in this environment.

chapter twelve

Nontest item acceptance methods coordination

12.1 Organizational responsibilities

The whole acceptance process could be accomplished by a test and evaluation organization, manufacturing organization, or a quality engineering organization. In many organizations, this work is partitioned up such that two or even three of these organizations are involved. Design engineering could even get involved with some analyses of test results. Mass properties may require access to manufactured products to weigh them. Software acceptance may be accomplished by a software quality assurance (SQA) function independent of the QA organization. These dispersed responsibilities make it difficult to manage the overall acceptance process, assure accountability of all participants, and to acquire and retain good records of the results so that they are easily accessible by all.

In a common arrangement, the acceptance test procedures are prepared by engineering, and manufacturing is responsible for performing acceptance tests and producing test results. The tests are witnessed and data sheets stamped by QA. Demonstrations may be accomplished by manufacturing or engineering personnel and witnessed by quality. Some tests may require engineering support for analysis of the results. And, QA may be called upon to perform independent examinations in addition to their witnessing of the actions of others, which qualify as examinations in themselves.

An organizational arrangement discussed in Chapter 14 pools QA, test and evaluation, and system engineering into one organization focused on system requirements, risk, validation, verification, integration, and optimization. This organization is made responsible for all verification management actions, whether they be qualification or acceptance, and some implementation actions. This arrangement was conceived as a way of centralizing V&V process control that is very seldom fulfilled in practice. Commonly, acceptance test is well developed and implemented, once again, because of a strong and competent test and evaluation organization. The acceptance verification tasks accomplished by other methods are, however, easily lost in the paperwork. Our goal in this chapter is to find ways to bring the other methods of acceptance verification into the light and under the same scrutiny as test.

12.2 The coordination task

No matter the way your company has chosen to organize to produce product and verify product acceptance, the overall verification program coordination fundamentals cannot be ignored. Figure 12-1 illustrates one way to establish order in this area through a matrix. All of the acceptance tasks are listed on one axis of the matrix and the responsible organizations on the other axis. ATP, commonly an acronym for acceptance test plan, has the meaning acceptance task plan/procedure here to extend the kinds of tasks to include all four methods that may be applied.

In the matrix, we indicate the kinds of responsibilities required of these organizations for the indicated tasks. One example is shown. We could arrange these responsibilities in many different ways. This figure is not intended to support or encourage the application of serial methods in the context of a strong functional management axis of the matrix. The intent is that the product teams formed on the program would have membership by the indicated specialists (from these functional organizations) for the indicated reasons. In the case of manufacturing and related activities that dominate a program at the point where acceptance work is being accomplished, these teams may be organized around the manufacturing facilities and/or led by manufacturing people. During the time the acceptance task plans/procedures are being developed and approved, the program may still be functioning with engineering-dominated teams built around the product at some level of indenture.

12.3 Acceptance task matrix

Just as in qualification and test acceptance verification, we should have a master list of all acceptance verification tasks that must be applied to the product. This matrix will tell what method is involved and establish responsibility for accomplishing the work described in the acceptance portion of the integrated verification plan. There will be a tendency to focus only on the acceptance test tasks and we have to consciously consider all of the methods to be applied in the acceptance process.

12.4 Examination cases

12.4.1 Quality acceptance examinations

Two kinds of quality acceptance examinations exist: (1) witnessing of the actions of others to provide convincing evidence that those actions were in

ACCEPTANCE TASK	ENGINEERING			PRODUCTION	QUALITY
	D&A	SE	T&E		
PREPARE ATP	P		S		
APPROVE ATP		P	R	R	R
IMPLEMENT ATP				P	
ASSURE COMPLIANCE WITH ATP					P

P PRIMARY RESPONSIBLITY
R REVIEWER
S SUPPORTING ROLE
ATP ACCEPTANCE TASK PLAN/PROCEDURE
D&A DESIGN AND ANALYSIS
SE SYSTEM ENGINEERING
T&E TEST AND EVALUATION

Figure 12-1 Nontest acceptance verification method responsibilities coordination.

fact accomplished and accomplished correctly in accordance with written procedures and (2) specific examination actions performed on the product by quality personnel.

In the first case, a quality engineer observes the work performed by others and compared it with a work standard at specific points in the process or in a continuous stream from beginning to end. The quality engineer reaches a conclusion about the correctness of the performance and the recordings of the results throughout the work activity. The action observed could be a test, examination, demonstration, or analysis although tests are the most commonly witnessed activities. In order for the results to be disbelieved a conspiracy involving at least two people would have to be in place, a situation generally accepted as unlikely. Some very critical integration actions, especially in space applications may require a two-man rule to guard against the potential for one person to purposely or inadvertently introduce faults. It may be possible to convince a customer that one of these people is performing a quality function but the two-man action will likely have to be witnessed by a quality engineer as well.

The work of the quality technician or engineer, as a person who monitors correct performance of work done by others, is often implemented by the use of an acceptance test data sheet. The person performing the test enters test results on the sheet and it is witnessed by the quality engineer. Acceptance of quality is signified by a signature or initials and a quality stamp registered to that inspector. The resultant data sheet becomes part of the acceptance verification evidence.

The acceptance process for a product may also include direct quality assurance action as well as these kinds of indirect actions. Examples of direct actions are

a. At some point in the production process, when a particular feature of the product is best exposed and the earliest in the production process that feature is in its final, unalterable configuration, a quality engineer may be called upon to make a measurement or observation of that feature against a clearly defined standard. This could be done by unaided visual means, use of a simple tool like a caliper and ruler or gauge blocks, application of a tool designed to give a clear pass/fail indication visually or by other means, or comparison of the feature with the related engineering drawings.
b. An inaccessible feature of the product is subjected to an X-ray inspection and observed results compared with pass/fail criteria.
c. A multiple-ply fiberglass structure is examined by a quality engineer for freedom from voids by striking the surface at various places with a quarter (U.S. 25-cent piece) and listening to the resultant sound. One experienced in this examination technique can detect voids with great reliability at very low examination apparatus cost.

It is granted that the difference between the two kinds of QA described above is slim, and that the two kinds merge together as you close the time between the original manufacturing action and the quality examination. In the first case the action is being observed real time. In the latter case, the observation is delayed in time from the action being inspected. Generally, quality examinations should be made in real time, but this is not always possible or, at least, practical. Some examples follow:

a. *Work Space Constraint.* It may not be possible for two people to locate themselves such that the manufacturing person can do the work while the quality person is observing.
b. *Safety Constraint.* During the time the action is being performed there is some danger for the person performing the work whereas the article is relatively safe after the act is complete and the effects of the work can be clearly observed after the fact.
c. *Cost Constraint.* Each examination costs money, and the more we introduce, the more money they cost. Also, each examination is a potential source of discontinuity in the production process if it is necessary for manufacturing work on the product to cease during the examination. If the examination drags out in time, manufacturing labor can be wasted as workers remain idle. The cost of making sure it is right is reflected in the manufacturing cost and may, in the aggregate, show up as a manufacturing overrun in the extreme case for which manufacturing must answer. This is a common area of conflict between quality and manufacturing and a sound reason why quality must be independent of the manufacturing management hierarchy.

Where the manufacturing action and examination action are close coupled, it is generally obvious what manufacturing action or actions are covered by the examination. Where some time does elapse between a manufacturing action and the related quality examination, no matter the cause, it should be noted in the quality examination procedures or in some form of map between these data sets precisely to which manufacturing action or actions the examination traces.

Clearly, these examination points must be very carefully chosen in the interest of cost and effectiveness. Their location in the production process should be selected for a lot of the same reasons used when deciding where to place equipment test points for troubleshooting or instrumentation pickup points for flight test data collection in a new aircraft. We are, in effect, instrumenting the production process for QA measurements. Ideally, these measurements should be capable of statement in numerical form such that a metric approach can be applied over time or per unit of articles and these numbers used to chart process integrity using statistical process control techniques.

12.4.2 *Engineering participation in acceptance examination*

The need for routine participation of design engineers in the acceptance examination process should be viewed as a symptom of failure in the development process. A example of where this can happen is the development of the engineering for an item using sketches rather than formal engineering drawings in the interest of reduced cost in the early program steps with the intent to upgrade the engineering between completion of qualification and high-rate production. Should it later develop that there are no funds to upgrade the engineering but the customer is enthusiastically encouraging you to proceed into production with full appreciation of the potential risk, it may be possible for very qualified manufacturing people to build or assemble the items incorrectly just because of ambiguities in the engineering. Without engineering participation, this kind of problem can lead to unexplained production line problems that are finally traced to a particular person being out sick, retired, or promoted. That person had determined how to interpret the engineering and that knowledge did not get passed along through any written media such that the replacement could unfailingly do the work correctly.

Engineers could be brought into the examination process but it should be recognized as a short-term solution to a problem driven by poor engineering documentation. As these problems are exposed, engineering should work those off as budget permits such that at some time in the future, manufacturing and quality people can properly do their jobs unaided by direct engineering support.

12.4.3 *Software acceptance examination*

Computer software is a very different entity than hardware and its acceptance is done in a very different way. If we thoroughly test a software entity in its qualification process and can conclude that it is satisfactory for the intended application, it would seem that we have but to copy it as many times as we wish to manufacture it. Each copy of the software should have precisely the same content as the original proven in qualification. The reliability of the copying process is very high and most people would accept that the copied article is identical to the original. In order to make absolutely certain that this is the case, the copied article could be subjected to a comparison with the original. Any failures to compare perfectly should be a cause for concern in that the two files should be identical. Alternatively, the copied article could be subjected to a checksum routine and the results compared with the original checksum.

The knowing customer's concern for software acceptance might be more closely related to uncertainties in the integrity of the developer's process than the simple correlation between master and delivered product. The question becomes, "Is the current software master the same code that passed qualification?" This is especially a concern after one or more changes to the product subsequent to its initial development, qualification, and distribution. These concerns can best be answered by the developer applying sound software configuration management principles in the context of a well-managed software library with access rigorously controlled and histories well kept. There should be a believable, traceable history of the product configuration and actions taken to the master during its life cycle.

12.5 Demonstration cases

12.5.1 Logistics demonstrations

One of the key elements in many systems is the technical data supplied to the customer to support operation and maintenance of the system. This data is generally accepted as a lot where it is delivered in paper document form. The customer may require 100 copies of each manual and accept them as a lot based on the technical data having been verified and all problems discovered during verification having been resolved. The words validation and verification are used a little differently here. The validation process often occurs at the contractor's facility and the procedures called for in the data are accomplished by contractor people, perhaps using qualification and system test articles. The verification work is then done at a customer facility when sufficient product has been delivered in the final configuration subsequent to FCA/PCA to support the work.

It is entirely possible and very common that the verification work will uncover problems not observed during validation. One of the reasons for this is that the people doing the work in verification are engineers very familiar with the product as a result of their work over a period of what may have been several years. These people may see things in the procedures that are not there while someone from the customer's ranks during verification will not be able to complete the same procedure because the equivalent knowledge level has not been achieved. The data may have to be changed in validation to make it more specific and complete so that anyone who has completed some prescribed level of training can complete the work.

The author discovered this lesson the hard way as a field engineer with the Strategic Air Command at Bien Hoa Air Base in South Vietnam. He had been deployed in support of a new model of the Ryan series of unmanned aircraft called a model 147H. During flight test he had found a problem in the compartment pressurization test procedure in technical data and rewrote the whole procedure. Apparently, it was never validated other than mentally or analytically because of the pressure of many more serious problems. Apparently, it was never verified at Davis Mounthan Air Force Base either where the unmanned aircraft group was headquartered and from which the vehicle had been deployed. During the buildup of the first vehicle one evening after deployment, one of the troops came into the tech rep trailer and very irately asked who the idiot was that had written the pressurization test procedure. The author owned up that he was the idiot when he found that in order to attach the compartment pressurization test source pressure hose one of the compartment doors had to be open violating the pressure integrity of the compartments. It took the rest of the night to rewrite the procedure and run the test finally verifying it.

12.5.2 Flight demonstration

When an aircraft manufacturer ships its product it flies to its destination. This could be considered a final demonstration of the airworthiness of the aircraft. Each aircraft would have previously have had to complete a series of flights to prove this, of course. The first flight of the model will have to go through a complete flight test and FAA certification process involving instrumented flights but the routine delivery of ship 56 will not have to undergo this whole battery of tests. The author would consider these kinds of flights demonstrations rather than tests since the do not involve extensive instrumentation, rather just the product being used in accordance with a predetermined plan. The pilot is in effect demonstrating the capabilities of the aircraft and that they match the required capabilities.

12.6 Analysis cases

In general, test results are much preferred to analysis results for customer acceptance. However, some of the tests that are applied may produce results that are not immediately obvious as to whether or not the product satisfies the requirements. In these cases analysis of the data may be required. The analysis tasks, just like the verification tasks mapping to the other methods, should be listed in our task matrix and identified in terms of who or what organization is responsible. The analysis task should be described in the acceptance analysis portion of our integrated verification plan.

Our work scheduling for the manufacturing process should identify these acceptance analyses causing the assigned person(s) to accomplish them producing a conclusion documented in some way for inclusion in the acceptance item record.

chapter thirteen

Product verification management and audit

13.1 The second stage of verification

The qualification process is intended to prove that the design synthesized from the item development requirements satisfies those requirements. The development requirements, captured in a development, Part I, or performance specification, were defined prior to the design effort so as to be design independent permitting the person or team responsible for synthesis to consider alternative competing design solutions and make a selection of the best overall solution through a trade study or other organized alternative evaluation process. This series of verification actions is only accomplished once to prove that the design is sound, and the evidence supporting this conclusion is audited on a DoD program in what is called a functional configuration audit (FCA). The tests accomplished during this work commonly stress the item beyond its capabilities to expose the extent of capabilities and extent of margins available in normal operation of the item and to validate the models used to create the designs. These stresses reduce the life of the unit and may even make them immediately unusable as a result of a successful qualification series. Therefore, these same inspections cannot be applied to every item produced for delivery since the delivered items would not have their planned service life at time of delivery.

What form of inspection might then be appropriate for determining whether or not the product is acceptable for delivery? This represents the second stage of verification, to prove that the product is a good manufacturing replication of the design that was proved sound through qualification. The acceptance process is accomplished on every article produced, on every product lot, or through some form of sampling of the production output. The process leading to the first manufactured item is subject to audit on DoD programs in what is called a physical configuration audit (PCA). Some customers may refer to this as a first article inspection process. The word inspection is used in this chapter to denote any of the four methods already discussed: test, analysis, examination, and demonstration.

13.2 The beginning of acceptance verification

The acceptance process should have its beginning in the content of the Part II, product, or detail specifications. This kind of specification is created based on a particular design solution and gives requirements for the physical product. This may include specific voltage measurements at particular test points under conditions defined in terms of a set of inputs, for example. Measurements will have to be made during a test and compared with requirements. As an alternative to two-part specifications, you may choose to use one-part specifications. In this case the specification may have to have a dual set of values or tolerances, one for qualification and another for acceptance. This difference is very important for the reason noted above. Subsequent to testing, the item must have a full lifetime before it. In either specification case, the beginning of the acceptance process should, like qualification, be in specification content.

Unfortunately, this is not the case in many organizations. In many organizations, the acceptance test procedures are prepared independently of the specifications based on the experience of test and evaluation and QA department personnel and their observations of the design concept. Subsequently, if a Part II specification is required, the test procedures are used as the basis for it. This is exactly backwards. It is analogous to accomplishing the design of a product and then developing the specification. The Part I specification is the basis for the design of the product and the qualification process while the Part II specification is the basis of the design of the acceptance process. If you first design the acceptance test process, then it makes little sense to create the requirements for the process subsequently. So, a fundamental element of acceptance management should be to insist on acceptance verification requirements as a prerequisite to acceptance planning activity.

Figure 13-1 illustrates the overall qualification and acceptance processes. In this chapter we will focus on the blocks with the heavy border beginning with the development of the detailed (Part II or product) specification. This process includes an audit of the acceptance evidence called, in DoD parlance, a PCA, but the process continues as long as the product remains in production.

13.3 The basis of acceptance

When dealing with a large customer like DoD or NASA, the acceptance process should be driven by what provides customers with convincing evidence that each article or lot of the product is what they want, what they paid for. This evidence pool should also be minimized based on end-to-end parameters, system-level parameters, and parameters that provide a lot of insight into the condition of the product. So, of all of the things that can be inspected on a product, we have to be selective and not base our evaluation on the possible, rather on the minimum set of inspections that comprehensively discloses the quality and condition of the product.

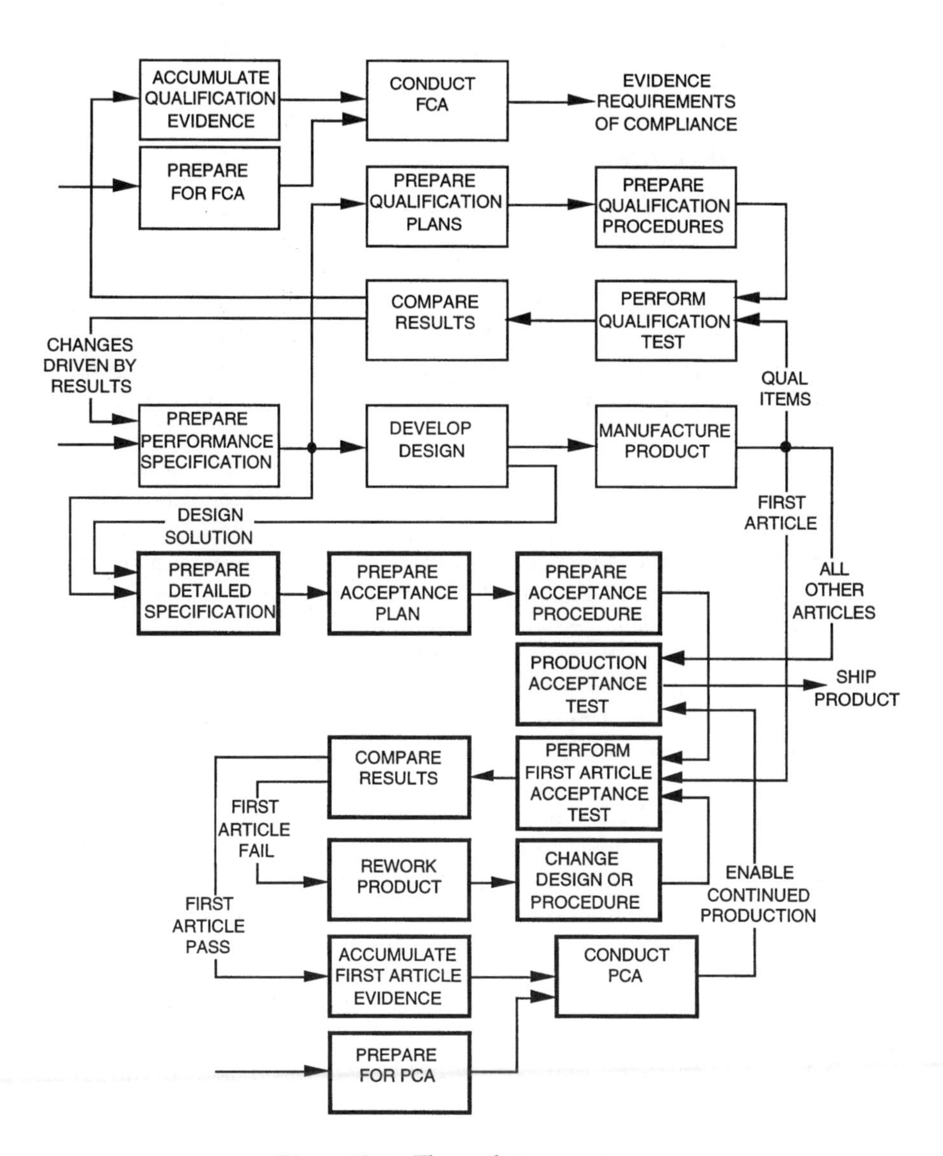

Figure 13-1 The path to acceptance.

Some people feel that the content of detail (Part II or product) specification should be determined prior to the design. That, frankly, is impossible. As shown in Figure 13-1, we develop the performance (Part I or development) specification as a prerequisite to the design. It should be design independent. The detail specification must contain design-specific information because it is focused on actual observations of the product features, voltage measurements, physical dimensions, motions, and so forth. Therefore, the content of detail specifications must be created subsequent to the development of at least the preliminary design with a possible update after the detailed design has been formulated.

13.4 Acceptance documentation

The suite of documentation needed as a basis for a sound item acceptance inspection includes the following:

a. *Detail Specification.* Table 11-1 suggests valid content of the detail specification relative to a performance specification. This content could be integrated into the same specification used for development. These specifications should be prepared during the detailed design phase.
b. *Quality Inspection Plan.* Quality Assurance should fashion an inspection plan based on the evolving design concept that identifies specific areas that should be inspected and the most appropriate times and conditions for those inspections in the evolving manufacturing plan. The goals of each inspection should be clearly defined.
c. *Acceptance Test Plan.* Assuming that the product has some functional capabilities rather than simply a physical existence, a test of some kind will be appropriate using instrumentation of some kind to convert reactions to a form suitable for human viewing. Each of the tests to be performed should be identified and defined in terms of goals, resources, budget, and schedule needs.
d. *Acceptance Procedures.* Each inspection identified in the plan should have developed a procedure telling how to accomplish the work. This information may be integrated into a combined plan and procedure as described in Chapters 6 and 11 also including all verification work.
e. *Acceptance Reports.* The acceptance reports may be as simple as sheets for recording values observed in tests and quality examinations or may entail computer recorded reading that can be reviewed on the screen of a computer terminal. The results should be collected for each article. The customer may require delivery of this data or require that the contractor maintain this data on file for a minimum period of time.

The program schedule should identify these documents and when they should be released. Responsible managers should then track the development of these documents and apply management skill as necessary to encourage on-time delivery of quality documents. Each document should pass through some form of review and approval.

13.5 Management of the work

All of the inspections work should appear on program planning and schedules for each article. Management should have available to them cost schedule control system and manufacturing tracking data indicating progress for each article or lot through the manufacturing process. As impediments are exposed, alternative corrective actions should be evaluated and a preferred approach implemented to encourage continued progress on planned work.

The aggregate of the work that must be managed in the acceptance process falls into two major components. First, the products examined at various stages of the production process within the producing company and at all of its suppliers. In these examinations, QA persons determine whether or not required manufacturing steps were performed and performed correctly to good result. If everything is in order in the traditional paper-oriented process, the QA person signs his or her name on or stamps a document to indicate acceptability. The other major inspection is in the form of acceptance testing where functional capabilities are subjected to input stimulus combinations and corresponding outputs compared with a standard. Commonly, checklists are used to record the results. The test technician records values observed and a QA inspector initials or signs the list to indicate compliance with written procedures to good results.

Management must encourage the development of clear manufacturing and quality planning data as well as a sound achievable manufacturing schedule that defines the work that must be done and when it must be done including the quality inspections and the acceptance tests. These activities have to be integrated into the manufacturing process and coordinated between the several parties. In some organizations manufacturing looks upon the inspection process as an impediment to their progress and this wrong-headed attitude can be encouraged where an attempt is made to butter the inspection process onto a previously crafted manufacturing plan that did not account for the time and cost of the inspection process.

The reader should know what is coming next at this point. Yes, we need a team effort. Here is another situation where several specialized persons must cooperate to achieve a goal that none of them can achieve individually. The overall manufacturing and inspection process (word used in the broadest sense to include test) should be constructed by a team of people involving manufacturing, quality, material, tooling, and engineering. Ideally, all of these people would have participated in the concurrent definition of mutually consistent product and process requirements followed by concurrent development of mutually consistent product and process designs. All of the planning data would have been assembled during the latter. Then, the management process is a matter of implementing the plans in accordance with planned budget and schedule constraints and adjusting for changing conditions and problems.

The teams effective in the production phase ought to be restructured from the development period as illustrated in Figure 13-2. Figure 13-2a shows the suggested structure in development with several integrated product teams (IPT) on each development program along with a program integration team (PIT) and a business integration team (BIT).

An enterprise integration team reporting to the executive (by whatever name) acts to optimize at the enterprise level providing the best balance of

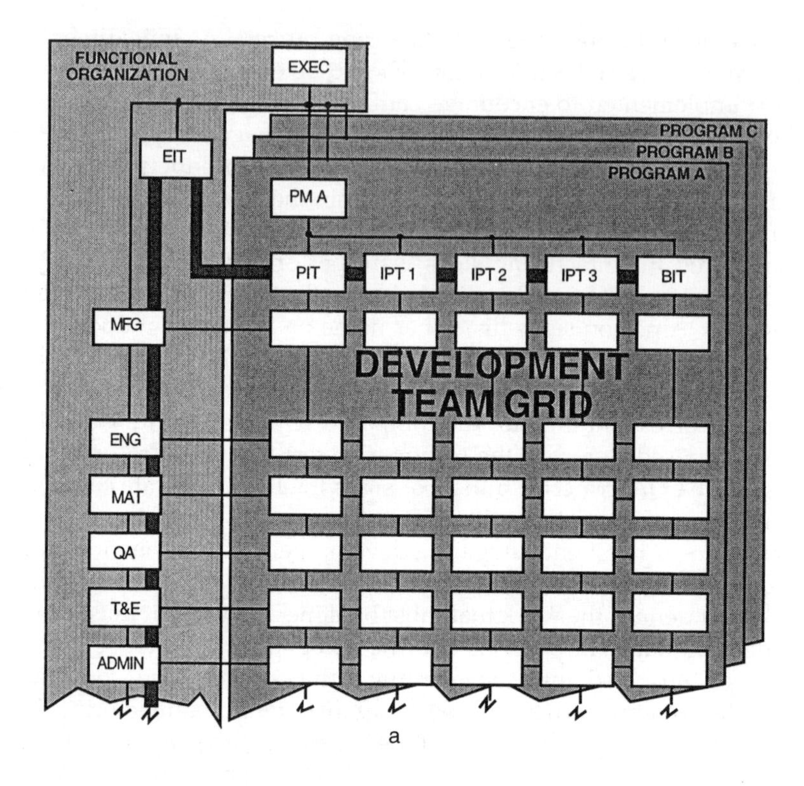

Figure 13-2 Development-to-production organizational transform. (a) Development phase; (b) production phase.

enterprise resources to the programs based on contractual obligations, available enterprise resources, and profit opportunities. This integration work is denoted by the heavy concurrent development bond joining the several activities. The PIT does integration work across the several programs through the PITs, and the PIT on each program carries on the integration work within the program. During development, the PIT acts as the system agent across the teams optimizing at the program product and process level. As the product maturity reaches a production-ready state, the IPTs can collapse into the PIT with continued BIT (contracts, legal, etc.) program support. Production is managed through manufacturing facility managers under the direction of an enterprise functional manufacturing manager with the enterprise integration team (EIT) optimizing for balanced support to satisfy program commitments. The PIT remains available for engineering in support of production changes and engineering change proposals. PIT specialists support facility teams as required by those teams.

In this arrangement, it was a program responsibility to develop suitable plans and procedures to accomplish acceptance examinations and tests in the manufacturing environment. The production responsibility for particular items was assigned to particular manufacturing facilities and vendors which

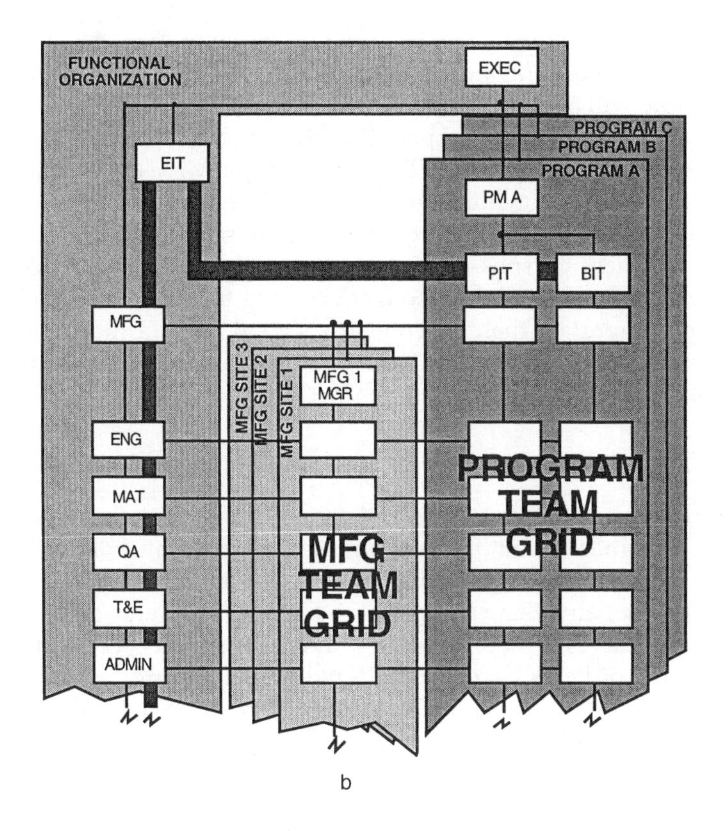

Figure 13-2 (continued)

are all treated as vendors by the program. These facilities must carry out the planned acceptance work as covered in the approved plans and procedures.

13.6 FRACAS

13.6.1 Acceptance ethics

A key element of the production process is a means to acquire information on failures to satisfy acceptance criteria. When items fail to satisfy inspection requirements, the item should not be passed through to the customer without some kind of corrective action. A few companies have gotten themselves in great difficulty with DoD in the past by treating failed items as having passed inspection requirements. In some cases, the tests called for were not even run in the interest of achieving schedule requirements. In other cases, failures were not reported and items shipped. Clearly, it is necessary to engage in deceit to put it mildly in order to perform in this fashion.

Management must make it clear to the workforce that integrity of acceptance data and resultant decisions is important to the company as well as to the people who are doing the work. A company that falsifies records not only damages its customers and its own reputation in time, it also damages those who work there and are required to perform in this fashion. An ethical person simply will not work in this environment. So, either a person's ethics are attacked or the person is replaced with someone who does not have a problem with this arrangement. Clearly, the company is a loser in either case. The moral fiber of the organization has been damaged.

The two causes for this kind of behavior are schedule and budget difficulty. Management persons throughout the chain dealing with acceptance should be alert for those doing the work coming to believe that delivery on time is more important than the company's good name. Unfortunately, this problem generally arises from management encouragement (intentional or otherwise) rather than grass roots creativity.

13.6.2 FRACAS implementation

The organization needs an effective FRACAS to provide management with accurate information on failures to satisfy acceptance requirements. Figure 13-3 illustrates a representative system for discussion. Failures can occur, of course, anywhere in the production stream, and the acceptance examination process should detect first that a failure has occurred and second what item or items have been affected by that failure. So, the first step in the FRACAS process is failure detection whereupon it must be reported in some formal way.

A failure can be caused by a number of different problems and we should first determine the root cause before settling upon a corrective course of action. We should no more decide a corrective action without understanding the problem here than we should seek to develop a design solution to a problem which we do not yet understand. A fishbone, cause-and-effect, or Ishikawa diagram is helpful in exhaustively searching for alternative causes. A Pareto diagram may be useful in prioritizing the corrective actions planned. Figure 13-4 illustrates these diagrammatic treatments and the reader is encouraged to consult a book on process reengineering, total quality management (TQM), or continuous process improvement for details.

Briefly, the Ishikawa diagram offers a disciplined approach for identifying the effect (failure) one wishes to investigate and possible causes for that problem while the Pareto diagram shows us that a relatively few probable causes generally drive the negative effect we wish to avoid. In that we seldom have all of the resources we wish were available, we are often forced to prioritize our actions to those areas that will produce the biggest positive effect.

Figure 13-3 suggests that there must be only a single cause of a failure and that is not always the case. It may be possible to correct for a given failure in a number of ways. It may be possible to implement a design change, a manufacturing process change, or a procedural change to correct for an observed failure or some combination of these. Our goal should be to select

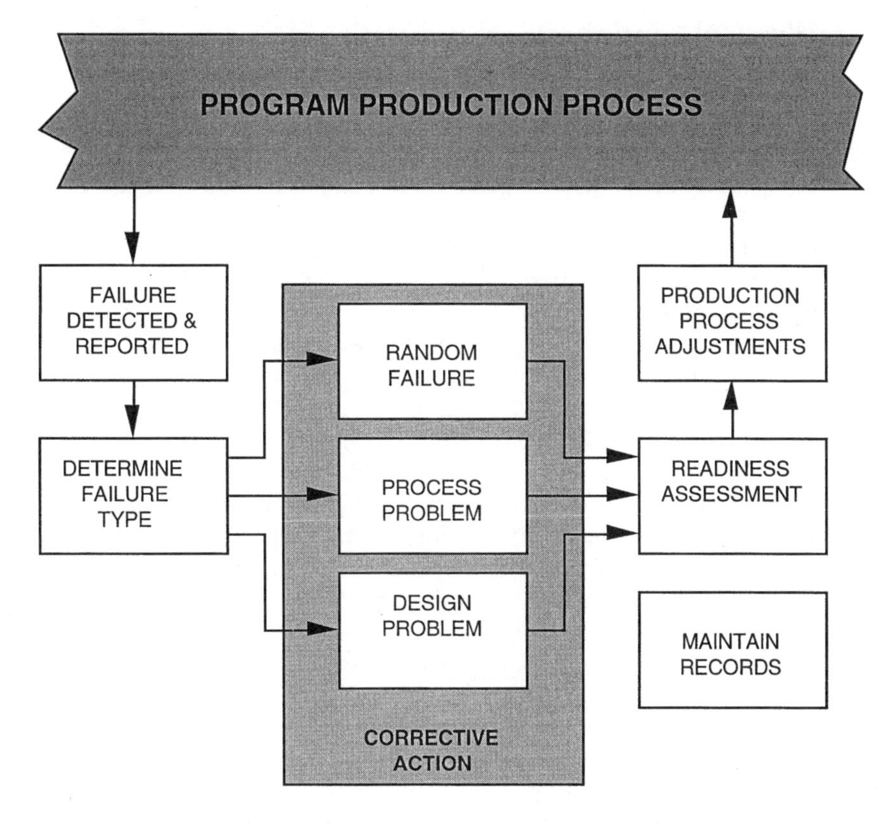

Figure 13-3 FRACAS flow.

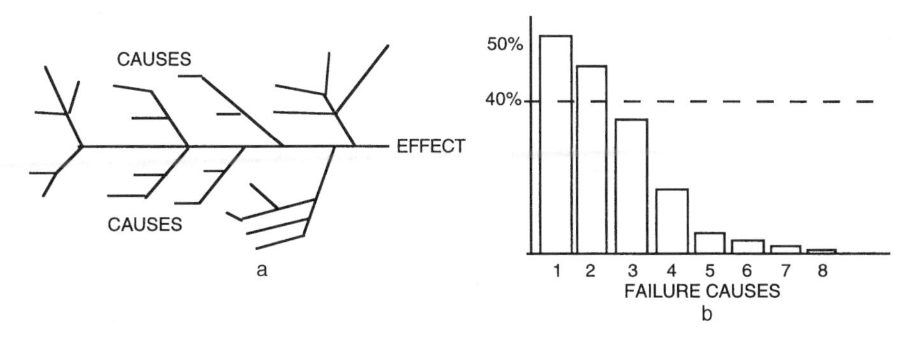

Figure 13-4 Supporting techniques. (a) Ishikawa (fishbone) diagram; (b) Pareto diagram.

the best corrective action in the aggregate generally in terms of life cycle cost effects but on some programs and in some phases near-term cost may be more important at the time.

13.7 Physical configuration audit (PCA)

The PCA provides a way to reach a conclusion about the quality of the manufacturing process as a prerequisite to committing to high-rate production using that process. If there are problems in that process that have not yet been exposed, it is better to expose them earlier than later. Since PCA is performed on the first production article, one cannot get any closer to the beginning of production. It should be pointed out that a sound development process up to the point of entering production of the first article will have a positive effect on reducing the risk that problems will occur in the production process. Another view of this translation was provided by W. J. Willoughby in his transition from development to production templates documented in NAVSO P-6071 and other places.

PCA attracts all of the evidence collected about the production of the first article into one place where it can be audited by a team of customer and contractor people against the requirements. The result desired by the contractor is that the evidence will be convincing that the product satisfies the detail requirements, conforms to the engineering drawings and lists, and was manufactured in accordance with the approved planning data.

13.7.1 PCA planning and preparation

As in the FCA, the contractor should develop an agenda subject to customer approval, and prepare and organize the materials to be used at the audit. In that the audit covers a physical reality rather than the intellectual entity audited at FCA, you will have to deal with things as well as paper in the audit. An audit space close to the production area will encourage the most-efficient use of time during the audit. The first article should be made available in this space for direct visible examination. All of the acceptance report data should be available in paper form or via computer terminals depending on the format of this data. The exact arrangements that the contractor makes should be based on the mode of audit selected by the customer.

13.7.2 PCA implementation

The customer may require any one of three different modes for audit presentation. The customer preference should be known to the contractor at the same time the agenda is crafted because the different modes encourage very different audit environments. The contractor should also develop and reach agreement with the customer on the criteria for successful completion of the audit which includes clear exit criteria.

13.7.2.1 The physically oriented audit

One of the three modes involves making the product available for customer examination and comparison with the engineering drawings. A U.S. Air Force customer may populate the audit with several staff noncommissioned officers with the right technical skills for the product line. They disassemble the product and compare the parts with the engineering drawings confirming that the product satisfies its engineering. They may also conduct or witness functional tests on the product to ensure that it satisfies test requirements.

13.7.2.2 The paper-dominated audit

The acceptance process generates a considerable amount of evidence in paper or computer media. If the customer elects to apply this mode of audit it is a totally different kind of people attracted to the audit. They will be people with good research and analytical skills rather than good practical skills. They will review all of the evidence in one of several patterns to assure themselves that the product accurately reflects its controlling documentation. The auditors may selectively evaluate the evidence looking for inconsistencies. If they find one, they will search deeper; if more are uncovered, they may audit the whole set of evidence. If, on the other hand, no problems are found in the first layer, the audit terminates with a good report. Alternatively, the audit may focus on only one aspect of the evidence such as qualification test reports using the progressive technique just discussed. A third approach is to simply plan to audit all of the evidence in some particular pattern.

13.7.2.3 Combined mode audit

A customer could bring in both populations discussed above and perform both kinds of audit. This audit could be coordinated between the two groups to apply a progressive approach or the plan could be to audit all data.

13.7.3 Post-PCA activity

The results of the PCA should be documented in minutes containing identification of any remaining problems and a timetable for resolving them. The precise mechanism for final closure should be included. These actions may be fairly complex entailing changes in specifications, changes in design, changes in acceptance test plans and procedures, and changes in manufacturing and quality planning. It will be necessary to track all of these changes while they are in work and coordinate the package for integration into subsequent production articles. Some of the changes may not be possible in time to influence near-term production unless production is halted for an extended period. The adverse effects of this on continuity of the program team is generally so negative that the decision will be made to continue production with one or more deviations accepted by the customer for a brief period of time until the changes can be fully integrated into the design and planning data. Those units produced under a deviation may have to be modified in the field subsequent to the incorporation of the complete fix in the production process. If so, modification kits will have to be produced, proofed, and installed in the field to bring these units up to the final configuration. If the changes are sufficiently extensive, it may even be necessary to run a delta PCA on the first article through the upgraded production line.

13.8 Software acceptance

Software follows essentially the same acceptance sequence described earlier. The fundamental difference is that software does not have a physical existence except to the extent that it resides on a medium that will be delivered. The software will be delivered on tape, disk, or other recording media and it is this media that must be accepted. Once the computer software has been proved to accomplish the planned functionality, the acceptance test may be as simple as a checksum or as complex as a special test involving operation of the software in its planned environment with actual interfaces and human operators at work in preplanned situations similar to the anticipated operating environment.

chapter fourteen

Process validation and verification

14.1 Is there a difference?

In Chapter 1 a Venn diagram was offered (Figure 1-11) suggesting that the program requirements universe included not only the product (or program-peculiar) specifications content but the process documentation as well. These requirements will be captured in documents called policies, plans, and procedures. Should we exempt these process requirements from validation and verification? It is suggested that the same situation is present here as in product requirements.

It is possible to create process controls such that they impose cost and schedule risks. It is very common that companies accept wildly unreasonable schedules offered in government requests for proposal (RFP) in order to get the contract. The author recalls that the advanced cruise missile program endured two proposal cycles in the early 1980s. An initial operating capability (IOC) date was defined in the first cycle and it was very optimistic and risky. After many months of study and proposal preparation, the competing contractors turned in their proposals and the Air Force decided not to accept any of them. Employees of General Dynamics Convair reasoned that they had won according to the selection criteria and Boeing was supposed to have won. The Air Force put another RFP on the street with the same IOC as the first one and, after several more months, GD Convair was selected. Never has there been such a rapid transform between elation and depression as among the GD Convair team members upon hearing the news of the win. You could see it on the faces of the people who had survived two proposal cycles over a period of many months of overwork. First, the eyes opened wide amid a face full of great satisfaction and an utterance, "We won!", followed almost immediately by a hollow look of despair and surrounded by the expanding inaudible wave, "Oh my God, we won." That program finally achieved its IOC quite a bit later than the contract required following a tremendous development effort that had to deal with multiple simultaneous technology challenges and overconstraining conditions leading to one or more situations where the technical and programmatic solution space was nearly a null condition.

We should avoid creating program plans and schedules that have not been validated, that have not been tested for credibility. It is true that customers, as in the case above, do not always offer prospective contractors good choices in terms of responding to their impossible schedule or losing the business. But, when a contractor insists on responding to such requirements, they should offer a proposal that gives the best possible chance of success and identify the risks attendant to that plan, and where possible, a means to mitigate the related risks. The machinery for validation of these plans is risk management. You could argue that this is the case for product requirements validation, as well, with few complaints from the author.

14.2 Process validation

As we prepare the program plans, schedules, budgets, and procedures as well as the supplier statements of work and other planning data, the content should pass through the same kind of filter used for product requirements validation in our search for program requirements that can be achieved and that will achieve program goals. Whoever writes plan content should have a responsibility of including in those plans not only what is required in the way of work but also the resources required, at some level of indenture, to accomplish that work (financial, time, personnel skills and numbers, material, equipment, facilities, etc.). Each task, at some level of indenture, should carry with it intended accomplishments and clear completion criteria telling how to identify when the task is complete. Realistic cost and schedule estimates, based on past performance, should be coordinated with these tasks. Satisfying these criteria encourages low-risk planning just as writing the verification requirements while writing the requirements in a specification results in better specifications. The purpose of validation, here as in the case of specifications, is risk identification and mitigation.

14.2.1 Completeness

Our process planning should be complete, free of voids. The best way to assure this is to create an overall process diagram that in Chapter 1 we referred to as a life cycle functional flow diagram, an example of which is given new expression in Figure 14-1. For most companies, this diagram will be appropriate for all programs in which they will ever become involved. The differences come into play at lower levels of indenture suggested in the diagram. The program uniqueness in this diagram primarily enters into the details of the product operational and logistics support functions where we may apply functional flow diagramming to expand our understanding of the needed functionality. The development, manufacturing, and verification (test) functions include program-peculiar content in terms of the specific

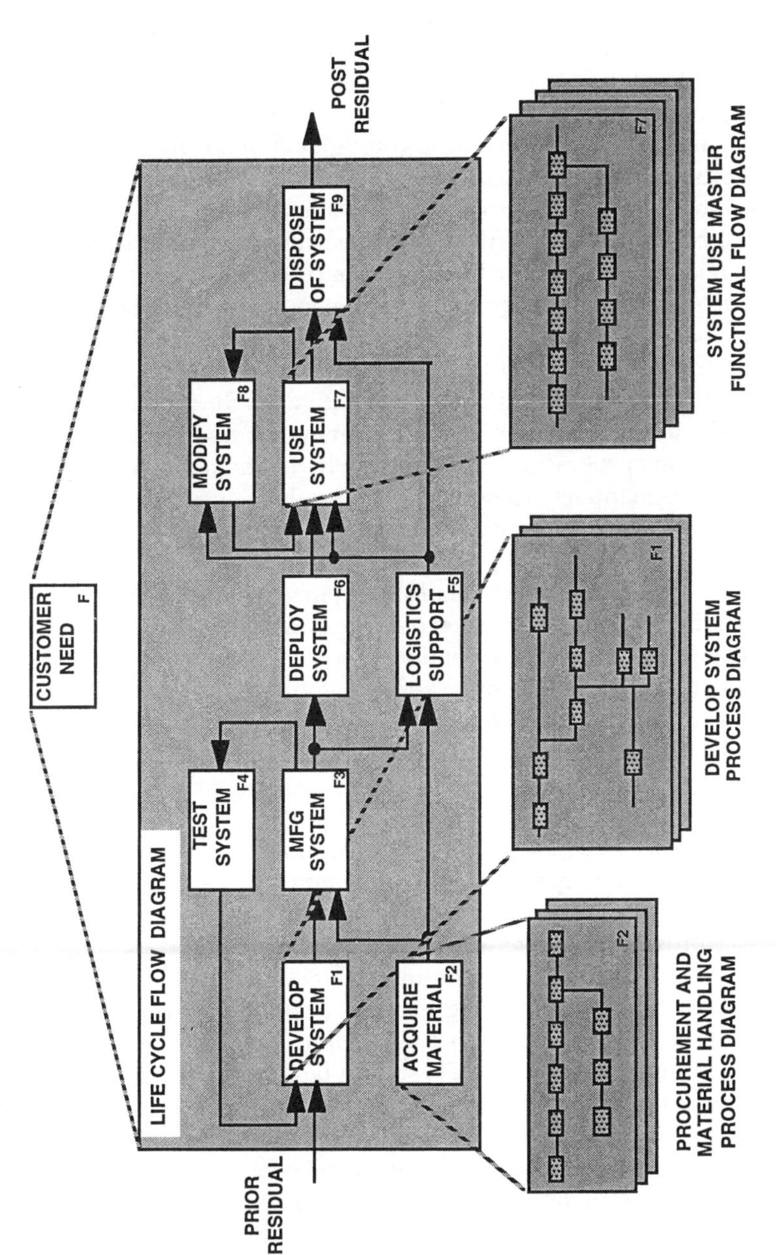

Figure 14-1 Life cycle functional flow diagram.

product entities that must be developed but generally follow a similar pattern of behavior that can be depicted in a generic process flow diagram. In support of this largely generic process, over time, the author believes, a successful company can evolve a generic set of planning data that will require relatively small adjustments from program to program. One part of the planning information that should be nearly identical for all programs is the how-to planning.

In an organization that is managed through a functional staff supplying personnel to programs organized into teams in accordance with the system architecture and led through strong program management, this information can be supplied in the form of functional department manuals that provide the generic information on how to perform particular tasks given that they are required on programs. These manuals are supplemented on programs with program-specific plans that tell what work must be done, when it must be done, and by whom it must be done. The how-to coverage for all tasks called out in the program-peculiar planning is given in the functional department manuals, the whole content of which may not necessarily be implemented on any one program. On programs where it is necessary to deviate from the generic manuals, sections should be developed for a program-unique program directives manual that contains instructions specific to that program.

Company or division management should manage this program-unique manual-building process by exception offering criticism of the program manager where this manual takes exception to too large a volume of generic how-to content. The enterprise must encourage programs to apply the task repetition scenario to the maximum extent possible to maximize the continued drive toward perfection on the part of the staff. At the same time, it is through program experimentation that the enterprise will try new things and create new generic content. Once again, we run into an interest in a condition of balance between extremes. In the author's context, this is one of the things that an enterprise integration team (EIT), reporting to the top enterprise manager, would pay attention to. As discussed elsewhere in this book, the EIT provides the enterprise integration and optimization function adjudicating the conflicts across programs for functional management. Every organizational entity needs an integration and optimization agent and the EIT, by whatever name you prefer, is this agent for the whole enterprise just as the program integration team (PIT) should be for each program.

During the proposal process, the EIT should formally review the planned program procedures and offer alternatives where the program attempts to run too far afield from the norm yet encourage a few experiments in process. The danger is that too many programs will have in progress too many such initiatives shredding the possibilities for repetition. Process change must be relentless but incremental and small. The program-peculiar excursions into process experimentation should be orchestrated based on past program lessons learned derived from metrics put in place just for this purpose and evaluated for improvement priorities. New proposals and programs offer laboratories within which to experiment with new techniques. Otherwise, we may have to introduce changes while programs are in progress.

This arrangement encourages capability improvement through repetition and progressive incremental improvement through continuous process improvement of the generic department planning data, effective continuing training of the staff in new techniques learned through program experimentation, and improvement of the functional department toolboxes consistent with the practices. All of these resources are made available to all programs providing a foundation of excellence upon which to build the program-peculiar planning documentation.

Most aerospace divisions, companies, or business units deal with relatively few different customers and a relatively narrow product line permitting some heritage in the program-peculiar planning documentation as well. But, for any program, this planning data should expand within the context of a structured top-down process to encourage completeness. This process should start with a functionally derived architecture overlaid with a product work breakdown structure (WBS) augmented by process or services WBS elements that do not clearly map to any one product element. The WBS elements should then be expanded into statement of work (SOW) paragraphs giving high-level work definition. This is a relatively simple transform. For each architecture element X at some level of indenture we simply identify the following kinds of work statements:

X.1 Perform requirements analysis yielding an item specification reviewed and approved at an item requirements review for item level or system requirements review (SRR) for system level as a precursor of design work. Where necessary, perform validation work to reduce risks identified in the transform between requirements and designs. Define verification requirements, plan verification work, and establish the verification management structure and database. Completion criteria are the release of the approved specification with customer approval where required or approval by in-house management at least one level higher that that immediately responsible for the specification.

X.2 Accomplish synthesis of the requirements yielding a preliminary design reviewed and approved at a preliminary design review (PDR) and a detailed design reviewed and approved at a critical design review (CDR). Perform analyses necessary to support and confirm an adequate design. Completion criteria are release of at least 95% of all planned engineering drawings, release of all planned analysis reports, customer and/or company management approval of the CDR presentation, and closure of all action items related to the CDR.

X.3 Perform integration, assembly, and test (IAT) actions to create the initial articles needed for engineering qualification analysis and test purposes. Complete all planned verification actions on the engineering articles, produce written evidence of compliance with the item performance specification requirements, and present that evidence for customer and/or in-house management review and approval at a functional configuration audit (FCA). Closure criteria require completion of all qualification tasks and related reports, customer acceptance of the FCA, and closure of all FCA action items.

X.4 Manufacture a first article and subject it to acceptance verification actions driven by the content of the item detail specification. Present the results of the acceptance process of the first article at a physical configuration audit (PCA) and gain approval of the results from customer and/or company management. Closure criteria include customer acceptance of PCA closure and closure of all PCA action items.

These paragraphs in the SOW should then be expanded into the content of an integrated master plan by identifying the contributing functional departments and obtaining detailed work definition from each of them coordinated with the corresponding SOW paragraphs. The functional departments will bring into the plan particular intended accomplishments and associated detailed tasks that must be accomplished such that they can be mapped to the generic planning data content which tells how those tasks shall be accomplished. In an application of the U.S. Air Force integrated management system, the functional inputs to program or team tasks can be correlated with the significant accomplishments called for in that approach as covered in the author's earlier book titled *System Engineering Planning and Enterprise Identity*.[1]

This disciplined approach mirrors the same process discussed in Chapter 1 for product requirements analysis. And why not! The contents of the SOW are the process requirements for a program. The author argues that the planning and procedures data correspond to the program design and this design should be responsive and traceable to the corresponding requirements. We first understand the grand problem and progressively expand that definition downward into more detail in an organized fashion. It is possible to create program planning data in a bottom-up direction, but it is very difficult to integrate the resultant environment so as to encourage completeness.

For each of the tasks in the integrated master plan, we should identify the completion criteria, an appropriate budget and period of time in the program schedule when the task must be accomplished either in an event-driven or calendar-driven format, and its relationships with other tasks. By using a top-down disciplined planning process, there is a good chance that we will have satisfied the completeness criteria.

The documentation for the program planning strings could be included in hierarchically organized paper documents, placed in relational databases similar to those suggested for specifications and related verification data where the documents can be printed if necessary from the database records organized by paragraph, or organized into hypertext documents or multimedia data supplemented with video clips and voice. The latter approach, particularly, has a lot of possibilities for the generic how-to data. As noted previously, this information should be the basis for company recurrent training programs and in this format it could be used for just-in-time training of individuals or groups about to embark on a new program applying techniques for which individuals or groups need a refresher. This information would serve equally well, coupled with a good find function, as a lookup source on the job for specific concerns.

Figure 14-2 offers one possible view of the aggregate enterprise documentation approach. The enterprise mission statement has been expanded into a high-level practices manual giving global requirements defined by top management including the functional organization structure definition and charters for those organizations. All of the how-to documentation expands from this base under functional management responsibility. The program planning documentation is expanded from a customer need expressed in a top-level product architecture defined though a careful mission analysis overlaid by a product and process WBS expanded in turn into an SOW (not shown in either case). The work defined in the SOW is summarized into a simple, top-level program plan and further expanded into an integrated master plan expressed also in terms of timing in an integrated master schedule.

The program manual included in Figure 14-2 offers the program-peculiar procedures discussed above. In general, this manual simply refers to the functional manuals as the source of detailed instructions in all cases where an exception is not presented in some detail in the program manual. Integrated master plan (IMP) content refers to the generic how-to manuals for details unless an exception is included in the program manual in which case the alternative procedure should be included in total.

Figure 14-2 also encourages the use of the generic how-to manuals as a source of textbooks for in-house courses. These may be delivered as organized one-day or multiple-day classes, fragmented in lunchtime classes, self-taught via the network in a just-in-time fashion, or simply read by the practitioners in paper form. Your company may not have the time or money to devote to generating the detailed information required to support the textbook application. An alternative is to prepare the plans so as to describe what to do and the sequence in which these activities should be done without providing great detail in the methods of accomplishing the work. To complete the story, you can refer in each section to standards, guides, or textbooks that do cover the way you prefer to do the work. The result is the same as if your company had devoted the time to write all of this material. In those few cases where you cannot find a suitable reference, you may have to prepare a preferred procedure.

14.2.2 Accounting for planning risks

As the content of our SOWs and program plans flows into reality, it should be screened for risks at some level of indenture just as we would perform validation on the content of our specifications as we create the content. Most of the risks that may become embedded in our planning documents will be of the cost or schedule variety rather than product performance, which are

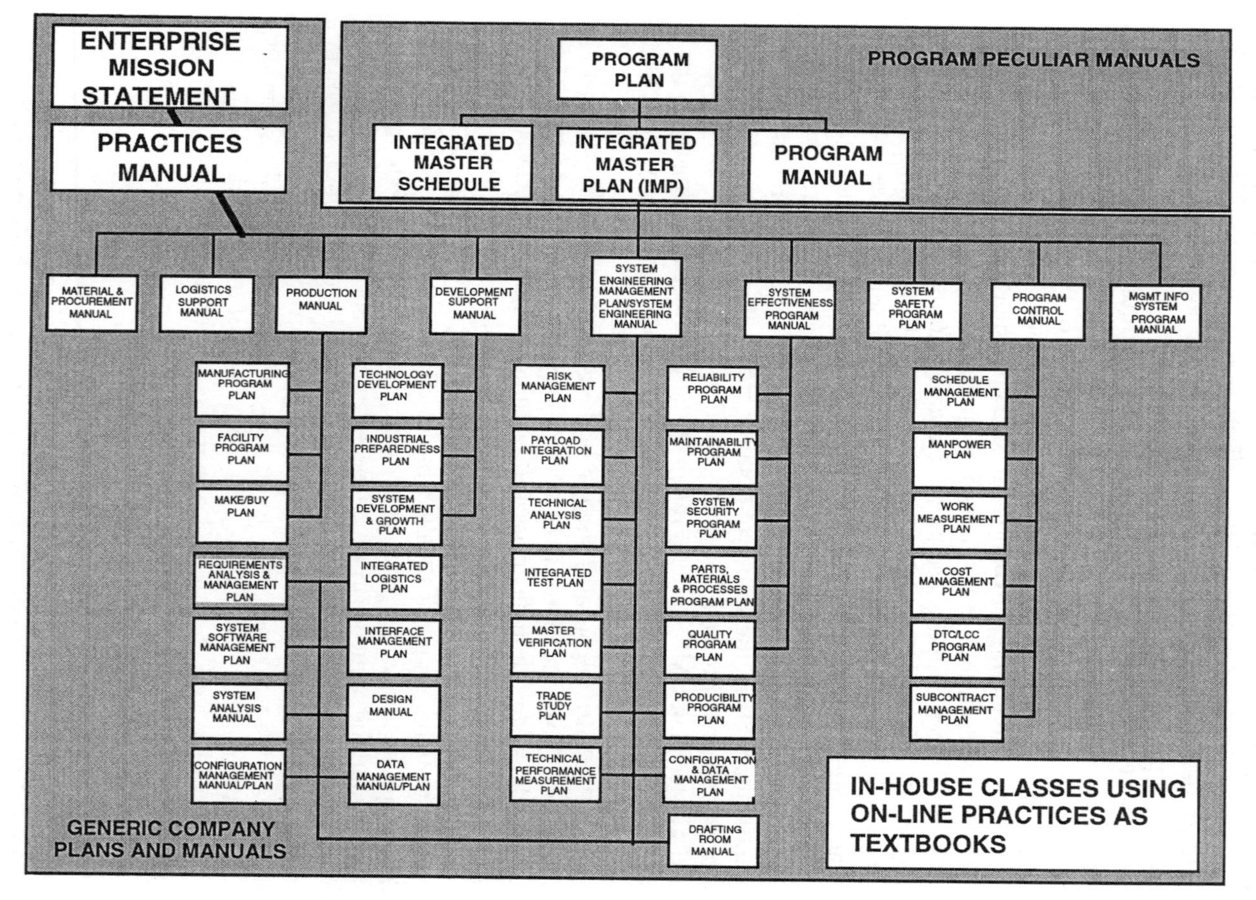

Figure 14-2 Program and functional planning documentation tree.

primarily driven by specifications content relative to our company's design capabilities and the technologies available to us.

We must ask whether the content of our plans is realistic in terms of the amount of budget and schedule available. Given that we have identified all of the tasks that must be accomplished at some level of indenture, budget and time should be allocated to them and combined in a process cost and schedule math model such that we can manipulate the model in a what-if fashion. There are many available project-planning software packages that are very effective in this application. Most large companies use one of these tools for manufacturing planning but may not make it available for more general planning purposes. If all else fails, adequate software of this kind can be acquired for operation on Mac and IBM-type desktop computers.

Margins should be included for cost figures at some level of indenture as a function of our perceived risk and the nature of the contract cost structure (cost plus incentive fee, fixed price, etc.) and float added to task schedule figures on the critical path as a function of risk as well. The resultant cost and schedule totals should be compared with what we know or think to be consistent with customer expectations and what we think to be a competitive position. If our cost or schedule figures are in excess of these, then we have to make a decision about whether to try to reallocate to close the gap or stick with our position and, if necessary, refuse to make a bid on the program because it poses too great a risk. In the former case, our proposal would have to include a clear and unmistakable statement of the program risk attendant to the customer's original cost and schedule figures in order to ensure that the customer fairly considers its proposal amid the other proposals of other bidders which may accept the risky position as a means to get in the door. Government customers are becoming more sophisticated than in years past when it comes to identification of risky proposals and will more often today give a truthfully defined program with attendant risks than many have observed from past experience.

It is possible to generate cost and schedule figures from the bottom up, also referred to as a grass roots estimate. The result can be the same as discussed here so long as we compute the totals and make the comparison with expectations and make the adjustments suggested by the comparison or decide not to bid. The danger in the bottom-up estimate and method of building the program task structure is the greater potential for omission of needed tasks. The top-down method encourages completeness. At the same time, the top-down decomposition method does run some risk of disconnecting important interfaces between tasks, as is the case in all applications of the decomposition scenario. Therefore, integration and optimization of the expanding task structure is necessary just as it is in the results of a functional decomposition applied to the product.

The author has in mind a simple process flow method of decomposition as suggested in Figure 14-1 but sees no fundamental flaw in applying well an object-oriented, data-augmented data flow diagramming, or Hatley–Pirbhai approach to name a few. All of these methods provide an environment within which we humans can seek to understand complexity in an organized fashion encouraging completeness.

The way estimates are commonly created in aerospace companies dealing with the government is that a grass roots input is requested from the contributing departments or functions. Seasoned veterans of this process build in a hidden man-hour margin because they know that management will skim a management reserve. This is all a charade and a very poor example of sound management techniques. A program organization is better served to encourage honest estimates and review the inputs and margin figures for credibility. On a program that is going to be managed by cross-functional teams, the task estimates should be built around the planned teams with inputs offered by the participants from the several participating functional departments and integrated by team leadership into an optimum estimate with margins and floats driven by perceived risks.

The aggregate margin and float figures become the program reserve controlled by the program office with some portion perhaps allocated to product team control. During program performance, the teams do their very best to accomplish their work without appealing to their margin and float. Depending on the risks that materialize, program and team management should allocate margin and float to resolve problems that cannot be resolved by other means. The cost figures and schedule demands included in the original estimates were based on perceived risks at the time they were determined, and over the run of the program some of those will materialize while others do not and still others never conceived will also come into play. This results in a mismatch between the basis for the margins and floats and the reality of program evolution, but it is the best that can be done at the time the margins and floats are determined because you are dealing with the future. The pluses and minuses will likely even out over the program duration and the aggregate margin and float will be consumed, though not in precisely the same way it was created.

This approach is identical to the use of margins adopted for the product requirements as a risk management technique. The requirements that present the greatest implementation risk are assigned margins and over the program development life these margins are generally consumed in the process of responding to risks that evolve. They commonly are not consumed in precisely the way they were originated. In some cases, a margin in one area such as weight or cost may be cashed in to improve the reliability of an item which is possible with a little more weight or a little higher design-to-cost figure. This is all part of the system optimization activity and it can be applied to process as well as product. What is more, the notion of margin class interchange just noted can be extended to the margin and float figures for process. That reliability problem might also have been solved if the schedule could be extended slightly (with other possible impacts) or if the budget could be expanded slightly.

In the latter case, we see a marriage between the design to cost (DTC) product requirements and the use of program task budget margins. The task budget margin very likely will provide the resources for the DTC program. It is important to manage the complete budget and float structure together rather than having engineering run a product requirements margin system and finance run a cost and schedule margin and float system. Whatever agent is globally responsible for program risk management, and in the author's view this would be the PIT, should be responsible for managing all of these margin systems.

14.3 Program process design

As in the case of product design, the process design should be responsive to the process requirements captured in our program planning documentation. The process design consists of detailed procedures to implement planning, personnel selection to satisfy predefined specialty engineering needs and establishment of the means to acquire these people, definition and acquisition of the program space or facility and association of the personnel and teams with this space, identification of the needed material resources and acquisition of them, clear assignment of previously defined responsibilities to persons or teams, and development of the information and communications resource infrastructure. Upon completion of the process design work we should be ready to light the program pilot light and turn up the gas. We should be ready for program implementation in accordance with a sound, validated plan.

14.4 Process verification

The verification of our process plans and procedures is accomplished in the implementation of them by monitoring preplanned program and functional metrics and using the results as the basis for actions taken to improve program performance. As in the case of product verification, process verification occurs subsequent to the design and implementation phase. There is a difference in that the results of the process verification work may be more useful on the next program rather than the current one. Metrics collected early in the program may yield improvement suggestions that can be applied prior to the end of the program, but metrics data collected later in the program may not be useful for that program. This should not lead the reader to conclude that program process verification is therefore a worthless waste of time and money. Quite the contrary, it is the connecting link for your continuous process improvement activities that will, over time, progressively improve company capability and performance. It provides a conduit for lessons learned from each program experience to the company's future of diminished faults.

14.4.1 Program and functional metrics

Metrics are measurements that permit us to measure our performance numerically and therefore make comparisons between programs and within

Table 14-1 Metrics List Representative Candidates

Task name	Metric candidates	Sense
Proposal Development	Red team quality rating	High
	Cost of preparation as a fraction of the program cost estimate	Low
Mission Analysis	Man-hours required	Low
	Number of requirements derived	High
Functional Analysis and Decomposition	Number of requirements derived	High
Requirements Analysis and Validation	Average man-hours per specification	Low
Specification Management	Number of requirements changed subsequent to specification release	Low
	Requirements per paragraph	Low
Requirements Verification	Average number of verification tasks per item	Low
Risk Management	Program risk figure	Low
	Number of active risks	Low
	Number of risks mitigated to null	High
Major Review Performance	Average number of action items received per review	Low
	Average number of days between review end and review closure	Low

programs at different times. Given that we have a generic process that can be applied to specific programs, that we have generic how-to documentation in the form of functional department manuals for all of the tasks expressed on the generic process diagram as blocks, and that we actually follow these practices on programs, we have but to determine where to install test points in our process diagram from which metric data will flow. A good place to start is to list the key tasks in our process life cycle such as the representative systems engineering tasks listed in Table 14-1.

The list in Table 14-1 is not intended to be comprehensive. The Sense column tell which value sense is desirable. Refer to the *Metrics Guidebook for Integrated Systems and Product Development* published by the International Council on Systems Engineering (INCOSE 1995) for a good source of specific metrics related to the systems engineering process.

The metrics selected should coordinate with specific tasks that are generally applied on all programs. Unusually good variations from expected values on specific programs should be investigated for cause as a potential source of process improvements whereas the opposite should be investigated for opportunities to improve the performance of specific personnel, reevaluate resource allocations, or correct personnel performance. Poor performance repeated from program to program should encourage us to select that task as a candidate for process improvement.

It is assumed that the reader is familiar with how to track these measurements in time graphically and to use the display of performance as a basis for management action to improve performance. Ideally, each organizational entity would have a series of metrics that is a good basis for determining how well their process is doing. The process-related ones will likely relate to cost and schedule performance primarily but may include other parameters. They should also be held accountable for a subset of the product metrics most often referred to as technical performance measurements discussed in Chapter 1.

14.4.2 Use of C/SCS in process verification

For each step in the program process, at some level of indenture, there should have been allocated appropriate schedule and budget through the WBS associated with the step. All program work should roll up through the IMP and SOW to the WBS. The contractor's cost/schedule control system (C/SCS) should produce reports telling the current cost and schedule status for each task at some level of indenture. The histories of these parameters should be retained and used as a basis for adjusting estimates for future programs. The C/SCS can become one element in a closed-loop control process driving the cost estimates toward low-risk values. This is a hard process to start because it only offers a delayed gratification rather than an immediate payoff. This same problem stands in the way of many improvements. The solution to this impediment is to simply accept that we will be in business for some time so we should work toward the long-term benefit of our enterprise.

14.4.3 Progressive planning improvements

The metrics data collected from programs should, over time, lead us to conclusions about tasks that our business unit does very well and those that are done very poorly. The latter lessons learned should be the object of study to determine how we can improve performance on future programs. First, we need to determine what tasks are done poorly, and then we need to prioritize the corrective actions needed to improve performance. These actions may entail improving our written practices (generic or program peculiar), computer tools and other supporting resources, facilities, or the training or skills base of our personnel. We need to determine which combination of these approaches would be most effective in correcting the high-priority problems determined from evaluation of our metrics collected on programs.

14.5 Organizational possibilities

Up to this point, the author's organizational framework has been a company with multiple programs organized into cross-functional teams with personnel and resources supplied by a traditional functional management structure. The author's past also has prejudiced his perspective in the direction of a strong independent test and evaluation organization functionally organized and run across several programs. This model also includes an independent quality organization and a system engineering function captured by engineering. In many companies, engineering management has been incapable of optimizing the application of the powerful force called system engineering to positively influence the product and process including the performance of engineering itself. The possibility exists that there may be a better way to relate these three organizational functions unveiled through an interest in verification.

Let us first extend the word *inspection* to embrace all four validation and verification methods we have discussed (test, analysis, demonstration, and examination) in keeping with MIL-STD-961D usage. Granted, some would use the term *test and evaluation* to extend across all four methods and the author has no quarrel with that term either. The application of the inspection process extends across the full length of the development and manufacturing process. Early in the development we apply it to validate the transform between requirements and product giving us confidence that we can create a compliant design. Subsequent to the design process, we manufacture product elements and subject them to qualification inspections, largely through the work of a test and evaluation function within an engineering function, proving that the design does satisfy the performance requirements. During production we subject the product to acceptance inspections to prove that the product properly represents the engineering and manufacturing product definition. The acceptance testing work is commonly staffed by manufacturing personnel and overseen by quality assurance persons.

Through all of this work there are those who do the related work, those who should coordinate, orchestrate, integrate, or manage it, and those who assure that it has been correctly accomplished. Table 14-2 lists all of the validation and verification possibilities and identifies the party who, in the author's view, should be responsible for coordinating or managing the work, for quality assurance, and who should be responsible for implementing the indicated work.

Manufacturing quality assurance is but one application of the verification process. It may make for a more efficient and technically sound process to scoop up the current test and evaluation and quality organizations into a V&V department by whatever name. The author has concluded from observation of the practice of systems engineering in many organizations that the engineering functional organization and systems engineering should perhaps have a relationship similar to that of manufacturing and quality assurance. When quality was under the manufacturing organization in many companies, it was compromised and unable to counter assertively management insistence on production over quality. By separating the function, it was empowered to participate on an equal footing.

It is too often the case that persons arriving at positions of responsibility for engineering organizations never had or mysteriously lose the good sense to encourage the systems approach in the organization they inherit. Efforts

Table 14-2 Verification Responsibilities

Level 1 task	Level 2 task	Method	Coordinator	Quality agent	Implementation
Validation		Test (DET)	T&E	Quality	T&E
		Analysis	T&E	Quality	As assigned
		Demonstration	T&E	Quality	As assigned
		Examination	T&E	Quality	As assigned
Verification	Qualification	Test	T&E	Quality	T&E
		Analysis	T&E	Quality	As assigned
		Demonstration	T&E	Quality	As assigned
		Examination	T&E	Quality	As assigned
	Acceptance	Test	T&E	Quality	Manufacturing
		Analysis	T&E	Quality	Manufacturing
		Demonstration	T&E	Quality	Manufacturing
		Examination	T&E	Quality	Manufacturing

to make improvements from below become career limiting, resulting in burnout for those who try. The organization suffers because it never places an emphasis on quality system engineering and it never achieves it. Figure 14-3 offers an alternative that will please no one except quality engineers, perhaps. It places the traditional systems engineering, quality assurance, and test and evaluation functions together in one functional organization called "Quality." The reader can pick a more-pleasing name if he or she likes. Persons who have come to love the term *total quality management* might select that term as the title of this organization and finally see it achieved.

The systems function in Figure 14-3 is responsible for risk management, mission analysis, system requirements analysis and management, specialty engineering (reliability, maintainability, logistics, safety, etc.), system analysis (aerodynamics, thermodynamics, mass properties, etc.), materials and properties, parts engineering, integration and interface development, system engineering administration (reviews, audits, action items, etc.), configuration management, and data management. Manufacturing quality is the traditional quality assurance organization responsible for verifying the quality of the process and product. The author would extend this responsibility to the complete process such that all functional organizations contributed practices, tools, and qualified personnel to programs and insist that their performance be audited on those programs in accordance with the written practices using metrics selected for that purpose. The new aggregate Quality organization should have global process quality responsibility.

The test and evaluation organization is responsible for the management of the product V&V process as described in this book as well as providing test laboratories, resources, and test engineering skills appropriate to the company's product line. People from this organization manage the process using the verification compliance and task matrices, they schedule the whole process, energize the responsible parties and resources at the appropriate time using planned budget, and track performance/status throughout.

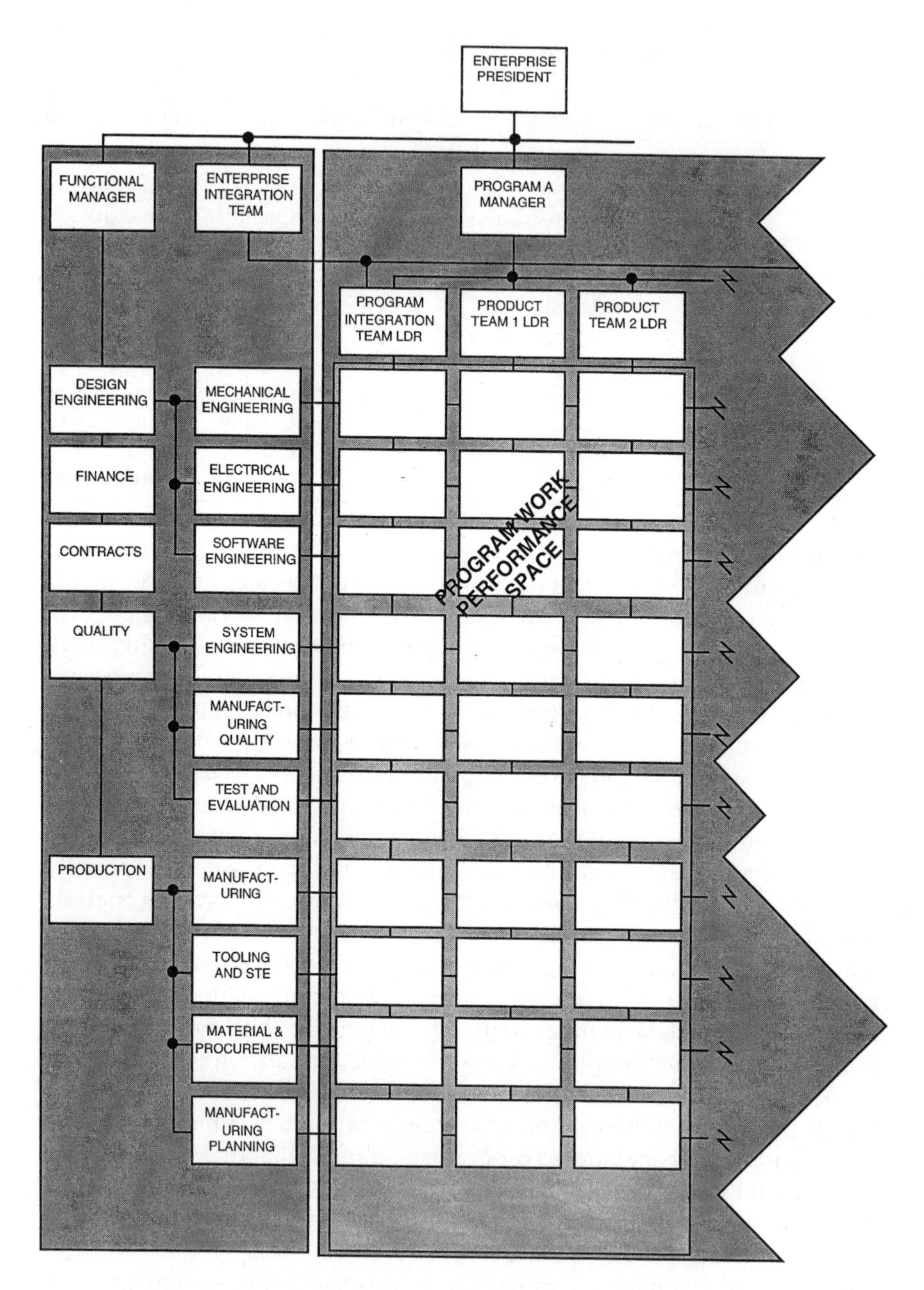

Figure 14-3 Matrix structure optimizing V&V functions.

Someone from this functional organization assigned to the PIT would lead the whole verification process composed of test, analysis, demonstration, and examination activities assigning responsibilities as appropriate.

The engineering organization in this structure is simply the creative engineering organization that solves problems defined by requirements in specifications. The results of their work in synthesizing the requirements is evaluated by the quality organization through validation work primarily driven by engineering concerns, verification work leading to qualification, and verification work leading to acceptance of specific product entities. The kinds of engineering specialties required are, of course, a function of the product line.

The Quality organization in the aggregate provides the program framework within which the people from all of the indicated organizations perform their work on each program in what is referred to in Figure 14-3 as the program work performance space. Within this space all of the people assigned to the program from the functional organizations are organized into teams oriented toward the product architecture for reasons exposed in Chapter 1. Only one of potentially several programs is illustrated.

The EIT works through the PITs to integrate and optimize at the enterprise level identifying conflicts between programs and working to adjudicate a good solution. A principal example of this is isolated program planning that calls for simultaneous use of enterprise resources beyond their current capability. One or more programs may have staffing peaks that exceed enterprise personnel availability, and it may be possible to level some of these peaks by shifting program goals in time without violating customer interests. Similarly, multiple programs may call for the same test or manufacturing capabilities in the same time frame. The EIT should also reach across the programs to encourage commonality of process by working with both program and functional organizational structures and managers.

References

1. Grady, J.O., *System Engineering Planning and Enterprise Identity,* CRC Press, Boca Raton, FL, 1995.

chapter fifteen

Postscript

15.1 Postscript plan

In the prior chapters we have explored the V words, offered clear (but not necessarily universally accepted) definitions, described the related work that must be done, and offered documentation templates for corresponding planning and reports. Generally, the content has recognized the system engineering and management concepts extant at the time the book was written (1995–1997). While writing the book, however, the author reached some conclusions about the status quo that he would prefer to share with others. These matters have a bearing on the performance of V&V work but some of them also spread out over the whole systems approach as well.

15.2 Closure on meanings

It is hoped that this book will bring attention to some of the vocabulary differences between hardware and software people and encourage closure toward a common vocabulary for systems. The development of products to satisfy our very complex needs continues to be a very difficult process and promises to become more so rather than less because our past successes encourage us to take on more and more complex problems. As we solve difficult problems we become aware of the fact that our past solutions were but subsets of grander problems encouraging us to merge multiple systems by automating their interfaces. In order to avoid development retarding forces, we should move to reach agreement on a common vocabulary across the widening hardware–software abyss. It is of absolutely no consequence which terms are favored and which are dropped. This is true of the V&V words which this book focuses upon as well as all of the other terms in conflict. As noted in the Foreword, this book was at least in part motivated by frustration over the different meanings assigned to these two words and it has been a fulfilling experience for the author to express his opinions on this matter. But, it would be far better for us all to agree on the meaning of these words even if it happened that we selected the opposite or a different meaning from that expressed in this book.

15.3 Hopes for balanced treatment

In the past, the existence of strong functionally oriented test and evaluation departments in industry, and the corresponding investment in costly test laboratories requiring good centralized management, has encouraged the evolution of a mind-set emphasizing test methods regardless of the situation. This is a positive trend but the other methods commonly have no single unifying functional mentor. It is hoped that the continuing interest in the cross-functional team concept will relax the relative strength of the lobby for test vs. other methods. The author is an advocate of matrix management with strong programs organized by cross-functional teams and supportive functional departments providing programs with the resources they require (qualified people, tools, and generic practices, as well as test laboratories). So the answer is not to reduce the excellence that test and evaluation organizations have achieved but to find a way to give the other methods a home that is every bit as effective in promoting excellence.

The organizational improvement suggested in Chapter 14 is to give the traditional functional test and evaluation department the full V&V management responsibility. It is not suggested that personnel of such a department actually accomplish all V&V work. Reliability engineers are going to have to perform reliability analyses, structural analysts will have to perform their specialized work, and logistics engineers will have to continue to perform maintenance demonstrations. What is suggested is that what we now call test engineers become more broadly qualified to embrace skills needed to understand and coordinate V&V work under all four methods. On a given program, we would find these kinds of engineers in each product team and on the system team. The V&V engineer would orchestrate the V&V process for items and the system as discussed in this book. Where test is appropriate, a pure test engineer would be assigned the responsibility to plan and manage the work possibly using a test laboratory managed by the functional V&V department and made available to all programs. V&V work corresponding to other methods would be similarly assigned to other experts in the appropriate field.

People assigned to programs from the V&V department would be responsible for management of all of the work embracing Section 4 of all of the specifications, V&V task planning, and implementation of this work. These people would act as project engineers working within their product teams across the program from the system team to orchestrate the complete process. Most of the work would be accomplished by specialists within the teams, but their performance would be guided by the V&V documentation described in earlier chapters to which they may have to contribute. On a large program a verification process leader in the PIT may have reporting to him or her validation, as well as qualification, acceptance, and system verification leaders. Each, in turn, manages strings of task leaders through method coordinators as depicted in Figure 7-1. On smaller programs, much of this overhead can be collapsed into one overall V&V leader/coordinator.

This suggestion is in keeping with the attitudes that some test and evaluation engineers have that their activity is really the superset of V&V and that the test and evaluation boundary is an overly constraining influence. Most other solutions, including continued dispersal, are not constructive. It makes no sense to create an analysis functional department or a demonstration department. Demonstration, examination, and analysis are appropriately done by people from many different disciplines and those disciplines correspond to the correct orientation for the functional organizational structure. Test should be centralized within a company because of the need to manage the test laboratory resources efficiently. The other three methods are appropriately distributed across all other disciplines but they should have central management or coordination.

Chapter 14 offered a more radical organizational change teaming the traditional system engineering, test and evaluation, and quality assurance organizations into a single functional department. The grouping of test and evaluation and quality might be more easily achieved as a first step on the road to achieving a true total quality organization. The engineering organization may offer a lot of resistance to separation of the systems function from their control even though in so many companies engineering management has failed to encourage the growth and effectiveness of its system function. The author believes that it is high time that we tried some other alternatives en route to system engineering excellence.

15.4 Information opportunities

Few companies have taken full advantage of available computer technology to improve their V&V capabilities. As discussed in earlier chapters, all of the content of all of the V&V documents can be captured in relational databases providing traceability throughout. On-line data in database format from which documents can be created as needed used in combination with computer projection for meetings can have an explosively positive effect on the speed with which one can accomplish this work and offers a significant cost reduction opportunity eliminating related reproduction, distribution, and lost time or inefficiency costs.

Most companies have already made the investment in machinery (computers, servers, networks, and software) and have only to apply those assets creatively to this very significant information management problem.

Unfortunately, the tool companies have not yet expanded their interest to embrace the full V&V problem. The author believes there is an unfulfilled market available to the toolmaker that opens their eyes to a broader view. It will not be sufficient simply to build the tables corresponding to the needed data content. The successful tool will permit distributed operation with assignment of fragments to particular persons or teams and central management of the expanding information base. These tools should also take advantage of some of the creative ideas put into play by McDonnell Douglas on their FA-18E/F program permitting access to the data in read-only mode via their intranet by a broad range of people who are not required to maintain skills in several tools each of which can be quite difficult to master and retain skill. There is a need for tool integration and optimization at a higher level of organization.

15.5 The role of the system engineer

Chapter 14 described an organizational arrangement that many professional system engineers and managers may feel attenuates the role of the system engineer in the development process. Quite the contrary, it expands the role by accepting test and evaluation engineers as system engineers and associates those who retain the job title of system engineer primarily with the development phase of a program beginning when there is only a customer need and moving through the program activity until CDR. There is a rich tapestry of work between these events that is broader than most of us can support. There will still be system engineers who focus on mission definition and others more proficient in engineering and manufacturing development. There is also and will always be a need for the people Eberhardt Rechtin calls system architects, people the author is comfortable referring to as grand system engineers, system engineers working at a higher plane of abstraction with very broad lateral boundaries.

Subsequent to CDR, the design should be relatively static yielding ideally only to engineering changes encouraged by the customer based on improved understanding of the solution to the problem and the new possibilities that product capabilities encourage. Most post-CDR work on a well-run program should be focused on integration, assembly, and test of real hardware and software. The tough development choices should have been made long before, and the program should have a sound foundation upon which to manufacture and code the product confidently. Yes, there is system engineering work remaining to be accomplished but the V&V engineers we have described are well suited to do that system engineering work.

The author has drawn criticism from fellow system engineers in the classroom and work environments about associating the QA function with system engineering. The criticism commonly offered is that Quality is generally staffed by less-qualified engineers and nondegree persons. Associating closely with Quality engineers will somehow bring down property values and ruin the neighborhood. If this is true, that the Quality organization is not staffed with quality people, that is a separate problem and industry should be doing everything possible to change that situation.

There is a much bigger problem in the application of a quality system engineering process in many organizations than others dragging the good name of system engineering down, however, if that premise is true. There are companies with a strong and respected system capability, but in many companies it would be hard to imagine the system engineering function sinking any lower in esteem within the organization. Too often, engineering management is less than enthusiastic about the system function. Functional managers are allowed to interfere with or even lead the daily workings of programs leading to serial work performance on programs. All too often, the systems charter has become a paperwork-dominated activity responsible for performing the residual work that no one in the design and analysis functions wants to do, such as documentation. The difficult product technical integration work either goes undone or it happens by chance between the design engineers who happen to stumble upon conflicts. The same criticism directed by some at Quality engineers is often valid for system engineers in some companies in that, over time, engineers who can't make it in design or analysis collect in the paperwork function.

Many engineers and engineering managers reacted very negatively to the requirements of Air Force Systems Command Manual (AFSCM) 375 series in the 1960s which required a very structured process complete with a specific set of forms coordinated with keypunch data entry from the forms into a mainframe computer. Each intercontinental ballistic missile program offered a little bit different version of this manual through the 1970s and 1980s but they were all dominated by 80-column Holerith card technology for mainframe computer input. Many engineers who experienced this environment despised the structure and the tedium associated with these processes. Some of these engineers matured into the engineering managers and chief engineers of the 1980s and 1990s, and in many of the organizations where these standards were applied by these managers the work was relegated to people called system engineers focused on documentation. The logical, artistic, and technical skills required for requirements analysis and technical integration fell into disuse in many of these companies.

While supporting a company that was developing a proposal for a space defense system, the author commented that they had used a hierarchical functional analysis process to define the system architecture and that it was likely that they had not clearly understood all of the needed functionality or concluded with the optimum architecture. It was proposed that a functional flow approach be tried to determine if a simpler system might result. The proposal manager did not want to impede proposal progress so he asked the author and one of his own company's system engineers to do that analysis in another city where their systems division was located. The analysis was done in a very few days yielding some ways to simplify the deployment system significantly by optimizing at a higher level of integration including the launch vehicle, but the proposal was not going to be changed at that point.

The author heard the same refrain from several chief engineers and proposal technical managers over the period of the 1980s, "I don't want my creative engineers damaged by that documentation mess." Other expressions of this feeling included "System requirements analysis only consumes a lot of trees"; "SRA has no beginning and no end, just an expensive middle"; and "Monkeys could fill out these forms and that's what I feel like while doing it."

More recently, companies sold on cross-functional teams eliminate their functional organizations, including system engineering, and projectize thereby losing the glue that holds the whole continuous improvement machine together. Some otherwise gifted managers believe that it is possible to assemble cross-functional teams that will flawlessly interact to develop common interfaces without the need of a system integration and optimization function. Apparently they feel that somehow people will magically start talking to one another simply because of an organizational change where this never happened before while the company was a matrix with strong functional management and programs dominated by serial work performance with all of its related problems.

Aerospace companies know that they must put on a good show in the system engineering area because their customers have come to respect the value that a sound system engineering activity should add to their product even though the members of the customer program office team may not clearly understand how this magic is supposed to happen. Too often, this is mostly show with little substance and the customer is unable to detect the difference. There are too many stories like the company that was developing a very advanced aircraft without delivering the required system engineering data items. After repeated warnings from the customer, the company hired some system engineers and assigned them the required work separated from the *real* engineers who were continuing with the design effort (although in the same town in this case). Once the data was completed and shown to an appreciative customer, the system engineers were separated and the program continued as before.

There is a lot of real system work to be done in companies and on programs, but the current relationship between design engineering and system engineering management is not optimum. The system engineer needs to be set free from the constraining power of design engineering management in order to help the whole organization, including design engineering, move toward perfection. System engineering needs a strong voice outside the design engineering channel supporting its proper activities on programs as the product system technical integration and optimization agent. In doing this work it is normal to offer constructive criticism of the design and analysis functions and their ongoing work. Those who must offer this criticism should not have to work within a regime where the top management of the design function has career power over them. An alternative is the selection of persons to manage engineering who have a broader view and an understanding of the fundamentals that drive specialization and need for system engineering.

Some examples of similar conditions that make it difficult to accomplish one's work with integrity include

a. The QA engineer whose reporting path includes manufacturing management.
b. Companies that manufacture commercial aircraft are required to have people on the staff called designated engineering representatives (DER). These people are on the company payroll but are responsible for assuring compliance with FAA regulations. In order to satisfy their

function, they have to have a large degree of immunity from company management retribution despite the very uncomfortable message that they may have to bring to the table from time to time.

c. While the author was a field engineer for Teledyne Ryan Aeronautical in the 1960s and 1970s, he often experienced a case of divided loyalties while in the field representing his company's product. The ultimate customer (Strategic Air Command for most of the author's field service work) was always rightly interested in the truth about the product and this was generally, but not always, the attitude of the company and the government acquisition agent. The author's conclusion was that no matter how painful at present a piece of information might be, immediate truth was a better selection in the long run. The results of this course of action will not always be appreciated on all sides. For example, the author found that another field engineer had told the troops how to adjust the Doppler pulse scale factor potentiometer based on prior mission inaccuracy to improve future mission guidance accuracy. There was no technical data to cover this adjustment and there was no guarantee that others who were briefed on the procedure verbally in the absence of the field engineer would use this capability correctly. One night while the author was socializing with the location commander he told the colonel what was going on and asked him to avoid the military solution. The next day the colonel applied the military solution making everyone else very angry with the author for spilling the beans.

The system engineer needs to be freed from these kinds of constraints so as to be able to contribute critically and forcefully to the evolving preferred solution. The organizational arrangement encouraged in Chapter 14 provides the needed degree of separation between the professional application of knowledge and career opportunities limitations. This is not to say that a system engineer should be free from all criticism or management influence, only that good work in the service of the program should be rewarded no matter the immediate gain or pain.

15.6 Singling up the work breakdown structure (WBS) lines

The good ship WBS has been secured to the dock by two lines in the past, the product and the process (or services) components. Commonly, test and evaluation, a principal part of the V&V process, is a separate services-oriented WBS. The author maintains this is unnecessary on development and production programs and that the added complexity is an impediment to effective and efficient program planning. The whole WBS for a development or production program should be product oriented in that services are accomplished in all cases to develop product at some level of indenture. The product and the process used to produce the product are two very different things deserving unique orthogonal identification.

Customers wish to have a product delivered. He who would create and deliver that product will have to expend work on every element of that product thus providing services. Today, the common practice is that work which can be easily correlated with the product entities will be scored against the WBS for those product entities. The work that cannot easily be associated with any of the entities selected as WBS elements or which must be associated with the system level will be scored against services WBS elements such as program management and system engineering.

If we recognize that work can be accomplished against the whole system as well as any element of the system, we shall have a workable model that will enable an efficient transform between generic enterprise work planning and the work needed for a specific program. The DoD customer has, in the past, been so concerned with their need to control cost and schedule that they have forced contractors to apply a planning model that denies them an efficient map or transform between their identity and the needed program work. Therefore, these contractors have generally failed to develop an identity through common written practices to be applied on all programs, sound internal education programs (on-the-job training and mentoring accepted as well as more formal methods) focused on that generic process, and efficient methods for building program-planning structures based on their process definition.

Figure 15-1 illustrates the proposed relationships among product items, functional departments, generic work commonly required on the enterprise's programs, and time. The understanding is that the enterprise is managed through a matrix with lean functional departments responsible for providing programs with qualified people skilled in standard practices found to be effective in the company's product line and customer base and good tools matched to the practices and people skills. Each of these functional departments has an approved charter of work it is responsible for performing on programs and which provides the basis for personnel acquisition and training.

The front vertical plane in Figure 15-1 illustrates the aggregate functional department charter by showing the correlation between all functional departments and all possible work they may be called upon to do on programs. The darkened blocks indicate this correlation. The crosshatched block in each column indicates the department that should supply the persons to lead that particular effort on programs in the product development teams assembled about the architecture. The indicated departments must prepare practices supportive of the practice defined by the task lead department (crosshatched block). The task lead department should also be primary in providing tools to do the work for that task on all programs and for providing any related training.

The reader should note that when an enterprise fully projectizes its workforce it loses the agent that can accomplish the generic work useful in providing the foundation for all programs. Each program becomes a new process factory. Therefore, there can be no central process, there can be no continuous improvement, there can be no progress, only a never-ending

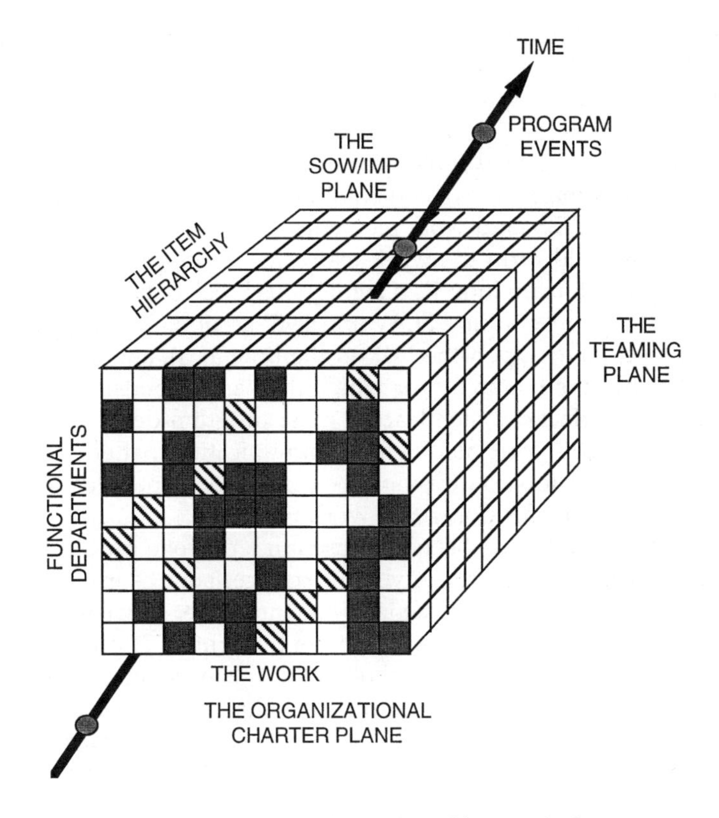

Figure 15-1 Product-only WBS enables good planning.

struggle, at best, at the lower fringes of excellence driven by the superhuman efforts of one or more heros.

In the suggested arrangement, the WBS and the architecture become one since there is no process component of the WBS. When planning a program composed of items identified on the item hierarchy (architecture or WBS) axis in Figure 15-1, we must select the work required from our generic work axis forming the statement of work (SOW) and integrated master plan (IMP) content (the SOW/IMP plane). This work selection leads to a clear identification of the functional departments from which personnel should be selected and the practices that must be applied on the program. Once we have decided the architecture nodes about which we will form cross-functional teams, the teaming plane tells us to which teams these people from the functional departments should be assigned.

The SOW that results from the first level of correlation between the product architecture and generic work must then be expanded into a more-detailed plan. For each SOW paragraph, we need to identify in a subparagraph in the IMP the work to be accomplished by each contributing department which should include the cross-functional, co-process integration work to be accomplished by the task lead department. All of the work thus selected must then be placed in a time context by mapping it to major events terminating that work and corresponding to the conduct of major reviews forming the foundation of the program integrated master schedule (IMS). We see the program plan as a series of nodes arranged in a series-parallel network mapped to trios in our matrix illustrated in Figure 15-1. These nodes, in turn, are traceable to a combination of contributing functional departments, product architecture or WBS, and work elements.

In that we should form the work teams focused on the product architecture or WBS, each team inherits a contiguous block of budget in the WBS and work in the corresponding SOW and IMP section, a clearly correlated schedule (IMS expands upon the IMP in the time axis), and a specification at the top level of the hierarchy for which they are responsible. The cross-organizational interfaces, which lead to most program development problems, become exactly correlated with communication patterns that must occur between the teams to develop those interfaces, and thus the people are clearly accountable for doing this work well.

This single-line WBS has been discussed in terms of a physical product, but most products also require deliverable services such as training, system test and evaluation work, site activation work, and other services. Figure 15-2 suggests that these services can be framed in the context of physical entities some of which may be delivered and others not. The training program may entail the services of teaching the courses developed initially or for the life cycle. Indeed, the whole system could be operated by the contractor throughout its life and this would clearly be a service. But, these services can be related to the things in the system. The author distinguishes between the services that would be performed on the delivered system and those required to create the system. The latter should all be coordinated with the product physical product.

Well, how about a contract that entails only services? The WBS for such a contract can still be formed of services to be delivered but there remains a distinction between the work that must be done to prepare for performing those services and the performance of those services. The former can be oriented about an architecture of services being designed. The latter constitutes the product that is delivered. The proposition is that the services performed to develop the deliverable services should be mapped to the deliverable services for WBS purposes.

There may still be an argument that there are some traditional work elements that will simply not map to the product structure shown in Figure 15-2, so the elements listed in the appendices of canceled MIL-STD-881A were reviewed for correlation. The traditional services-oriented (non-prime-item) elements are listed below in Table 15-1 correlated with the element of the product-only WBS illustrated in Figure 15-2 to which they would be attached. The author's preferred base 60 identification system has been used for WBS, but the reader may substitute a logistics numbering system, the common four-digit structure, or a decimal-delimited style if desired.

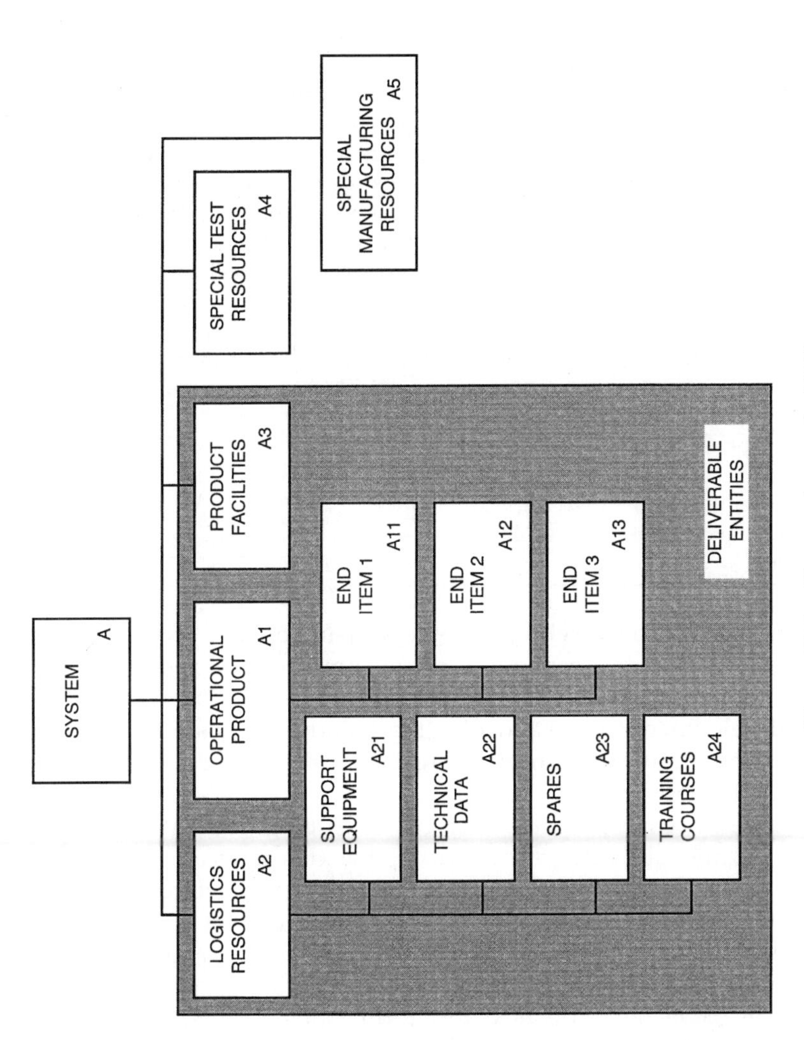

Figure 15-2 The product-only WBS.

Table 15-1 Non-product WBS Transform

Non-prime-item work element	Comment
Training	A24
Peculiar support equipment	A21
Systems test and evaluation	A1 elements as appropriate to the test; system testing may be mapped to architecture item A or A1 depending on the scope of the planned testing; otherwise, the work maps to the activities associated with items within architecture A1
System engineering/ program management	Elements of system A as appropriate; system engineering work accomplished at the system level would map to A; deliverable operational product element work would map to A1; this same pattern would be extended down to team management and the systems work accomplished within the context of those teams
Data	A22
Operational site activation	A3 as the work relates to facilitization and preparation of sites from which the product will be operated and maintained
Common support equipment	A21
Industrial facilities	A4, A5
Initial spares and initial repair parts	A24

In the product-only WBS all of the system services or processes are simply part of the enterprise product development process and they are all linked through the program-planning process to product entities where they pick up development budget and schedule. Figure 15-3 illustrates the overall generic-planning-to-program-planning transform that the product-only WBS fully enables.

The enterprise has prepared itself by developing an identity composed of an organizational charter plane showing the map between common tasks that must be applied on all programs and their functional department structure. Each functional department has developed written practices telling how to do the tasks mapped to them coordinated with the designated task lead department's practice and tools available to do the work. This how-to information could be in detailed text format or be provided in training materials traceable to the content of an abbreviated practice. This generic identity data is continuously improved based on lessons learned from ongoing programs and should be the basis for in-house system engineering training.

When a program comes into the house, the work is planned by first performing a functional decomposition and analysis on the customer need

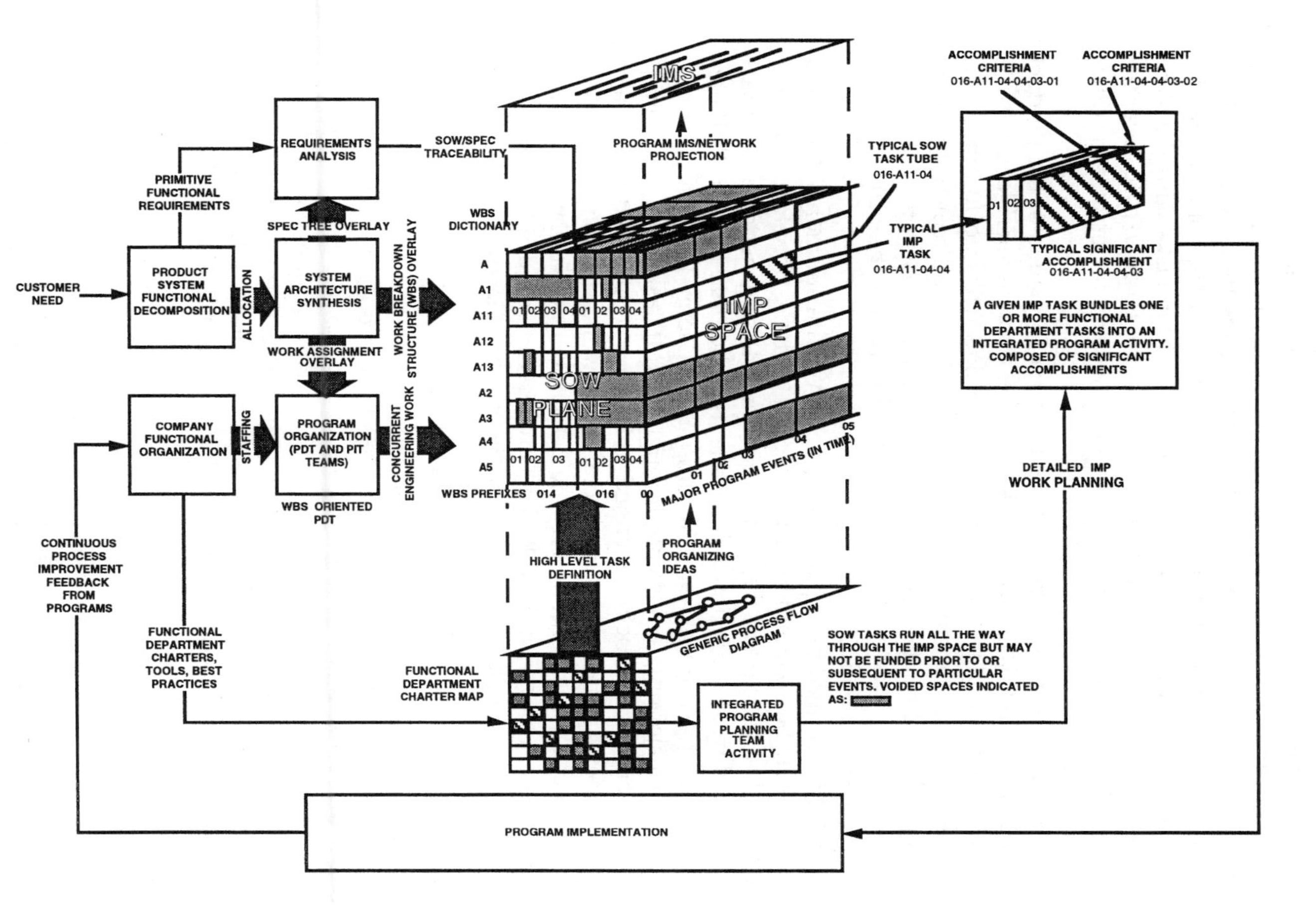

Figure 15-3 The grand planning environment using product-only WBS.

which yields a preferred architecture synthesized from the allocated functionality and the aggregate views of engineering, manufacturing, procurement and material, finance, logistics, and quality within the context of a program integration team (PIT). The product architecture becomes the WBS (base 60 identification used here can be substituted with any other system) which is the basis for development of the program statement of work (SOW). The SOW is created by the PIT by linking high-level tasks with the WBS elements as discussed earlier. Figure 15-3 includes WBS prefixes to cover the most-complicated case where one prefix may be for recurring and the other for nonrecurring.

The WBS elements also become the basis for formation of the product development teams (PDT) each of which receives a contiguous block of WBS, SOW, and related budget in a simple fashion that will lead to a need for relatively low management intensity. The continuing requirements analysis process performed by the PIT yields a specification for each team defining the design problem to be solved by the team within the context of the planning data and budget.

The IMP is formed by expansion of the SOW with additional detail that tells what has to be done and by whom it must be done. How-to information is provided to all programs by reference in the IMP to generic how-to practices provided by the functional departments rather that through IMP narratives unique to each program. All IMP tasks at some level on indenture are linked to major program events, through the IMP task-coding technique. The events are used to define planned milestones coordinated with major program reviews. Scheduling people craft the raw information into an IMS. Teams expand upon the IMP to provide details about their planned activity so as to be consistent with the IMS. This process can be aided by possession of a generic process flow diagram or network that places generic tasks at some level of indenture into a sequence context.

The work definition flows into the IMP by mapping functional department charters to SOW tasks and performing integration and optimization on the details to craft a least-cost, most-effective process for developing a product that satisfies the customer need. In the U.S. Air Force integrated management system, these functional tasks mapped into the IMP tasks become the significant accomplishments phrased in terms of the result of having performed that work. Each of these tasks includes at least one accomplishment criterion whereby one can tell when the task is finished.

15.7 Focusing on the whole

There are so many pieces of a successful V&V activity on a program and this book covers many of them. It is not enough for an organization to have people on the payroll who are gifted in the details of a process. It is necessary but not sufficient. The organization must be capable of fusing together the skills of the individuals to form a powerful communications network yielding a tremendous data rate between these individuals. It is through the application of gifted engineers in the right specialized fields on difficult jobs joined together through an effective communication process that we achieve great ends in the development of products to satisfy complex and demanding needs.

The proposition is offered that great strides in the performance of system engineering will seldom come from education alone (another necessary but not sufficient case). The organization that would perform this process well must have a shared corporate memory of product and process knowledge defined in source materials available to all that acts as a foundation for a tremendously effective conversation between the participants. In the ideal application, the participants retain their individuality but during working hours become elements of the equivalent of one great thinking machine driven by their common knowledge base and effective communications media that, aided by gifted system engineers, integrates their evolving specialized findings into the whole.

These conditions will not materialize from a cloud of stars off your company's fairy godmother's baton tip. They must be brought into being through the result of good management starting at the top and be moved down and throughout the organization by a determined management staff. Some of the principles noted in the prior chapters that encourage a desirable result are

a. An ultimate enterprise manager who understands the need for both specialization and integration as well as how the systems approach and function fits into this pattern of human behavior. This person must also have the force of personality to single-mindedly emphasize the necessity that everyone cooperate in implementing a documented process as well as contributing to continuous improvement of that process.
b. A matrix structure that requires functional management to deliver to programs good people, good tools, and good practices as well as provide service on program boards and teams in advisory and critical review functions. The matrix structure should also call upon programs to provide the strong day-to-day management function for the people assigned to the programs. These programs should be organized as cross-functional teams oriented about the product architecture.
c. The program teams identify and define requirements as a prerequisite to application of concurrent design principles to develop compliant design solutions. Requirements are reviewed and approved prior to authorizing detail design which, in turn, is reviewed and approved prior to authorizing manufacture of qualification articles. There is a good condition of traceability among the development requirements, verification requirements, and verification plans and procedures leading to well-executed verification work that discloses accurately the condition of compliance of the designs with the requirements.

d. Risk is minimized through a consciously applied program to identify potential problems and find ways to reduce the probability of occurrence and adverse consequences that may result.

The importance of a clear expression of management support for systems work cannot be overemphasized. The author encountered a good example of management support while completing a string of two 96-hour system engineering training programs at Hughes Defense Communications in Fort Wayne, Indiana while writing this book. Prior to the beginning of the first course a video tape of the president of the company (Magnavox at the time) explaining how important it was that the company perform system engineering work well was run for each class. At the end of the program a graduation exercise included a talk by the head of engineering encouraging the engineers to work to apply the principals covered in the program and to come see him if they ran into difficulty implementing those principles on programs.

Management cannot issue encouragement to the troops and then withdraw to the safe ground. This is like encouraging the people of Hungary to revolt back in the 1950s and then not helping them when the Russians brought in the tanks. The effect on your troops is similar. It is very career limiting for the troops and quickly leads to individual burnout and group failures in the systems approach. After each failed attempt it becomes all the more difficult to energize the process another time.

Excellence in system engineering, like so many things involving us humans, is simple but it is not easy. It requires the continuous application of energy, brains, and management strength and inspiration. Managers in companies that would do system engineering well must not depend on the troops to bring forth an excellent systems approach, they must lead the movement to continuously improve, not as a response to some new buzzword but as a permanent dedication to excellence no matter the intensity of the noise surrounding them. There will be times when others, subordinates, peers, and superiors, will do their best to avoid supporting your ideas with zeal and may even seek to damage your best efforts while doing their best to appear otherwise.

So, the most fundamental characteristic of a successful system engineering supporter is an inner unshakable strength of position based on the sure knowledge that in order to achieve great and complex engineering goals, one must craft a framework of product infrastructure married to teams of individual human beings each with the right specialized knowledge (and therefore limitations) welded into powerful engines of composite thought matched to the right resources (money, time, work definition, facilities, machines, and materials) and characterized by the simplest aggregate interface between product teams and product items in terms of work definition, budget, planning data, and degree of alignment between product interfaces and communication patterns that must exist between the teams. This message has to be propagated through the organization with steadiness over time while the systems capability is improved to deliver on its promise.

While a young Marine Sergeant, the author was asked by his wife to be, "What do you intend to do with your life?" The author answered, "Work with machines, that's where the complexity lies." The author has long since come to realize that the complexity resides in us humans. The infrastructure useful in the development of systems is much more complex than the product produced by it. So, it should be no wonder that it is seldom done to perfection and so often done poorly. The principles of doing system development work well and organizing the work of people to cooperate efficiently in so doing are publicly available now in many books, however, so programs are run poorly by program managers unnecessarily.

Index

A

D

E

F

N

O

P

R

U

V

DATE DE RETOUR L.-Brault
07 DEC. 1998